T0122316

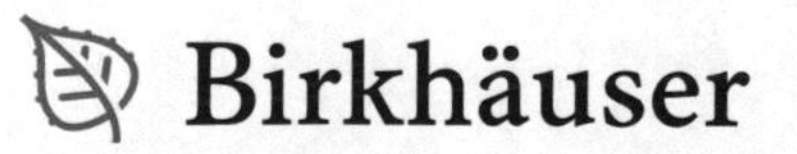
Birkhäuser

ISNM

International Series of Numerical Mathematics

Volume 169

More information about this series at http://www.springer.com/series/4819

Volker Schulz • Diaraf Seck
Editors

Shape Optimization, Homogenization and Optimal Control

DFG-AIMS workshop held at the AIMS Center Senegal, March 13-16, 2017

Editors
Volker Schulz
FB IV Abt. Mathematik
Universität Trier
Trier, Germany

Diaraf Seck
Département de Mathématiques de la
Université Cheikh Anta Diop de Daka
Dakar-Fann, Senegal

ISSN 0373-3149 ISSN 2296-6072 (electronic)
International Series of Numerical Mathematics
ISBN 978-3-030-08023-5 ISBN 978-3-319-90469-6 (eBook)
https://doi.org/10.1007/978-3-319-90469-6

Mathematics Subject Classification (2010): 49Q10, 35B27, 49J15, 49J20, 65K10

Softcover re-print of the Hardcover 1st edition 2018

Printed on acid-free paper

This book is published under the imprint Birkhäuser, www.birkhauser-science.com by the registered company Springer Nature Switzerland AG part of Springer Nature.
The registered company address is: Gewerbestrasse 11, 6330 Cham, Switzerland

Preface

This volume contains the proceedings of the DFG-AIMS Workshop on shape optimization, homogenization and control, held at AIMS Sénégal in Mbour, Sénégal, on March 13–16, 2017. This workshop joined mathematicians from Sénégal, Germany and various other countries to discuss the topics of shape optimization, homogenization and control and to explore mutual interests and options of further joint research.

The contributions in this volume give an insight into current research activities in Africa, Germany and internationally in those fields. Seeds for collaboration can be found in the first four papers in the field of homogenization. Modelling and optimal control in partial differential equations is the topic of the next six papers, again mixed from Africa and Germany. Finally, new results in the field of shape optimization are discussed in the final international three papers.

This workshop has been supported by the Deutsche Forschungsgemeinschaft (DFG), which is the largest self-governing funding organisation for basic research in Germany and in Europe, and by the African Institute for Mathematical Sciences (AIMS) in Senegal, which is one of six centres of a pan-African network of centres of excellence for postgraduate education, research and outreach in mathematical sciences.

We wish to thank all the contributors to this workshop for making it an outstanding scientific and intellectual event. We are convinced that this workshop and the contributions collected in this volume show that future collaborations in the field of mathematics between Africa and Germany are already visible on the horizon.

Trier, Germany — Volker Schulz
Dakar-Fann, Senegal — Diaraf Seck

Contents

Homogenization in Magnetic-Shape-Memory Polymer Composites

Sergio Conti, Martin Lenz, Matthäus Pawelczyk, and Martin Rumpf

Abstract Magnetic-shape-memory materials (e.g. specific NiMnGa alloys) react with a large change of shape to the presence of an external magnetic field. As an alternative for the difficult to manufacture single crystal of these alloys we study composite materials in which small magnetic-shape-memory particles are embedded in a polymer matrix. The macroscopic properties of the composite depend strongly on the geometry of the microstructure and on the characteristics of the particles and the polymer.

We present a variational model based on micromagnetism and elasticity, and derive via homogenization an effective macroscopic model under the assumption that the microstructure is periodic. We then study numerically the resulting cell problem, and discuss the effect of the microstructure on the macroscopic material behavior. Our results may be used to optimize the shape of the particles and the microstructure.

Keywords Homogenization · Magnetic shape memory · Micromagnetism · Calculus of variations · Shape optimization

S. Conti (✉)
Institut für Angewandte Mathematik, Universität Bonn, Bonn, Germany
e-mail: sergio.conti@uni-bonn.de

M. Lenz
Institut für Numerische Simulation, Universität Bonn, Bonn, Germany
e-mail: martin.lenz@ins.uni-bonn.de

M. Pawelczyk
Institut für Geometrie, Technische Universität Dresden, Dresden, Germany
e-mail: Matthaeus.Pawelczyk@tu-dresden.de

M. Rumpf
Institut für Numerische Simulation, Universität Bonn, Bonn, Germany
e-mail: martin.rumpf@uni-bonn.de

V. Schulz, D. Seck (eds.), *Shape Optimization, Homogenization and Optimal Control*, International Series of Numerical Mathematics 169,
https://doi.org/10.1007/978-3-319-90469-6_1

Mathematics Subject Classification (2010). Primary 74Q05; Secondary 74F15

1 Introduction

The peculiar mechanical behavior of magnetic shape memory (MSM) alloys renders them interesting for many applications. At the same time, they give rise to a new class of models, whose mathematical study is challenging. MSM alloys are multiferroic materials, which combine a magnetic phase transition with a shape-memory one. The deformation driven by a magnetic field deformation is due to a coupling between the two order parameters, which arises from the difference in magnetic anisotropy between the elastic variants. A typical example is a NiMnGa alloy [15, 21, 26, 29, 30].

Shape-memory metals are characterized by a phase transformation from a high-temperature, high-symmetry phase, called austenite, to a low-temperature, low-symmetry phase, called martensite. Since the point group of the martensite lattice is smaller than in the austenite phase, there are a number of different variants of the martensitic phase, which are obtained from the austenite via different but symmetry-related eigenstrains. Whereas in ceramic materials elastic phase transformations typically generate spontaneous strains of less than 1%, in alloys the deformations can be significant. For example in NiMnGa single crystals the deformation between two martensitic variants is about 11%. The practical exploitation of the shape-memory effect for actuation has been hindered by the fact that the transition between austenite and martensite can often be induced only by slow heating and cooling processes. Furthermore, it is difficult to select one variant over the other. This leads in typical situations to low-temperature states characterized by a fine mixture of the different variants, in which martensite appears without significant macroscopic deformation.

The coupling of the structural phase transformation to a magnetic field renders the materials much easier to control. The difference in anisotropy between the different variants of the martensitic phase leads to a coupling between the magnetization and the eigenstrain and provides a simple mechanism for selecting one of the variants over the others. At least in clean single crystals, large strains up to 11% have been induced by the application of external magnetic fields [21, 26, 29, 30].

It is however practically very difficult to produce single crystals of good quality. Additionally, single crystals tend to be brittle. At the same time, in polycrystals the deformation is blocked by incompatibilities at the grain boundaries. It has therefore been proposed to use small single-crystal MSM particles embedded in a softer matrix [12, 13, 16, 25, 27], although magnetically induced strains are typically smaller than in single crystals [17, 18, 28]. The composite geometry opens the way for optimization of the particle shape and in general of the microstructure.

In this paper we review recent mathematical progress and extend the analysis of MSM-polymer composites. A model based on micromagnetism and elasticity was developed in [7], and will be presented in Sect. 2 below. Under the assumption of periodicity of the microstructure, a rigorous homogenization result was then derived in [22]. This has a posteriori given a justification to the heuristic cell-problem computations that had been performed in [7, 8] to study numerically the influence of the shape of the particles on the macroscopic material behavior both in composites and in polycrystals. Some of these results are discussed, also in view of the new homogenization result, in Sect. 4. The relaxation of the model, which is appropriate for situations where the particles are much larger than the domain size, has been addressed in [9]. Our model was extended to the study of time-dependent problems, in a setting in which each particle changes gradually from one phase to the other, in [10].

2 The Model

In this section we describe the general physical model, while the precise mathematical assumptions needed for the homogenization result are presented in the next section.

We work in the framework of continuum mechanics and micromagnetism. For simplicity we discuss only the two-dimensional case, the extension of the model to three dimensions is immediate. We refer to Brown [5], Hubert and Schäfer [14] for background on micromagnetism, and to Ball and James [4], Pitteri and Zanzotto [24] and Bhattacharya [3] for the treatment of diffusionless solid-solid phase transitions. The model we use here was first presented in [7], to which we refer for further motivation and details.

Let $\Omega \subset \mathbb{R}^2$ describe the reference configuration for the composite material, and let $\omega \subset \Omega$ be the part occupied by the magnetic-shape-memory material (MSM). We model the magnetization in the MSM-particles as a measurable function $m : \omega \to \mathbb{R}^2$ with $|m| = m_s$. Here $m_s \in (0, \infty)$ is a parameter representing the saturation magnetization, which depends on the temperature and the specific choice of material. For the analytical treatment we assume the temperature and material to be fixed, and thus we may assume, after normalization, $m_s = 1$. Denoting by S^1 the set of unit vectors in $\mathbb{R}^2$, we get $m \in S^1$ on ω. For notational convenience we extend m to $\mathbb{R}^2 \setminus \omega$ by 0, and introduce $S^1_0 := S^1 \cup \{0\}$.

The austenite–martensite phase transition leads to the presence of d ($d = 2$ in our concrete application) symmetry-related variants of the martensitic phase. Magnetically, each of them has a preferred direction for the magnetization, called easy axis; we denote those directions by $f_1, \ldots, f_d \in S^1$. The phase index $p(x)$ represents the variant at point $x \in \omega$. It is convenient to view it as an element of $\mathbb{R}^d$, taking values in a discrete set $\mathcal{B} := \{e_1, e_2, \ldots, e_d\} \subset \mathbb{R}^d$. This way any $\overline{p}$ in the convex hull of $\mathcal{B}$, denoted by conv $\mathcal{B}$, is a unique convex combination of vectors in $\mathcal{B}$, allowing the tracking of the contribution of different variants. As before, we extend p by 0 on $\mathbb{R}^2 \setminus \omega$. Thus p takes values in $\mathcal{B}_0 := \mathcal{B} \cup \{0\} \subset \mathbb{R}^d$.

The coupling between the phase variable p and the magnetization m is expressed via the anisotropy energy, which is obtained by integrating a density φ depending on $x, m(x)$, and $p(x)$ (in practice, if $x \in \omega$ then φ can be seen as a function of $m(x) \cdot f_{p(x)}$). It is minimized if $x \notin \omega$ or $m(x)$, $p(x)$ are compatible with each other, in the sense that $m(x)$ and $f_{p(x)}$ are parallel, see [14]. The global anisotropy energy $E_{\text{aniso}} : L^\infty(\Omega, S_0^1) \times L^\infty(\Omega, \mathcal{B}_0) \to [0, \infty)$ then reads

$$E_{\text{aniso}}[m, p] := \int_{\mathbb{R}^2} \varphi(x, m(x), p(x))\, \mathrm{d}x. \tag{2.1}$$

A specific expression for φ is, at this level of modeling, not needed.

The magnetization m in turn induces a magnetic field, called the demagnetization field. By Maxwell's equation this field is given by ∇u, where u solves

$$\Delta u + \operatorname{div} m = 0 \quad \text{on } \mathbb{R}^2. \tag{2.2}$$

Here, (2.2) is understood distributionally. Since we extended m by zero outside ω, the boundary terms are automatically included in the distributional derivatives. The demagnetization energy $E_{\text{demag}} : L^\infty(\Omega, S_0^1) \to [0, \infty)$ can be computed by

$$E_{\text{demag}}[m] := \frac{1}{2} \int_{\mathbb{R}^2} |\nabla u_m(x)|^2\, \mathrm{d}x, \tag{2.3}$$

where u_m solves $\Delta u_m + \operatorname{div} m = 0$ [14, subsection 3.2.5] with suitable boundary conditions at infinity. The well-posedness of this problem will be discussed in the next section.

Next, we model the external magnetic field applied to the composite material as in [14, subsection 3.2.4]. If the external field is given by $h_e \in L^1(\Omega, \mathbb{R}^2)$, then the external energy $E_{\text{ext}} : L^\infty(\Omega, S_0^1) \to \mathbb{R}$ is

$$E_{\text{ext}}[m] := -\int_{\Omega} h_e(x) \cdot m(x)\, \mathrm{d}x. \tag{2.4}$$

This energy is minimal for m oriented in the direction of h_e.

In ferromagnets the microscopic magnetization m tends to be aligned in neighbouring cells, leading to a macroscopic magnetization and to an energy cost for fluctuations in the magnetization itself. This behaviour if often modelled by the exchange energy $E_{\text{ex}} : L^\infty(\Omega, S_0^1) \to [0, \infty]$, given by

$$E_{\text{ex}}[m] := \varepsilon^2 \int_{\omega} |\nabla m(x)|^2 \mathrm{d}x$$

for $m \in W^{1,2}(\omega, S^1)$ and ∞ otherwise (see [14, section 3.2.2]). The parameter ε measures the exchange length, which can be understood as the (small) length scale over which the exchange term between overlapping atomic orbitals is significant.

This concludes the description of the different components of the magnetic energy $\widetilde{E^{\mathrm{mag}}} : L^\infty(\Omega, \mathbb{R}^2) \times L^\infty(\Omega, \mathcal{B}_0) \to [0, \infty]$, given by

$$\widetilde{E^{\mathrm{mag}}}[m, p] := E_{\mathrm{aniso}}[m, p] + E_{\mathrm{demag}}[m] + E_{\mathrm{ext}}[m] + E_{\mathrm{ex}}[m].$$

If ω consists only of sufficiently 'small' grains, as we will assume in the homogenization process, then it is reasonable to assume that every grain is a single crystal, and consists only of a single domain, in which m is constant. Indeed, in this case the exchange energy is negligible, as is shown in [11]. Thus the final magnetic energy $E^{\mathrm{mag}} : L^\infty(\Omega, \mathbb{R}^2) \times L^\infty(\Omega, \mathcal{B}_0) \to [0, \infty)$ is given by

$$E^{\mathrm{mag}}[m, p] := E_{\mathrm{aniso}}[m, p] + E_{\mathrm{demag}}[m] + E_{\mathrm{ext}}[m].$$

We finally address the elastic properties of the composite material. Let $\widetilde{v} \in W^{1,2}(\Omega, \mathbb{R}^2)$ be the deformation of the composite material, which describes the spontaneous stretching of the MSM material in response to the external field and the subsequent deformation of the polymer. Let $\widetilde{W} = W(x, F, p)$ be the (nonlinear) elastic energy density, where $F(x) = \nabla v(x)$ is the deformation gradient at x, and $p(x) \in \{0, e_1, \dots, e_d\}$ denotes the martensitic variant (as usual, $p = 0$ in the polymer). A possible choice would be $\widetilde{W}(x, F, p) = \mathrm{dist}^2(FR(x), \mathrm{SO}(2)(\mathrm{Id} - A_p))$ if $x \in \omega$ and $\widetilde{W}(x, F, p) = \mathrm{dist}^2(F, \mathrm{SO}(2))$ else. Here the rotation $R \in L^\infty(\omega, \mathrm{SO}(2))$ encodes the local orientation of the crystal lattice, and A_p the eigenstrain of the phase p. The magnetization in the deformed configuration, where Maxwell's equation has to hold, is then $m \circ \widetilde{v}$. Thus, we now would need to solve $\Delta u + \mathrm{div}\,(m \circ \widetilde{v}) = 0$ on $\mathbb{R}^2$. In addition, the anisotropy energy φ has to take into account the fact that the local magnetization $m(x)$ needs to be compared with the easy axis in the deformed (and, possibly, rotated) crystal.

However, experiments show that the displacements are moderate, and thus we approximate $\widetilde{v}(x) = x + v(x)$ with a small displacement v. Consistently expanding all terms to the first nontrivial order in m and v, we consider the magnetization m instead of $m \circ \widetilde{v}$ and ignore elastic rotations in the anisotropy energy. Therefore we will work with the energy

$$E[v, m, p] := E_{\mathrm{def}}[v, m, p] + E^{\mathrm{mag}}[m, p],$$

where $E_{\mathrm{def}}[v, m, p]$ is the linear elastic energy stored in the displacement v and obtained integrating a quadratic form W corresponding to the linearization of $\widetilde{W}$.

3 Analytical Homogenization

The MSM–polymer composite is a finely microstructured material, and therefore very difficult to simulate directly. A direct discretization would need to resolve explicitly all scales from the one of the MSM particles, of the order of a few microns, to the one of the macroscopic sample, which may be in the centimeter range. The theory of homogenization permits both a simpler qualitative understanding of the material behavior and an efficient simulation. The key idea is to exploit the difference in length scales to address the two scales separately. Under specific assumptions of the microstructure (periodicity in the present case) one derives an effective macroscopic material model, which describes the behavior of the composite on a length scale much larger than the one of the microstructure, by solving suitable cell problems. At this level one does not need to address the macroscopic shape of the sample, boundary conditions or applied forces, but instead studies only the microstructure. One then uses, in a separate process, the effective material model to perform macroscopic simulations, which takes into account boundary conditions and external forces. Typical examples of homogenization include the process leading from atomistic models of matter to continuum elasticity and the treatment of materials obtained by mixing two different components on a very fine scale. Whereas the general theory for composites of two materials with different elastic properties is well developed, we are here interested in a situation in which the magnetic properties are also different, and coupled to an elastic phase transition. A purely magnetic problem was studied in [23]. The homogenization result for our system does not follow from the general theory, but needs to be proven specifically for the case at hand. For a comprehensive overview over the subject, we refer to Braides and Defranceschi [2], Cioranescu and Donato [6], Allaire [1] and Milton [20].

Let $\Omega \subset \mathbb{R}^2$ be an open, bounded Lipschitz domain, and let $\omega \subset Q := (0,1)^2$ be measurable. For every $k \in \mathbb{N}$ let $\omega_k := \Omega \cap \big(\frac{1}{k}(\omega + \mathbb{Z}^2)\big)$, see Fig. 1 for an illustration. It is easily seen that $\lim_{k\to\infty} \mathcal{L}^2(\omega_k) = \mathcal{L}^2(\omega)\,\mathcal{L}^2(\Omega)$ is the volume fraction of the composite occupied by the MSM particles. Here, $\mathcal{L}^2$ is the Lebesgue measure. The

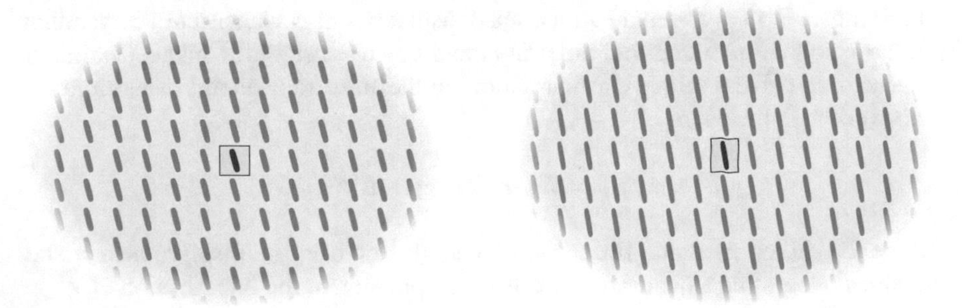

Fig. 1 Sketch of the two-scale approach to the behavior of composites. The periodic cell is emphasized in the center. Left picture: reference configuration. Right picture: deformed configuration. Both the particle and the shape of the unit cell are deformed

set of admissible magnetization fields and phase indices for ω_k is given by

$$S_k^{\text{mag}} := \left\{(m, p) \in L^\infty(\mathbb{R}^2, \mathbb{R}^2) \times L^\infty(\mathbb{R}^2, \mathbb{R}^d) : \; |m| = |p| = \chi_{\omega_k}\right\},$$

where χ_{ω_k} is the characteristic function of ω_k.

In order to compute the demagnetization energy we need to show the well-posedness of (2.2). By Lax-Milgram's Theorem there exists for any measurable m with $|m| = \chi_{\omega_k}$ a unique solution u of (2.2) in the space

$$W_*^{1,2}(\mathbb{R}^2) := \left\{u \in W_{\text{loc}}^{1,2}(\mathbb{R}^2) : \int_{\{|x|\le 1\}} u(x)\text{d}x = 0 \text{ and } \nabla u \in L^2(\mathbb{R}^2, \mathbb{R}^2)\right\}.$$

The condition on the average of u on the unit ball is an arbitrarily chosen criterion used to make the solution unique, since both the equation defining u and the energy only depend on the gradient of u. Alternatively one could work in the space of curl-free L^2 vector fields. The explicit appearance of the potential u in the definition of the space is useful for truncation and interpolation. We also remark that, due to the fact that the fundamental solution of Laplace's equation is logarithmic in two dimensions, the potential u is not in $L^2(\mathbb{R}^2)$. In $\mathbb{R}^3$ instead one would obtain a unique solution $u \in W^{1,2}(\mathbb{R}^3)$. Given a magnetization field m, we denote by u_m the solution of (2.2) in $W_*^{1,2}(\mathbb{R}^2)$. Let the anisotropic energy density $\varphi : \mathbb{R}^2 \times S_0^1 \times \mathcal{B}_0 \to [0, \infty)$ in (2.1) satisfy

1. φ is a Carathéodory function, i.e.,
   (a) $\varphi(\cdot, m, p)$ is measurable for all $(m, p) \in S_0^1 \times \mathcal{B}_0$.
   (b) $\varphi(y, \cdot, \cdot)$ is continuous for all $y \in \mathbb{R}^2$.
2. The map $y \mapsto \varphi(y, m, p)$ is Q-periodic for any $m \in S_0^1$, $p \in \mathcal{B}_0$.
3. For any $y \in \mathbb{R}^2 \setminus \omega$ we have $\varphi(y, \cdot, \cdot) = 0$.
4. There exists $M > 0$, such that for all $m, m' \in S^1$, $p \in \mathcal{B}$, $y \in \omega$ we have

$$|\varphi(y, m, p) - \varphi(y, m', p)| \le M|m - m'|.$$

For the external magnetic field we assume $h_e \in L^1(\Omega, \mathbb{R}^2)$. We define the magnetic energy $E_k^{\text{mag}} : L^\infty(\mathbb{R}^2, \mathbb{R}^2) \times L^\infty(\mathbb{R}^2, \mathbb{R}^d) \to (-\infty, \infty]$ by

$$E_k^{\text{mag}}[m, p] := \int_\Omega \varphi(kx, m(x), p(x))\text{d}x + \frac{1}{2}\int_{\mathbb{R}^2} |\nabla u_m|^2 \text{d}x - \int_\Omega m(x) \cdot h_e(x)\text{d}x$$

if $(m, p) \in S_k$, and $E_k^{\text{mag}}[m, p] = \infty$ otherwise.

The homogenization limit corresponds to taking weak limits of p_k and m_k, which are L^∞ fields. It is therefore not surprising that the conditions $|p_k| = |m_k| = \chi_{\omega_k}$ will be relaxed in the limit, with χ_{ω_k} converging weakly to the function $\mathcal{L}^2(\omega)\chi_\Omega$, and the equality conditions being replaced by inequalities representing the convexification of the conditions. Indeed, we will show that the space of the possible limiting values m and p is

$$S_\infty^{\mathrm{mag}} :=\{(m,p) \in L^\infty(\mathbb{R}^2,\mathbb{R}^2) \times L^\infty(\mathbb{R}^2, \operatorname{conv}\mathcal{B}_0) : \\ |m| \le \mathcal{L}^2(\omega)\chi_\Omega,\ |p|_{\ell^1} = \mathcal{L}^2(\omega)\chi_\Omega\},$$

where $|p|_{\ell^1} = \sum_{k=1}^d |p_k|$ and $\operatorname{conv}\mathcal{B}_0$ is the convex hull of $\mathcal{B}_0$.

The effective energy $E_\infty^{\mathrm{mag}} : L^\infty(\mathbb{R}^2,\mathbb{R}^2) \times L^\infty(\mathbb{R}^2,\mathbb{R}^d) \to (-\infty,\infty]$ on S_∞^{mag} is then given by

$$E_\infty^{\mathrm{mag}}[m,p] := \int_\Omega f_{\mathrm{hom}}^{\mathrm{mag}}(m(x),p(x))\mathrm{d}x + \frac{1}{2}\int_{\mathbb{R}^2} |\nabla u_m|^2 \mathrm{d}x - \int_\Omega m(x)\cdot h_e(x)\mathrm{d}x,$$

and $E_\infty^{\mathrm{mag}} = \infty$ otherwise. Here $f_{\mathrm{hom}}^{\mathrm{mag}} : \mathbb{R}^2 \times \operatorname{conv}\mathcal{B}_0 \to [0,\infty]$ is given by

$$f_{\mathrm{hom}}^{\mathrm{mag}}(\mu,\rho) = \inf_{((M_k,P_k))_{k\in\mathbb{N}} \in \mathcal{A}_{\mu,\rho}} \liminf_{k\to\infty} \left(\int_Q \varphi(kx, M_k(x), P_k(x)) + \frac{1}{2}|\nabla U_{M_k}|^2 \mathrm{d}x \right) \tag{3.1}$$

where $U_M \in W^{1,2}_{\mathrm{loc}}(\mathbb{R}^2,\mathbb{R}^2)$ is the unique function satisfying

$$U_M \text{ is } Q\text{-periodic},\quad \int_Q U_M(x)\mathrm{d}x = 0 \quad\text{and}\quad \Delta U_M + \operatorname{div} M = 0 \ \text{ on } Q,$$

and

$$\mathcal{A}_{\mu,\rho} := \Big\{ ((M_k,P_k))_{k\in\mathbb{N}} \subset L^\infty(Q,S_0^1) \times L^\infty(Q,\mathcal{B}_0) : \\ \forall k \in \mathbb{N} : |M_k| = |P_k| = \chi_\omega,\quad \int_Q M_k \mathrm{d}x = \mu \quad \text{and} \\ M_k \overset{*}{\rightharpoonup} \mu \text{ on } L^\infty(Q,\mathbb{R}^2),\quad P_k \overset{*}{\rightharpoonup} \rho \text{ on } L^\infty(Q,\mathbb{R}^d) \Big\}.$$

We now can state the homogenization theorem for the purely magnetic energy. This and the next results have been proven in [22] extending the classical homogenization theory for elastic materials [1, 2, 6, 20] and the homogenization results for three-dimensional purely magnetic materials in [23]; a brief summary of the proof is given below.

Theorem 3.1 *With the above assumptions on φ we have the following results:*

(i) *Let $((m_k,p_k))_{k\in\mathbb{N}} \subset L^\infty(\mathbb{R}^2,\mathbb{R}^2) \times L^\infty(\mathbb{R}^2,\mathbb{R}^d)$ with*

$$\liminf_{k\to\infty} E_k^{\mathrm{mag}}[m_k,p_k] < \infty.$$

Then there exist a subsequence (not relabeled), and $(m, p) \in S_\infty^{\mathrm{mag}}$ *such that*

$$m_k \overset{*}{\rightharpoonup} m \text{ in } L^\infty(\Omega, \mathbb{R}^2), \quad p_k \overset{*}{\rightharpoonup} p \text{ in } L^\infty(\Omega, \mathbb{R}^d),$$
$$\nabla u_{m_k} \rightharpoonup \nabla u_m \text{ in } L^2(\mathbb{R}^2), \quad \nabla u_{m_k} \to \nabla u_m \text{ in } L^2(\mathbb{R}^2 \setminus \Omega),$$

where $u_m, u_{m_k} \in W_*^{1,2}(\mathbb{R}^2)$ *are solutions to*

$$\Delta u_{m_k} + \operatorname{div} m_k = \Delta u_m + \operatorname{div} m = 0 \text{ on } \mathbb{R}^2.$$

(ii) *For any* $(m, p) \in L^\infty(\mathbb{R}^2, \mathbb{R}^2) \times L^\infty(\mathbb{R}^2, \mathbb{R}^d)$ *there exists a sequence* $((m_k, p_k)) \subset L^\infty(\mathbb{R}^2, \mathbb{R}^2) \times L^\infty(\mathbb{R}^2, \mathbb{R}^d)$ *such that*

$$\limsup_{k\to\infty} E_k^{\mathrm{mag}}[m_k, p_k] = E_\infty^{\mathrm{mag}}[m, p].$$

(iii) *Let* $((m_k, p_k)) \subset L^\infty(\mathbb{R}^2, \mathbb{R}^2) \times L^\infty(\mathbb{R}^2, \mathbb{R}^d)$ *with* $m_k \overset{*}{\rightharpoonup} m$, $p_k \overset{*}{\rightharpoonup} p$ *in* $L^\infty(\Omega, \mathbb{R}^2)$, $L^\infty(\Omega, \mathbb{R}^d)$ *resp. Then*

$$\liminf_{k\to\infty} E_k^{\mathrm{mag}}[m_k, p_k] \geq E_\infty^{\mathrm{mag}}[m, p].$$

As usual, this result can be easily stated in terms of Γ-convergence.

The compactness result (i) for m_k, p_k follows easily, since the uniform bound on the energy implies that there is a subsequence with $|m_k| = |p_k| = \chi_{\omega_k}$ almost everywhere. Thus the L^∞-norm of m_k, p_k is uniformly bounded by 1, and we can extract a converging subsequence. The convergence of the demagnetization field then follows from [22, Lemma A.5.4].

A key ingredient in the proof is the fact that $f_{\mathrm{hom}}^{\mathrm{mag}}$ can be equivalently defined by minimizing over other sets than $\mathcal{A}_{\mu,\rho}$. Specifically, we can either drop the condition of having a fixed mean value on m_k on Q, or drop the condition that $M_k \overset{*}{\rightharpoonup} \mu$ without changing the value of $f_{\mathrm{hom}}^{\mathrm{mag}}(\mu, \rho)$. Also instead of solving $\Delta U + \operatorname{div} M = 0$ on Q with periodic boundary, we can solve it on Q with vanishing Dirichlet boundary condition. Furthermore the density $f_{\mathrm{hom}}^{\mathrm{mag}}$ does not change if U is replaced by the solution $U \in W_*^{1,2}(\mathbb{R}^2)$ with $\Delta U + \operatorname{div} \chi_Q M = 0$ and the term $\int_Q |\nabla U|^2$ in (3.1) replaced by the integral $\int_{\mathbb{R}^2} |\nabla U|^2$. We refer to Pawelczyk [22, Proposition 2.2.1] for a precise statement of these facts and a proof.

This equivalence is crucial to relate the macroscopic demagnetization field u to the microscopic fields U, which are the result of oscillations in the magnetic charges. More precisely, we first prove that $f_{\mathrm{hom}}^{\mathrm{mag}}$ is a continuous function and E_∞^{mag} restricted to the limiting function S_∞^{mag} is continuous w.r.t. the L^1-topology. Thus it suffices to prove (ii) for step functions m, p. This is done by covering Ω by squares, and assuming m, p are constant on each of them. On each square we can explicitly construct a recovery sequence of periodic functions, and by glueing them together we obtain a global recovery sequence. Details are given in [22, Theorem 2.3.1 and Theorem 2.3.2].

We now include the elastic energy in the energy in terms of a displacement v. Let $W : \mathbb{R}^2 \times \mathbb{R}^{2\times 2} \times \mathcal{B}_0 \to \mathbb{R}$ be a Carathéodory function, such that

1. W is Q-periodic in the first component.
2. There exists $r : [0,\infty) \to [0,\infty)$ with $r(\delta) \searrow 0$ for $\delta \searrow 0$, such that for any $x \in \mathbb{R}^2,\ F, G \in \mathbb{R}^{2\times 2},\ p \in \mathcal{B}_0$ it holds
$$W(x, F+G, p) \leq (1 + r(|G|))W(x, F, p) + r(|G|).$$
3. There exists $c > 0$, such that for any $x \in \mathbb{R}^2,\ F, G \in \mathbb{R}^{2\times 2},\ \rho \in \mathcal{B}_0$ it holds
$$\frac{|\mathrm{sym}\, F|^2}{c} - c \leq W(x, F, p) \leq c|\mathrm{sym}\, F|^2 + c,$$
where $\mathrm{sym}\, F = \frac{1}{2}(F + F^T)$.

We define the set of admissible triples (v, m, p) by
$$S_k := \left\{(v, m, p) \in W^{1,2}(\Omega, \mathbb{R}^2) \times S_k^{\mathrm{mag}}\right\},$$
and the energy $E_k : W^{1,2}(\Omega, \mathbb{R}^2) \times L^\infty(\mathbb{R}^2, \mathbb{R}^2) \times L^\infty(\mathbb{R}^2, \mathbb{R}^d) \to (-\infty, \infty]$ by
$$E_k[v, m, p] := E_k^{\mathrm{mag}}[m, p] + \int_\Omega W(kx, \nabla v(x), p(x))\mathrm{d}x$$
if $(v, m, p) \in S_k$, and $E_k[v, m, p] = \infty$ otherwise. The set of limiting configurations is given by $S_\infty := W^{1,2}(\Omega, \mathbb{R}^2) \times S_\infty^{\mathrm{mag}}$, and the effective energy is
$$E_\infty[v, m, p] := \int_\Omega f_{\mathrm{hom}}(\nabla v(x), m(x), p(x))\mathrm{d}x + \frac{1}{2}\int_{\mathbb{R}^2} |\nabla u_m|^2 \mathrm{d}x - \int_\Omega m \cdot h_e \mathrm{d}x,$$
where
$$f_{\mathrm{hom}}(\beta, \mu, \rho) := \inf_{((V_k, M_k, P_k))\in\mathcal{A}_{\beta,\mu,\rho}} \liminf_{k\to\infty}$$
$$\left(\int_Q W(kx, \nabla V_k(x), P(x)) + \varphi(kx, M_k(x), P_k(x)) + \frac{1}{2}|\nabla U_{M_k}|^2 \mathrm{d}x\right),$$
and
$$\mathcal{A}_{\beta,\mu,\rho} := \Big\{((V_k, M_k, P_k))_{k\in\mathbb{N}} \subset W^{1,2}(Q, \mathbb{R}^2) \times L^\infty(Q, S_0^1) \times L^\infty(Q, \mathcal{B}_0) :$$
$$((M_k, P_k)) \in \mathcal{A}_{\mu,\rho},\ \ \nabla V_k \rightharpoonup \beta \text{ in } L^2(Q, \mathbb{R}^2)\Big\}.$$

Theorem 3.2 *With the above assumptions on φ and W we have the following results:*

(i) *Let $((v_k, m_k, p_k))_{k\in\mathbb{N}} \subset W^{1,2}(\Omega, \mathbb{R}^2) \times L^\infty(\mathbb{R}^2, \mathbb{R}^2) \times L^\infty(\mathbb{R}^2, \mathbb{R}^d)$ with $\liminf_{k\to\infty} E_k[m_k, p_k] < \infty$. Then there exist a subsequence (not relabeled), $(v, m, p) \in S_\infty$, and affine maps $a_k : \mathbb{R}^2 \to \mathbb{R}^2$ such that*

$$
\begin{aligned}
&m_k \overset{*}{\rightharpoonup} m \text{ in } L^\infty(\Omega, \mathbb{R}^2), \quad p_k \overset{*}{\rightharpoonup} p \text{ in } L^\infty(\Omega, \mathbb{R}^d),\\
&\nabla u_{m_k} \rightharpoonup \nabla u_m \text{ in } L^2(\mathbb{R}^2), \quad \nabla u_{m_k} \to \nabla u_m \text{ in } L^2(\mathbb{R}^2 \setminus \Omega),\\
&v_k - a_k \rightharpoonup v \text{ in } W^{1,2}(\Omega, \mathbb{R}^2),
\end{aligned}
$$

where $u_m, u_{m_k} \in W^{1,2}_*(\mathbb{R}^2)$ are solutions of

$$
\Delta u_{m_k} + \operatorname{div} m_k = \Delta u_m + \operatorname{div} m = 0 \text{ on } \mathbb{R}^2.
$$

(ii) *For any $(v, m, p) \in W^{1,2}(\Omega, \mathbb{R}^2) \times L^\infty(\mathbb{R}^2, \mathbb{R}^2) \times L^\infty(\mathbb{R}^2, \mathbb{R}^d)$ there exists a sequence $((v_k, m_k, p_k)) \subset W^{1,2}(\Omega, \mathbb{R}^2) \times L^\infty(\mathbb{R}^2, \mathbb{R}^2) \times L^\infty(\mathbb{R}^2, \mathbb{R}^d)$ such that*

$$
\limsup_{k\to\infty} E_k[v_k, m_k, p_k] = E_\infty[v, m, p].
$$

(iii) *Let $((v_k, m_k, p_k)) \subset W^{1,2}(\Omega, \mathbb{R}^2) \times L^\infty(\mathbb{R}^2, \mathbb{R}^2) \times L^\infty(\mathbb{R}^2, \mathbb{R}^d)$ with $m_k \overset{*}{\rightharpoonup} m$, $p_k \overset{*}{\rightharpoonup} p$ in $L^\infty(\Omega, \mathbb{R}^2)$, $L^\infty(\Omega, \mathbb{R}^d)$ resp., and $v_k \rightharpoonup v$ in $W^{1,2}(\Omega, \mathbb{R}^2)$. Then*

$$
\liminf_{k\to\infty} E_k[v_k, m_k, p_k] \geq E_\infty[v, m, p].
$$

The proof is similar to the previous one. For (i) it only remains to prove compactness for the sequence (v_k). The coercivity of W yields an uniform L^2-bound on the sequence $(\operatorname{sym} \nabla v_k)$. An application of Korn's and Poincaré's inequality yields then compactness.

While the proof of (iii) is analogous to the previous one, we need to be more careful in (ii). To approximate v in $W^{1,2}$ we use piecewise linear functions on triangles. This is, however, not the natural decomposition for the Q-periodic microstructure. This can be dealt with by covering each triangle by squares once more. Detailed proofs can be found in [22, Theorem 3.3.1 and Theorem 3.3.2].

4 Numerical Study of the Cell Problem

We illustrate the significance of the homogenization result by studying numerically the cell problem for a few selected microstructures. Our results characterize the dependence of the macroscopic material properties on the microstructure. From the viewpoint of applications, the key interest lies in devising microstructures which can be practically produced and which lead to (near) optimal macroscopic properties, such as for example a large transformation strain in response to an external magnetic field or a large work output. This constitutes a shape-optimization problem: we are interested in devising the shape and location of the particles which optimizes some scalar quantity. It is however different from most classical shape-optimization problems, in that it completely takes place at the microstructural level, and in that only a very restricted set of shapes is practically relevant. Indeed, the control on the shape and the ordering of the MSM particles is only very indirect, as they are obtained for example by mechanically grinding thin ribbons of MSM material [19]. Their orientation can, up to a certain point, be controlled by applying a magnetic field during the solidification of the polymer. At the same time, the elastic properties of the polymer can be, up to a certain degree, tuned by choosing different compositions. We therefore select a few geometric and material parameters (particle elongation, Young modulus of the polymer, etc.) and investigate their influence on the macroscopic material behavior. Our results can serve as a guide for the experimental search for production techniques which lead to composites with successively improved properties.

As customary in the theory of homogenization, we assume that the corrector fields (V_k, M_k, P_k) entering $f_{\text{hom}}(\beta, \mu, \rho)$ have the same periodicity of the microstructure (that is, $(0, 1/k)^2$) and that V_k obeys affine-periodic boundary conditions. In practice, we approximate $f_{\text{hom}}(\beta, \mu, \rho)$ by $f^{(1)}_{\text{hom}}(\beta, \mu, \rho)$, which (after rescaling) is defined as the infimum of

$$\int_Q W(x, \nabla V(x), P(x)) + \varphi(x, M(x), P(x)) + \frac{1}{2}|\nabla U_M|^2 \mathrm{d}x$$

over all $V \in W^{1,2}(Q, \mathbb{R}^2)$ which obey $V(x+e_i) = V(x)+\beta e_i$ in the sense of traces for $x \in \partial Q$, all $M \in L^\infty(Q, S^1_0)$ which obey $|M| = \chi_\omega$ and $\int_Q M dx = \mu$, and all $P \in L^\infty(Q, \mathcal{B}_0)$ with $|P| = \chi_\omega$. As usual in nonconvex homogenization, the usage of larger unit cells may lead to spontaneous symmetry breaking and formation of microstructures on intermediate scales, as it was discussed in [7, Sect. 6.3 and Fig. 13]. For the sake of simplicity we do not address this issue here. We instead include the magnetostrictive effect by solving Maxwell's equations in the deformed configuration (an effect that is not included in the first-order model used in the homogenization).

We use a boundary-element method for solving both the linear elastic problem (with piecewise affine ansatz functions) and the magnetic problem (with piecewise constant ansatz functions). Both the particle-matrix interfaces and the boundary of the unit cell are approximated by polygons. Each particle is assumed to have a uniform magnetization and to deform affinely; the boundary conditions on $\partial(0,1)^2$ are affine-periodic, as usual in the theory of homogenization. Our numerical algorithm is based on computing the (discrete) gradient of the energy in the relevant variables, performing a line search in the descent direction, and then updating the direction as in the conjugate gradient scheme.

We choose parameters that match the experimentally known values for NiMnGa particles, as in [7, 10]. In the magnetic energy we use $\frac{M_s}{\mu_0} = 0.50\frac{\text{MPa}}{\text{T}}$, $\frac{M_s^2}{\mu_0} = 0.31\,\text{MPa}$, $K_u = 0.13\,\text{MPa}$. We assume an external magnetic field of 1 T. The elastic constants taken into account for NiMnGa are $\epsilon_0 = 0.058$, $C_{11} = 160\,\text{GPa}$, $C_{44} = 40\,\text{GPa}$, $C_{11} - C_{12} = 4\,\text{GPa}$. The elastic modulus of the polymer varies between $E = 0.03\,\text{MPa}$ and $E = 80\,\text{MPa}$, its Poisson ratio is assumed fixed at $\nu = 0.45$.

We assume here that the polymer is solidified with the MSM particles in one of the two martensitic variants. This introduces a clear preference for one of the two, and renders transformation more difficult. At the same time it introduces a natural mechanism for transforming back to the original state after the external field is removed. From the viewpoint of material production, the current setting corresponds to the assumption that the solidification of the polymer occurs below the critical temperature for the solid-solid phase transformation (which is about 70 °C). We remark that in our previous work [7] we had instead assumed that solidification occurs in the austenitic phase, rendering the two martensitic variants symmetric.

Our numerical results are illustrated in Figs. 2, 3 and 4. We systematically display energies and spontaneous strains, assuming affine macroscopic deformations, in dependence on the polymer elasticity; our aim being to understand which type of

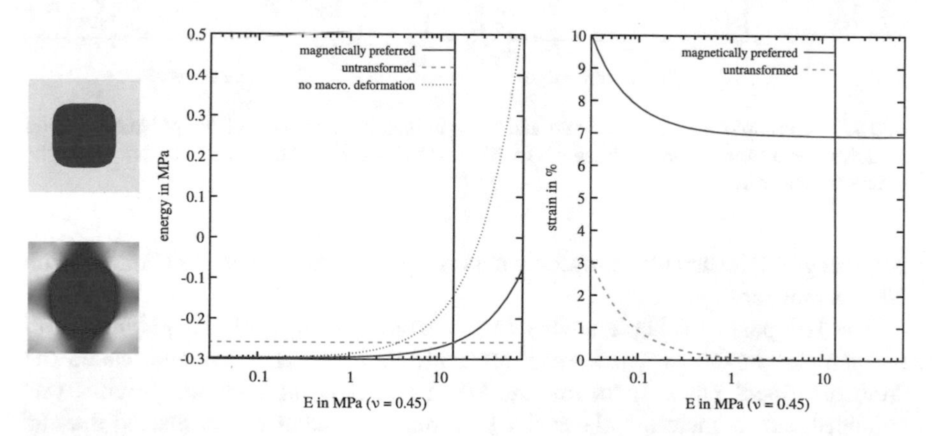

Fig. 2 Energy and macroscopic spontaneous strain as a function of the polymer's elastic modulus, for the two martensitic variants (solid line: magnetically preferred variant; dashed line: untransformed). The left plot depicts in addition the energy if no macroscopic deformation is allowed (dotted line). The sketches on the far left show the geometric configuration of the particle and the elastic energy density in the polymer (yellow: 0 kPa, red: 0.1 kPa, for an elastic modulus of 1 MPa)

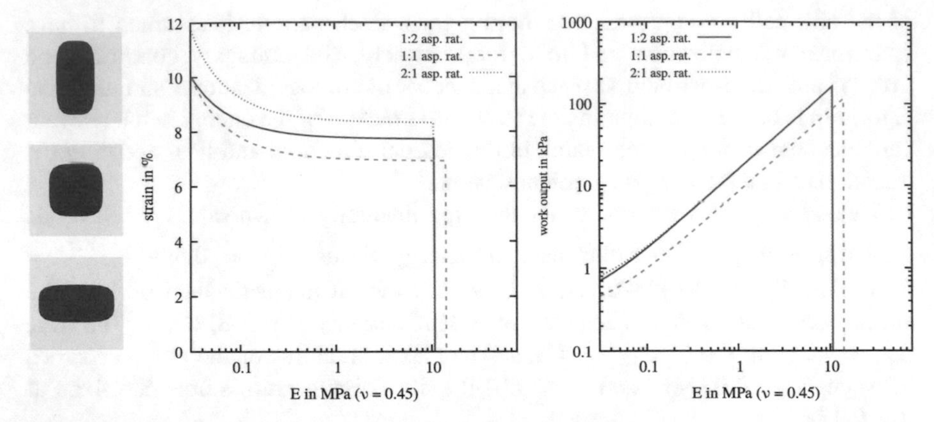

Fig. 3 Macroscopic spontaneous strain and work output, as a function of the polymer's elastic modulus, for different aspect ratios of the MSM particle. The sketches on the left show the different particle shapes under consideration. The curves for 1:2 and 2:1 aspect ratios in the left panel are almost undistinguishable. The external magnetic field is horizontal

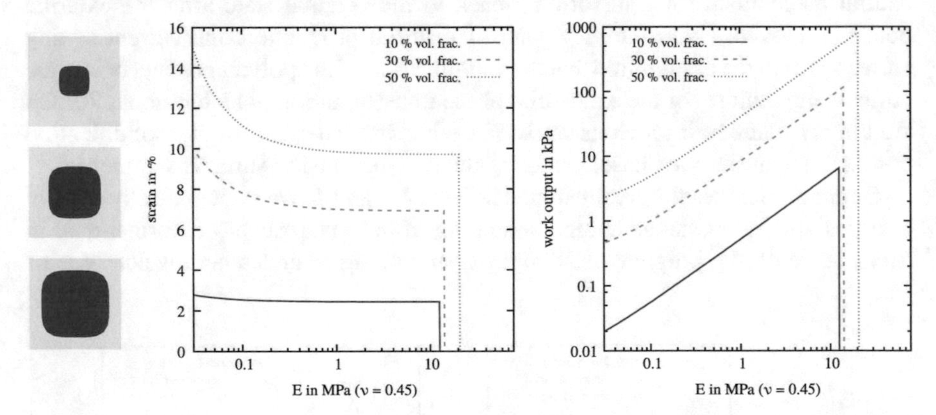

Fig. 4 Macroscopic spontaneous strain and work output, as a function of the polymer's elastic modulus, for different volume fractions of MSM material. The sketches on the left show the different volume fractions

polymer gives the better macroscopic material properties, for various choices of the other parameters.

The left panel of Fig. 2 compares the internal energy of the polymer with the particles in the two different martensitic variants. The first one, called the "untransformed" phase, is the one the MSM particles had when the polymer was solidified, and is therefore elastically preferred. The other one is instead the one which is magnetically preferred, with the easy axis aligned with the external field. Comparing the two it is clear that for the parameters considered here the phase transformation will only occur for soft polymers, with a Young modulus below around 14 MPa (black vertical line). We also plot the energy that the transformed phase would have if no macroscopic deformation would have been allowed: the difference between the two is a measure of the maximal work output that we can obtain for this material. The right panel illustrates the spontaneous strain corresponding to the two phases; we recall that the transformed one is only relevant for E below the critical value. Both phases have a significant transformation for very soft polymers, this is a magnetostrictive effect arising from the interaction between the magnetic dipoles of the particles, and does not depend on the phase transition. It is present only for extremely soft polymer.

Figure 3 shows the resulting spontaneous strain curve and the work output for three different shapes of the particles. Here, it is apparent that, while very soft polymers give the largest spontaneous strains, they are practically not very relevant, since the work output is minimal. The best polymers seem to be those with an intermediate Young modulus, between 1 and 10 MPa: on the one hand the transition is expected to occur reliably, since the difference between the energies of the two phases is significant, on the other hand the phase transition will lead to a macroscopic deformation, since the energy difference between the macroscopically undeformed and the macroscopically deformed state (the work output) is significant.

Elongated particles result in a somewhat larger macroscopic strain and work output. However, the critical elastic modulus of the polymer (above which no transformation takes place) is lower for longer particles, assuming constant volume fraction. The macroscopic deformation of a composite where the particles are aligned to the magnetic field is even larger than if the particles are aligned perpendicularly, but the work output for these two configuration is nearly identical.

Finally, Fig. 4 illustrates the dependence on the volume fraction. As intuitively expected, a larger amount of MSM material in the mixture leads to a larger spontaneous strain and larger work output. It is interesting to observe that the optimal choice for the polymer does not depend very strongly on the volume fraction. This is an important observation, since in material production high volume fractions are difficult to realize, and often turn out not to be uniform across the sample.

5 Conclusions and Outlook

In summary, we have discussed a model for composite materials, in which one component consists of magnetic-shape-memory particles, and the other of a polymer matrix. We discussed analytical homogenization, and rigorously derived a macroscopic effective model, whose energy density is obtained solving a suitable cell problem. We then addressed the cell problem numerically, and investigated how the microstructure can be optimized to obtain the composite with the best material properties.

In closing, we observe that the assumption that the particles are affinely deformed, which has a strong influence on the results presented and is only appropriate for very small particles, can be relaxed. In particular, in [9] we discussed a variant of this model in which the particles are assumed to be large with respect to the scale of the individual domains, so that three scales are present: the scale of a microstructure inside a single particle, the scale of the individual particles interacting with the polymer, and the scale on which macroscopic material properties are observed and measured. Furthermore, in [10] a time-dependent extension of the model was developed, assuming that two elastic phases are present inside each particles; the phase boundaries then move in response to the applied magnetic field. The formulation of a rate-independent model for the motion of phase boundaries permits in this case the study of hysteresis in the composite. The much richer picture that most likely will arise in the intermediate regimes, and the extension to three spatial dimensions, still remain unexplored.

Acknowledgements This work was partially supported by the Deutsche Forschungsgemeinschaft through Schwerpunktprogramm 1239 *Änderung von Mikrostruktur und Form fester Werkstoffe durch äußere Magnetfelder* and through Sonderforschungsbereich 1060 *Die Mathematik der emergenten Effekte*.

References

1. G. Allaire. *Shape optimization by the homogenization method*, volume 146 of *Applied Mathematical Sciences*. Springer, New York, 2002.
2. A. Braides and A. Defranceschi. *Homogenization of multiple integrals*. Claredon Press, Oxford, 1998.
3. K. Bhattacharya. *Microstructure of Martensite: Why it forms and how it gives rise to the Shape-Memory Effect*. Oxford University Press, Oxford, 2003.
4. J. Ball and R. D. James. Fine phase mixtures as minimizers of the energy. *Arch. Ration. Mech. Analysis*, 100:13, 1987.
5. W. Brown. *Micromagnetics*. Wiley, New York, 1963.
6. D. Cioranescu and P. Donato. *An introduction to homogenization*. Oxford University Press, Oxford, 1999.
7. S. Conti, M. Lenz, and M. Rumpf. Modeling and simulation of magnetic shape-memory polymer composites. *J. Mech. Phys. Solids*, 55:1462, 2007.
8. S. Conti, M. Lenz, and M. Rumpf. Macroscopic behaviour of magnetic shape-memory polycrystals and polymer composites. *Mat. Sci. Engrg. A*, 481–482:351, 2008.
9. S. Conti, M. Lenz, and M. Rumpf. Modeling and simulation of large microstructured particles in magnetic-shape-memory. *Adv. Eng. Mater.*, 14:582–588, 2012.
10. S. Conti, M. Lenz, and M. Rumpf. Hysteresis in magnetic shape memory composites: modeling and simulation. *J. Mech. Phys. Solids*, 89:272–286, 2016.
11. A. Desimone. Energy minimizers for large ferromagnetic bodies. *Arch. Rat. Mech. Anal.*, 125:99, 1993.
12. J. Feuchtwanger et al. Energy absorpion in Ni-Mn-Ga polymer composites. *J. Appl. Phys.*, 93:8528, 2003.
13. J. Feuchtwanger et al. Large energy absorpion in Ni-Mn-Ga polymer composites. *J. Appl. Phys.*, 97:10M319, 2005.
14. A. Hubert and R. Schäfer. *Magnetic domains*. Springer, Berlin, 1998.
15. O. Heczko, N. Scheerbaum, and O. Gutfleisch. Magnetic shape memory phenomena. In *Nanoscale magnetic materials and applications*, pp. 399–439. Springer, 2009.

16. H. Hosoda, S. Takeuchi, T. Inamura, and K. Wakashima. Material design and shape memory properties of smart composites composed of polymer and ferromagnetic shape memory alloy particles. *Sci. Technol. Adv. Mater.*, 5:503, 2004.
17. S. Kauffmann-Weiss, N. Scheerbaum, J. Liu, H. Klauss, L. Schultz, E. Mäder, R. Häßler, G. Heinrich, and O. Gutfleisch. Reversible magnetic field induced strain in Ni_2MnGa-polymer-composites. *Adv. Eng. Mater.*, 14:20–27, 2012.
18. J. Liu, N. Scheerbaum, S. Kauffmann-Weiss, and O. Gutfleisch. NiMn-based alloys and composites for magnetically controlled dampers and actuators. *Adv. Eng. Mater.*, 8:653–667, 2012.
19. J. Liu, N. Scheerbaum, S. Weiß, and O. Gutfleisch. Ni–Mn–In–Co single-crystalline particles for magnetic shape memory composites. *Applied Physics Letters*, 95:152503, 2009.
20. G. W. Milton. *The theory of composites*. Cambridge University Press, Cambridge, 2002.
21. S. J. Murray, M. Marioni, S. M. Allen, R. C. O'Handley, and T. A. Lograsso. 6% magnetic-field-induced strain by twin-boundary motion in ferromagnetic Ni-Mn-Ga. *Appl. Phys. Lett.*, 77:886, 2000.
22. M. Pawelczyk. Homogenization for magnetic shape memory materials. *Master's thesis, Universität Bonn*, 2014.
23. G. Pisante. Homogenization of micromagnetics large bodies. *ESAIM: control, optimisation and calculus of variations*, 10:295–314, 2004.
24. M. Pitteri and G. Zanzotto. *Continuum models for phase transitions and twinning in crystals*. CRC/Chapman & Hall, London, 2003.
25. N. Scheerbaum, D. Hinz, O. Gutfleisch, K.-H. Müller, and L. Schultz. Textured polymer bonded composites with NiMnGa magnetic shape memory particles. *Acta Mater.*, 55:2707, 2007.
26. A. Sozinov, A. A. Likhachev, N. Lanska, and K. Ullako. Giant magnetic-field-induced strain in NiMnGa seven-layered martensitic phase. *Appl. Phys. Lett.*, 80:1746, 2002.
27. B. Tian, F. Chen, Y. Tong, L. Li, and Y. Zheng. Bending properties of epoxy resin matrix composites filled with Ni-Mn-Ga ferromagnetic shape memory alloy powders. *Materials Letters*, 63:1729–1732, 2009.
28. B. Tian, F. Chen, Y. Tong, L. Li, and Y. Zheng. Magnetic field induced strain and damping behavior of Ni-Mn-Ga particles/epoxy resin composite. *Journal of Alloys and Compounds*, 604:137–141, 2014.
29. R. Tickle, R. James, T. Shield, M. Wuttig, and V. Kokorin. Ferromagnetic shape memory in the NiMnGa system. *IEEE Trans. Magn.*, 35:4301–4310, 1999.
30. K. Ullakko, J. K. Huang, C. Kantner, R. C. O'Handley, and V. V. Kokorin. Large magnetic-field-induced strains in Ni_2MnGa single crystals. *Appl. Phys. Lett.*, 69:1966, 1996.

Homogenization of Stochastic Parabolic Equations in Varying Domains

Mogtaba A. Y. Mohammed and Mamadou Sango

Abstract We present homogenization results for stochastic semilinear parabolic equations in varying domains which are stochastic counterparts of the some fundamental results of Khruslov and Marchenko, Skrypnik, Cioranescu, Dal Maso and Murat.

Keywords Stochastic partial differential equation · Homogenization · Perforated domains

1 Introduction

In this paper, we shall be concerned with homogenization results for stochastic parabolic equations in varying domains of various complexity. The theory known in mathematics as homogenization theory studies physical (or other natural) processes taking place in highly heterogeneous media, for instance composite materials used in the manufacture of highly sophisticated industrial products such as aircraft, satellites, or more basic structures in nature such as soils, human tissues, just to cite a few. The deterministic version of the theory goes back to the pioneering works of Khruslov and Marchenko, in the 1960s in the Soviet Union, they were the object of their celebrated monograph [22] (see also the long awaited English version [23]); we refer to [43] where the nonlinear version of their theory was

M. A. Y. Mohammed
Department of Mathematics and Applied Mathematics, University of Pretoria, Pretoria, South Africa

Department of Mathematics, Sudan University of Science and Technology, Khartoum, Sudan

M. Sango (✉)
Department of Mathematics and Applied Mathematics, University of Pretoria, Pretoria, South Africa

V. Schulz, D. Seck (eds.), *Shape Optimization, Homogenization and Optimal Control*, International Series of Numerical Mathematics 169,
https://doi.org/10.1007/978-3-319-90469-6_2

developed by Skrypnik. Subsequent developments of the theory are due to the Italian school of mathematics, notably De Giorgi who introduced the concept of Gamma-convergence and Spagnolo to whom we owe the notion of G-convergence; of historical importance is the landmark work of Tartar and Murat on the compensated compactness and H-convergence. The work of the Moscow school under the leadership of Oleinik, Zhikov and Kozlov (see [21]) was also crucial in the evolution of the field. The invention of the two scale-convergence by Nguetseng [28] and its subsequent development by Allaire [1] are milestones for homogenization of partial differential equations in the periodic setting; recent extensions of the two-scale convergence to non periodic frameworks continue to push the envelop of the classes of problems that may be studied to new frontiers, we note in that direction the works [7, 8, 29, 30, 48]. It should however be noted that from the vantage point of asymptotic expansions, the source of homogenization theory is even more remote; indeed the use of multiple scale asymptotic expansions which also leads to homogenization procedures featured prominently in the work of Raleigh and Poincare. The achievements of asymptotic expansions methods in homogenization are the object of the fundamental monograph by Bensoussan et al. [3]; the two-scale convergence is in fact the rigorous mathematical foundation of that approach.

Most of the results in homogenization theory mainly cover deterministic problems or partial differential equations with randomly oscillating coefficients. Besides the already mentioned works, several influential works cover the achievements made, for instance the monographs [9, 15, 47] and the papers [5, 21, 31, 32] and [40] on random homogenization; just to cite a few.

The study of homogenization theory for stochastic partial differential differential equations, driven by random processes (a central topic nowadays in Mathematics) is very recent and most results obtained concern mainly stochastic parabolic problems taking place in a periodic setting, almost periodic and ergodic [2, 19, 20, 34, 35, 45] and [46]. The homogenization of stochastic hyperbolic equations with periodically oscillating coefficients is even more recent and has been initiated in the works [24–26] of the authors. With the exception of the works of the first author [36, 39], almost nothing has been done in the direction of stochastic problems in perforated domains without any assumptions of periodicity; namely a stochastic counterpart of Kruslov Marchenko and Skrypnik theories and their generalized versions developed by Cioranescu, Murat and Dal Maso in [11] and in [14]. The goal of this paper is to present some results in these directions.

We consider only linear and semilinear stochastic parabolic equations governing processes such as heat conduction, diffusion and reaction-diffusion processes and taking place in highly heterogeneous media under the influence of random factors. We can also view these problems as simplified stochastic models of the motions of fluids in turbulent regime filling regions with tiny obstacles.

We end this introduction by stating some semantics needed to formulate our results. Let Ω be a region in $\mathbb{R}^n$ and T a positive real. We shall use the following well-known Lebesgue and Sobolev spaces $L_p(\Omega)$, $W_p^1(\Omega)$, $\overset{o}{W}{}_p^1(\Omega)$ $(p \geq 1)$

(we denote $H^1(\Omega) =: W_2^1(\Omega)$, $H_0^1(\Omega) =: \overset{o}{W_2^1}(\Omega)$) as defined, for example in [16]; $H^{-1}(\Omega)$ stands for the dual of $H_0^1(\Omega)$, $(.,.)$ denotes the scalar product in $L_2(\cdot)$. $C_0^\infty(\Omega)$ is the space of infinitely differentiable functions with compact support in Ω.

Let X be any Banach space of functions defined on Ω and let $(\mathcal{S}, \mathcal{F}, \mathcal{P})$ be any probability space over which the expectation is denoted by E. By the symbol $L_p\left(\mathcal{S}, L_q(0, T, X)\right)$ $(1 \leq p, q)$ we denote the space of random variables $u = u(\omega, t, x)$ defined on $\mathcal{S} \times [0, T]$ with values in X, and such that

(a) $u(\omega, t, x)$ is measurable with respect to (ω, t) and for each t is $\mathcal{F}_t$-measurable in ω.

(b) $u(\omega, t, x) \in X$, for almost all (ω, t) and

$$||u||_{L_p(\mathcal{S}, L_q(0,T,X))} = \left(E\int_0^T \left(||u(t)||_X^q\, dt\right)^{p/q}\right)^{1/p} < \infty; \tag{1}$$

If $q = \infty$, we write

$$||u||_{L_p(\mathcal{S}, L_\infty(0,T,X))} = \left(E ess \sup_{0\leq t\leq T} ||u(t)||_X^p\right)^{1/p} < \infty. \tag{2}$$

Let Z be any Banach space of functions defined on $(0, T) \times \Omega$. We denote by $L_p(\mathcal{S}, Z)$ the space of random variables u defined on S with values in Z and such that

$$||u||_{L_p(\mathcal{S}, Z)} = \left(E\, ||u||_Z^p\right)^{1/p} < \infty.$$

We also need some useful probabilistic compactness results due to Prokhorov and Skorokhod.

Let E be a separable Banach space and let $\mathcal{B}(E)$ be its Borel σ-field.

Definition 1 A family of probability measures $\mathcal{P}$ on $(E, \mathcal{B}(E))$ is tight if for any $\varepsilon > 0$, there exists a compact set $K_\varepsilon \subset E$ such that

$$\mu(K_\varepsilon) \geq 1 - \varepsilon, \text{ for all } \mu \in \mathcal{P}.$$

A sequence of measures $\{\mu_n\}$ on $(E, \mathcal{B}(E))$ is weakly convergent to a measure μ if for all continuous and bounded functions φ on E

$$\lim_{n\to\infty} \int_E \varphi(x)\, \mu_n(dx) = \int_E \varphi(x)\, \mu(dx).$$

The following result due to Prokhorov [33] shows that the tightness property is a compactness criterion.

Lemma 2 *A sequence of measures $\{\mu_n\}$ on $(E, \mathcal{B}(E))$ is tight if and only if it is relatively compact, that is, there exists a subsequence $\{\mu_{n_k}\}$ which weakly converges to a probability measure μ.*

Skorokhod proves in [42] the next result which relates the weak convergence of probability measures with that of almost everywhere convergence of random variables.

Lemma 3 *For an arbitrary sequence of probability measures $\{\mu_n\}$ on $(E, \mathcal{B}(E))$ weakly convergent to a probability measure μ, there exists a probability space $(\mathcal{S}, \mathcal{F}, P)$ and random variables $\xi, \xi_1, \ldots, \xi_n, \ldots$ with values in E such that the probability law of ξ_n,*

$$\mathcal{L}(\xi_n)(A) = P\{\omega \in \mathcal{S} : \xi_n(\omega) \in A\}, \text{ for all } A \in \mathcal{F},$$

is μ_n, the probability law of ξ is μ, and

$$\lim_{n\to\infty} \xi_n = \xi, \; P\text{-a.s.}$$

We borrowed the presentation of these lemmas from [12].

The plan of the paper is as follows. In Sect. 2, we present homogenization results for semilinear stochastic parabolic equations in domains with fine grained boundaries in the framework of Khurslov, Marchenko and Skrypnik homogenization theories. In Sect. 3, we extend the results of Sect. 2 to a stochastic heat equation in perforated domains satisfying Cioranescu-Murat conditions and finally in Sect. 4, we elaborate on the corresponding stochastic counterpart of Dal Maso-Murat's approach.

2 Homogenization of Semilinear Stochastic Parabolic Equations in a Domain with Fine Grained Boundaries

In this section we present some homogenization results for a sequence of semilinear stochastic parabolic equations in a varying domain with a non periodic structure; Skrypnik studied in [44] a more general quasilinear deterministic parabolic equation. This class of domains was introduced by Khruslov and Marchenko in [22] in the framework of homogenization of linear elliptic problems.

Let Ω be a bounded domain in the Euclidean space $\mathbf{R}^n (n \geq 3)$ with boundary $\partial\Omega$. For $0 < T < \infty$, we denote by Q_T the cylinder $(0, T) \times \Omega$. We assume that for each natural number s there is defined a finite number of disjoint closed sets $F_i^{(s)} (i = 1, 2, \ldots, I(s), I(s)$ is a natural number) lying inside Ω. Let $\Omega^{(s)}$ be the sequence of perforated domains obtained by removing the set $F^{(s)} = \cup_{i=1}^{I(s)} F_i^{(s)}$ from Ω, i.e., $\Omega^{(s)} = \Omega \backslash F^{(s)}$; the boundary $\partial\Omega^{(s)}$ of $\Omega^{(s)}$ is assumed to be sufficiently smooth (e.g., of class C^2). In the sequel we shall formulate some conditions on $F_i^{(s)}$ from which it follows in particular that as $s \to \infty$ the diameters of $F_i^{(s)}$ tend to zero. We consider the sequence of cylindrical domains $Q_T^{(s)} = (0, T) \times \Omega^{(s)}$. In $Q_T^{(s)}$, we consider the initial boundary value problem for the stochastic partial differential equation

$$du^{(s)} = \left[\Delta u^{(s)} + f\left(t, u^{(s)}\right)\right] dt + \sum_{k=1}^{d} g_k\left(t, u^{(s)}\right) dw_t^{(s)k}, \text{ in } Q_T^{(s)}, \tag{3}$$

$$u^{(s)}(t, x) = 0 \text{ on } \partial Q_T^{(s)}, \tag{4}$$

$$u^{(s)}(0, x) = u_0(x) \text{ in } \Omega^{(s)}, \tag{5}$$

where $\partial Q_T^{(s)} = (0, T) \times \partial\Omega^{(s)}$, f, g_k are real valued functions taking values in some Hilbert spaces, $u_0(x)$ is a deterministic real valued function defined on $\Omega^{(s)}$, Δ is the Laplace operator, i.e., $\Delta = \sum_{i=1}^{n} \partial^2/\partial x_i^2$, $W^{(s)}(t) = \left(w_t^{(s)1}, \ldots, w_t^{(s)d}\right)$ is a standard d-dimensional Wiener process.

Due to the presence of forcing terms which do not satisfy the Lipschitz condition (this precludes any possibility of uniqueness of the solution), the notion of solution for the problem (3)–(5) is that of probabilistic weak solution

$$\left(\mathcal{S}^{(s)}, \mathcal{F}^{(s)}, \left\{\mathcal{F}_t^{(s)}\right\}_{0\le t\le T}, \mathcal{P}^{(s)}, W^{(s)}, u^{(s)}\right),$$

which will be defined shortly.

We are concerned with the asymptotic behaviour of the sequence

$$\left(\mathcal{S}^{(s)}, \mathcal{F}^{(s)}, \left\{\mathcal{F}_t^{(s)}\right\}_{0\le t\le T}, \mathcal{P}^{(s)}, W^{(s)}, u^{(s)}\right), \quad s = 1, 2, \ldots$$

of solutions of problem (3)–(5) under some appropriate conditions on the data and on the perforated domain $\Omega^{(s)}$ that we make precise later on. We shall prove that under those conditions the sequence of solutions converge in suitable topologies to a probabilistic system $\left(\Xi, G, \{G_t\}_{0\le t\le T}, P, W, u\right)$ which is a weak solution of a homogenized stochastic evolution problem in the cylinder Q_T. Namely

$$du = [\Delta u - c(x)u + f(t, u)]\,dt + \sum_{k=1}^{d} g_k(t, u)\,dw_t^k \text{ in } Q_T$$

$$u(t, x) = 0 \text{ on } \partial Q_T,$$

$$u(0, x) = u_0(x) \text{ in } \Omega.$$

The function $c(x)$ is defined in terms of the geometry of the sets $F_i^{(s)}$ and is responsible for the additional term $c(x)\,u(t,x)$ in the above limit problem. The appearance of this term is the main feature of Khruslov-Marchenko's theory [22] mentioned earlier. It was extended to a more general framework by Cioranescu and Murat in [11] and also by Dal Maso and Murat in [14]. We shall present the corresponding results for the stochastic heat equation in forthcoming sections.

2.1 *Our Notion of Solution and Its Existence*

The notion of solution of interest to us for problem (3)–(5) is made precise in the following

Definition 4 By a weak solution of problem (3)–(5), we mean a system

$$\left(\mathcal{S}^{(s)}, \mathcal{F}^{(s)}, \left\{\mathcal{F}_t^{(s)}\right\}_{0\leq t\leq T}, \mathcal{P}^{(s)}, W^{(s)}, u^{(s)}\right)$$

such that for each s,

- $\left(\mathcal{S}^{(s)}, \mathcal{F}^{(s)}, \left\{\mathcal{F}_t^{(s)}\right\}_{0\leq t\leq T}, \mathcal{P}^{(s)}\right)$ is a probability space, $\left\{\mathcal{F}_t^{(s)}\right\}_{0\leq t\leq T}$ is a filtration on $\left(\mathcal{S}^{(s)}, \mathcal{F}^{(s)}, \mathcal{P}^{(s)}\right)$,
- $W^{(s)}(t)$ is sequence of $d-$dimensional $\mathcal{F}_t^{(s)}$ standard Wiener process,
- $u^{(s)}(t) \in L_p\left(\mathcal{S}^{(s)}, L_\infty\left(0,T,L_2\left(\Omega^{(s)}\right)\right)\right) \cap L_{p/2}\left(\mathcal{S}^{(s)}, L_2\left(0,T,H_0^1\left(\Omega^{(s)}\right)\right)\right)$, for any $p \in [2,\infty)$,
- for any $\eta \in H_0^1\left(\Omega^{(s)}\right)$ the integral identity

$$\left(u^{(s)}(t),\eta\right) = (u_0,\eta) - \int_0^t \sum_{i=1}^n \left(\frac{\partial u^{(s)}(\tau)}{\partial x_i}, \frac{\partial \eta}{\partial x_i}\right) d\tau + \int_0^t (f,\eta)\,d\tau + \sum_{k=1}^d \int_0^t (g_k,\eta)\,dw_\tau^{(s)k} \tag{6}$$

 holds for almost all ω and $t \in [0,T]$; $(\cdot,\cdot)$ denotes the inner product in L_2.

We assume the following conditions on the data :
$f : (0,T) \times L_2\left(\Omega^{(s)}\right) \to L_2\left(\Omega^{(s)}\right)$ is a nonlinear mapping

- measurable for almost every t,
- $v \to f(t,v)$ is continuous from $L_2\left(\Omega^{(s)}\right)$ into $L_2\left(\Omega^{(s)}\right)$,

$$\|f(t,v)\|_{L_2(\Omega^{(s)})} \leq C\left(1 + \|v\|_{L_2(\Omega^{(s)})}\right), \tag{7}$$

$g_k : (0, T) \times L_2\left(\Omega^{(s)}\right) \to L_2\left(\Omega^{(s)}\right)$ are nonlinear mappings

- measurable for almost every t,
- $v \to g_k(t, v)$ is continuous from $L_2\left(\Omega^{(s)}\right)$ into $L_2\left(\Omega^{(s)}\right)$,
-

$$||g_k(t, v)||_{L_2(\Omega^{(s)})} \leq C\left(1 + ||v||_{L_2(\Omega^{(s)})}\right). \tag{8}$$

-

$$u_0(x) \in L_2\left(\Omega^{(s)}\right). \tag{9}$$

We shall need appropriate functions spaces that we now introduce. Let $Z_\theta^{(s)}$ (indexed by $\theta > 0$ such that $T - \theta > 0$) be the space of functions $v(t, x)$ defined and measurable on $Q_T^{(s)}$ and such that

$$\sup_{0 \leq t \leq T} ||v||^2_{L_2(\Omega^{(s)})} \leq C_1, \int_0^T ||v(t)||^2_{H_0^1(\Omega^{(s)})} dt \leq C_2,$$

$$\int_0^{T-\theta} ||v(t+\theta) - v(t)||^2_{H^{-1}(\Omega^{(s)})} dt \leq C_3 \theta.$$

We endow $Z_\theta^{(s)}$ with the norm

$$\begin{aligned} &||v||_Z \\ &= \sup_{0 \leq t \leq T} ||v||_{L_2(\Omega^{(s)})} + \left(\int_0^T ||v(t)||^2_{H_0^1(\Omega^{(s)})} dt\right)^{1/2} \\ &+ \left(\sup_{\theta > 0} \frac{1}{\theta} \int_0^{T-\theta} ||v(t+\theta) - v(t)||^2_{H^{-1}(\Omega^{(s)})} dt\right)^{1/2}. \end{aligned}$$

We also consider the space $\mathcal{X}_{p,\theta}^{(s)}$ $(1 \leq p < \infty)$ of random variables v defined on $\mathcal{S}^{(s)} \times Q_T^{(s)}$ such that

$$E^{(s)} \sup_{0 \leq t \leq T} ||v||^{2p}_{L_2(\Omega^{(s)})} < \infty, \ E^{(s)} \left(\int_0^T ||v||^2_{H_0^1(\Omega^{(s)})} dt\right)^{p/2} < \infty,$$

$$E^{(s)} \int_0^{T-\theta} ||v(t+\theta) - v(t)||^2_{H^{-1}(\Omega^{(s)})} dt \leq C\theta,$$

$E^{(s)}$ denotes the mathematical expectation on $\left(\mathcal{S}^{(s)}, \mathcal{F}^{(s)}, \mathcal{P}^{(s)}\right)$. Endowed with the norm

$$||v||_{\mathcal{X}_{p,\theta}^{(s)}} = \left(E^{(s)} \sup_{0\le t\le T} ||v||_{L_2(\Omega^{(s)})}^{2p}\right)^{1/2p} + \left(E^{(s)} \left(\int_0^T ||v||_{H_0^1(\Omega^{(s)})}^2 dt\right)^{p/2}\right)^{2/p}$$
$$+E^{(s)} \left(\sup_{\theta>0} \frac{1}{\theta} \int_0^{T-\theta} ||v(t+\theta) - v(t)||_{H^{-1}(\Omega^{(s)})}^2 dt\right)^{1/2},$$

$\mathcal{X}_{p,\theta}^{(s)}$ is a Banach space. We denote by $\tilde{Z}_\theta$ and $\tilde{\mathcal{X}}_{p,\theta}$ the corresponding spaces with $\Omega^{(s)}$ replaced by Ω.

We have

Lemma 5 *The space $\tilde{Z}_\theta$ is compactly embedded in to $L_2(0, T, L_2(\Omega))$.*

This result is of Aubin-Lions-Dubibnsky-Simon type (see [41]) and due to Bensoussan [4].

The following existence result holds (see e.g.,[4, 27]).

Theorem 6 *Under the above conditions, for each s, the problem (3)–(5) has a weak solution*

$$\left(\mathcal{S}^{(s)}, \mathcal{F}^{(s)}, \left\{\mathcal{F}_t^{(s)}\right\}_{0\le t\le T}, \mathcal{P}^{(s)}, W^{(s)}, u^{(s)}\right)$$

such that $u^{(s)}$ satisfies

$$\left|\left|u^{(s)}\right|\right|_{\mathcal{X}_{p,\theta}^{(s)}} \le C, \tag{10}$$

with a constant C independent of s, for all $p \ge 1$ and for sufficiently small $\theta > 0$.

Note that Theorem 6 implies, among other results, that almost surely

$$u^{(s)}(\cdot) \in C_w\left(0, T, L_2\left(\Omega^{(s)}\right)\right),$$

that is $\lim_{t\to t'} \left(u^{(s)}(t), \varphi\right)_{\Omega^{(s)}} = \left(u^{(s)}(t'), \varphi\right)_{\Omega^{(s)}}$ for any $\varphi \in L_2\left(\Omega^{(s)}\right)$. This shows that the initial condition (5) is meaningful.

2.2 Probability Measures Generated by the Sequence $(W^{(s)}, u^{(s)})$ and Their Tightness

A crucial step in the homogenization of stochastic partial differential equation is the compactness of the family of probability measures generated by the sequence of weak solutions $\left(\mathcal{S}^{(s)}, \mathcal{F}^{(s)}, \left\{\mathcal{F}_t^{(s)}\right\}_{0\leq t\leq T}, \mathcal{P}^{(s)}, W^{(s)}, u^{(s)}\right)$. This constitute a major difference from the deterministic realm.

Let $\tilde{u}^{(s)}$, $\tilde{u}_0$ be the extension by zero to Q_T and to Ω of the functions $u^{(s)}$ and u_0, respectively such that $\mathcal{P}^{(s)}$-almost surely

$$\tilde{u}^{(s)}(x,t) = u^{(s)}(x,t) \ \ \forall\ (x,t) \in Q_T^{(s)}, \tilde{u}^{(s)}(x,t) = 0 \ \forall\ (x,t) \notin Q_T^{(s)}$$

$$\tilde{u}_0(x) = u_0(x) \ \text{ in } \Omega^{(s)} \text{ and } \tilde{u}_0(x) = 0 \text{ outside } \Omega^{(s)}.$$

Under the smoothness conditions on $\Omega^{(s)}$ and the fact that $u^{(s)}$ satisfies the Dirichlet boundary conditions such extensions always exist.

These extensions induce extensions $\tilde{f}$ and $\tilde{g}_k$ of f and g_k from $(0,T) \times L_2\left(\Omega^{(s)}\right) \to L_2\left(\Omega^{(s)}\right)$ to $(0,T) \times L_2(\Omega) \to L_2(\Omega)$ preserving the inequalities (7) and (8) with obvious changes of $\Omega^{(s)}$ by Ω; $\tilde{u}_0(x)$ is bounded in $L_2(\Omega)$. Therefore from (10) it follows that the function $\tilde{u}^{(s)}$ satisfies the estimate

$$\left\|\tilde{u}^{(s)}\right\|_{\tilde{\mathcal{X}}_{p,\theta}} \leq C, \tag{11}$$

with the constant C independent of s.

Now we consider the set

$$K = C\left(0, T, \mathbb{R}^d\right) \times L_2(0, T, L_2(\Omega))$$

and $\mathcal{B}(K)$ the σ-algebra of the Borel sets of K. For each s, let Φ be the map

$$\Phi : \mathcal{S} \to K \ : \omega \mapsto \left(W^{(s)}(\omega,\cdot), \tilde{u}^{(s)}(\omega,\cdot)\right),$$

For each s, we introduce the probability measure π_s on $(K, \mathcal{B}(K))$ defined by

$$\pi_s(A) = \mathcal{P}^{(s)}\left(\Phi^{-1}(A)\right) \ \text{ for all } A \in \mathcal{B}(K).$$

We have

Theorem 7 *The family of probability measures $\{\pi_s : s \in \mathbb{N}\}$ is tight.*

Thus we have by Prokhorov's compactness result Lemma 2, that there exists a subsequence $\left\{\pi_{s_j}\right\}$ and a measure π such that

$$\pi_{s_j} \to \pi$$

weakly. By Skorokhod's result Lemma 3, there exist a probability space (Ξ, G, P) and random variables $\left(W^{(s_j)}, u^{(s_j)}\right)$, (W, u) on (Ξ, G, P) with values in K such that the probability law of $\left(W^{(s_j)}, u^{(s_j)}\right)$ is π_{s_j}; hence $\left\{W^{(s_j)}\right\}$ is a sequence of d-dimensional Wiener processes. Furthermore

$$\left(W^{(s_j)}, u^{(s_j)}\right) \to (W, u) \text{ in } K,\ P\text{-a.s} \tag{12}$$

and the probability law of (W, u) is π.

Set

$$G_t = \sigma \left\{W(\tau), u(\tau)\right\}_{\tau \in [0,t]}.$$

It turns out that $W(t)$ is a G_t-standard Wiener process.

A crucial step is to ensure that the pair $(u^{(s_j)}, W^{(s_j)})$ satisfies the problem (3)–(5); we note that this a not a subsequence of $(u^{(s)}, W^{(s)})$; they only share at this stage the same law. This fact may seem obvious but in view of the nonlinearity of the intensity of the noise, the proof requires some work. Therefore

$$du^{(s_j)} = \left[\Delta u^{(s_j)} + f\left(t, u^{(s_j)}\right)\right] dt + \sum_{k=1}^{d} g_k\left(t, u^{(s_j)}\right) dw_t^{(s_j)k}, \text{ in } Q_T^{(s_j)}, \tag{13}$$

$$u^{(s_j)}(t, x) = 0 \text{ on } \partial Q_T^{(s_j)}, \tag{14}$$

$$u^{(s_j)}(0, x) = u_0(x) \text{ in } \Omega^{(s_j)}. \tag{15}$$

Next we have that the extension of $u^{(s_j)}$ to Q_T which we denote again by $u^{(s_j)}$ satisfies the estimate (11).

As a consequence, the following convergences hold: for any $p \in [1, \infty)$

$$u^{(s_j)}(\cdot) \rightharpoonup u(\cdot) \text{ weakly-star in } L_\infty\left(0, T, L_2(\Omega)\right),\ P - a.s. \tag{16}$$

$$u^{(s_j)} \rightharpoonup u \text{ weakly in } L_{p/2}\left(\Xi, L_2\left(0, T, H_0^1(\Omega)\right)\right). \tag{17}$$

Furthermore

$$u(\cdot) \in L_\infty(0, T, L_2(\Omega)) \ P - a.s., u \in L_{p/2}\left(\Xi, L_2\left(0, T, H_0^1(\Omega)\right)\right) \quad (18)$$

and almost surely

$$u(\cdot) \in C_w(0, T, L_2(\Omega)).$$

Let $p = 4$, we have that

$$u^{(s_j)} \in L_4(\Xi, L_\infty(0, T, L_2(\Omega))).$$

Combining this fact with (12) we deduce via Vitali's convergence theorem that

$$u^{(s_j)}(\cdot) \to u(\cdot) \text{ strongly in } L_2(0, T, L_2(\Omega)), \ P - a.s. \quad (19)$$

These results are crucial for the construction of the homogenized problem through appropriate passage to the limit.

2.3 Cell Problem

Our aim is to determine the initial boundary value problem satisfied by the function u and estimate its closeness to $u^{(s)}$ in suitable topologies finer than the above weak convergence. For this purpose we need some suitable assumptions on the domains $\Omega^{(s)}$ which are formulated using the solutions of suitable cell problems that we now introduce.

Let $B(x, \rho)$ denote the ball of radius ρ centered at x. Let $d_i^{(s)}$ be the radius of the smallest ball containing $F_i^{(s)}$ and $x_i^{(s)}$ the center of that ball. We denote by $r_i^{(s)}$ the distance between $\overline{B\left(x_i^{(s)}, d_i^{(s)}\right)}$ and the set $\overline{\cup_{i \neq j} B\left(x_j^{(s)}, d_j^{(s)}\right)} \cup \partial\Omega$; we shall impose some conditions (forthcoming Hypothesis 1) which guaranty that the sequence $\left(r_i^{(s)}\right)$ is positive, converges to zero not rapidly. In particular this condition prevent the ball $B\left(x_i^{(s)}, d_i^{(s)}\right)$ from crossing $\partial\Omega$; hence it lies inside Ω.

Since $B\left(x_i^{(s)}, d_i^{(s)}\right)$ lies inside Ω and Ω is open, then there exists a ball $B\left(x_i^{(s)}, a\right)$ ($a > d_i^{(s)}$ for sufficiently large s) inside Ω which contains $B\left(x_i^{(s)}, d_i^{(s)}\right)$.

We set $B_i^{(s)} = B\left(x_i^{(s)}, a\right) \setminus F_i^{(s)}$ and consider the deterministic functions $v_i^{(s)}(x) \in H^1\left(B_i^{(s)}\right)$ solutions of the boundary value problems

$$\begin{aligned} \Delta v_i^{(s)}(x) &= 0 \text{ in } B_i^{(s)}, \\ v_i^{(s)}(x) &= 1 \text{ on } \partial F_i^{(s)}, \\ v_i^{(s)}(x) &= 0 \text{ on } \partial B\left(x_i^{(s)}, a\right). \end{aligned} \tag{20}$$

We set $v_i^{(s)}(x) = 1$ on $F_i^{(s)}$ and $v_i^{(s)}(x) = 0$ outside $B\left(x_i^{(s)}, a\right)$. It is well-known that (20) is uniquely solvable. In particular $v_i^{(s)}(x)$ is the solution of the variational problem

$$\inf\left\{\int_{B\left(x_i^{(s)},a\right)} \left|\frac{\partial \varphi}{\partial x}\right|^2 dx : \varphi(x) \in H_0^1\left(B\left(x_i^{(s)}, a\right)\right), \varphi(x) = 1 \text{ in } F_i^{(s)}\right\}.$$

We note that this quantity is the capacity $Cap\left(F_i^{(s)}\right)$ of the set $F_i^{(s)}$(see e.g. Evans-Gariepy [17] for a definition), i.e.,

$$Cap\left(F_i^{(s)}\right) = \int_{B\left(x_i^{(s)},a\right)} \left|\frac{\partial v_i^{(s)}}{\partial x}\right|^2 dx.$$

Of crucial importance are the following pointwise a priori estimates.

By the maximum principle (see e.g., [18, Chap. 2]) and the extension of $v_i^{(s)}(x)$ by 1 in $F_i^{(s)}$, we get that

$$0 \le v_i^{(s)}(x) \le 1 \text{ in } \overline{B\left(x_i^{(s)}, a\right)}. \tag{21}$$

The following pointwise a priori estimates are well-known (see e.g., [22] for proof in the case of higher-order elliptic equations and [43], and [37, 38] for quasilinear elliptic equations and systems of equations):

$$\int_{B\left(x_i^{(s)},a\right)} \left|\frac{\partial v_i^{(s)}}{\partial x}\right|^2 dx \le C_1 \left[d_i^{(s)}\right]^{n-2}, \tag{22}$$

$$\left|D^\alpha v_i^{(s)}(x)\right| \le C_2 \frac{\left[d_i^{(s)}\right]^{n-2}}{\left|x - x_i^{(s)}\right|^{n-2+|\alpha|}} \text{ if } d_i^{(s)} < \left|x - x_i^{(s)}\right| < a, \tag{23}$$

for any $\alpha = (\alpha_1, \ldots, \alpha_n)$ with non negative integer components such that $|\alpha| \leq 1$, $D^\alpha = D_1^{\alpha_1} \cdots D_n^{\alpha_n}$, $D_i = \partial/\partial x_i$; the constants C_1 and C_2 are independent of s.

Combining the last estimate with the regularity results of the gradient of the solution of Dirichlet problem for elliptic equations (see [18]), we have

$$\left\| \frac{\partial v_i^{(s)}}{\partial x} \right\|_{L_\infty\left(B\left(x_i^{(s)},1\right)\right)} \leq \frac{C}{d_i^{(s)}}, \tag{24}$$

with the constant C independent on s. It is easy to check the sharpness of these estimates when $F_i^{(s)}$ is a ball.

Our conditions on the perforated domain $\Omega^{(s)}$ now follow,

We assume that there exist some positive constants A_1 and A_2 independent of s such that

H1.

$$d_i^{(s)} \leq A_1 r_i^{(s)}, \lim_{s\to\infty} \max_{1\leq i\leq I(s)} \left\{ r_i^{(s)} \right\} = 0. \tag{25}$$

H2.

$$\sum_{i=1}^{I(s)} \frac{\left[d_i^{(s)}\right]^{2(n-2)}}{\left[r_i^{(s)}\right]^n} < A_2. \tag{26}$$

H3. There exists a bounded function $c(x)$ such that for any open set $G \subset \Omega$

$$\lim_{s\to\infty} \sum_{i\in I_s(G)} Cap\left(F_i^{(s)}\right) = \int_G c(x)\,dx, \tag{27}$$

where $I_s(G)$ is the set of indices $i \in \{1, 2, \ldots, I(s)\}$ such that $F_i^{(s)} \subset G$.

2.4 Main Result and Ideas of Its Proof

We now in the position to state the main result of this section.

Theorem 8 *Let the conditions on the data (7),(8), (9) hold and assume that the hypotheses H1, H2 and H3 are satisfied. Given a sequence of weak solutions*

$$(\mathcal{S}^{(s)}, \mathcal{F}^{(s)}, \left\{\mathcal{F}_t^{(s)}\right\}_{0\leq t\leq T}, \mathcal{P}^{(s)}, W^{(s)}, u^{(s)})$$

of problem (3)–(5), there exist a probability space $\left(\Xi, \{G_t\}_{0\le t\le T}, G, P\right)$ *and stochastic processes* $\left(W^{(s_j)}, u^{(s_j)}\right)$, (W, u) $(W = \left(w_t^1, \ldots, w_t^d\right)$ *is a d-dimensional Wiener process) on this probability space such that* $\left(W^{(s_j)}, u^{(s_j)}\right)$ *has the same distribution as* $\left(W^{(s)}, u^{(s)}\right)$ *and converges to* (W, u) *in the sense of (12). Furthermore u satisfies almost surely in the sense of distributions the stochastic initial boundary value problem*

$$du = [\Delta u - c(x)u + f(t,u)]\,dt + \sum_{k=1}^{d} g_k(t,u)\,dw_t^k \text{ in } Q_T \tag{28}$$

$$u(t,x) = 0 \text{ on } \partial Q_T \tag{29}$$

$$u(0,x) = u_0(x) \text{ in } \Omega, \tag{30}$$

where $\partial Q_T = (0,T) \times \partial\Omega$. *In addition* $u^{(s_j)}$ *strongly converges to u in* $L_p\left(\Xi, L_p\left(0, T, \overset{o}{W}{}_p^1(\Omega)\right)\right)$ *for all* $p \in (1,2)$.

2.4.1 Ideas of the Proof of Theorem 8

Asymptotic Representation of $u^{(s)}$ and Corrector Result

Following Skrypnik [43], we construct appropriate test functions which will be used to establish a crucial corrector result, as well as the proof of Theorem 8.

We consider the numbers

$$\rho_i^{(s)} = \max\left\{\left(1 + \frac{1}{2A_1}\right)d_i^{(s)}, \frac{1}{2A_3}\left[r_i^{(s)}\right]^{\frac{n}{n-2}} \ln^2 r_i^{(s)}\right\}, \tag{31}$$

where A_1 is the constant from hypothesis H1 and

$$A_3 = \sup_{0<\tau\le diam\Omega}\left\{\tau^{\frac{2}{n-2}} \ln^2 \tau\right\}.$$

By the definition of $\rho_i^{(s)}$ and the hypothesis H1, we see that

$$\rho_i^{(s)} \le d_i^{(s)} + \frac{r_i^{(s)}}{2}.$$

Furthermore for $i \ne j$ we have that $B\left(x_i^{(s)}, d_i^{(s)} + \frac{r_i^{(s)}}{2}\right) \cap B\left(x_j^{(s)}, d_j^{(s)} + \frac{r_j^{(s)}}{2}\right) = \emptyset$.

Let θ_1 and θ_2 be such that $0 < \theta_2 < \theta_1 < 1$. We consider the functions $\psi_i^{(s)} \in C_o^\infty(\mathbf{R}^n)$, such that $0 \le \psi_i^{(s)}(x) \le 1$, $\psi_i^{(s)}(x) = 0$ for $\left|x - x_i^{(s)}\right| \ge \theta_1 \rho_i^{(s)}$, $\psi_i^{(s)}(x) = 1$ for $\left|x - x_i^{(s)}\right| \le \theta_2 \rho_i^{(s)}$, and

$$\left|\frac{\partial \psi_i^{(s)}}{\partial x}\right| \le \frac{C}{\rho_i^{(s)}},$$

with the constant C independent of s. We note that $\psi_i^{(s)}(x)\,\psi_j^{(s)}(x) = 0$ for $i \ne j$. Let us consider the following sets of indices

$$I_s' = \left\{ i = 1, 2, \ldots, I(s) : \left(1 + \frac{1}{2A_1}\right) d_i^{(s)} \ge \frac{1}{2A_3}\left[r_i^{(s)}\right]^{\frac{n}{n-2}} \ln^2 r_i^{(s)} \right\},$$

$$I_s'' = \left\{ i = 1, 2, \ldots, I(s) : \left(1 + \frac{1}{2A_1}\right) d_i^{(s)} < \frac{1}{2A_3}\left[r_i^{(s)}\right]^{\frac{n}{n-2}} \ln^2 r_i^{(s)} \right\}.$$

It is clear that $I_s' \cap I_s'' = \emptyset$.

We have

Lemma 9 *If the conditions H1 and H2 are satisfied, then*

$$\lim_{s\to\infty} \sum_{i\in I_s'} \left[d_i^{(s)}\right]^{n-2} = 0, \tag{32}$$

$$\lim_{s\to\infty} \sum_{i\in I_s''} \left[\rho_i^{(s)}\right]^{n} = 0. \tag{33}$$

Next we proceed to the construction of the corrector formula for $u^{(s)}$. Let $p = 4$ in (18), so that $u \in L_4(\Xi, L_\infty(0, T, L_2(\Omega))) \cap L_2\left(\Xi, L_2\left(0, T, H_0^1(\Omega)\right)\right)$. By density, there exists a sequence $(u_m)_{m=1,2,\ldots} \in L_4\left(\Xi, C_o^\infty(Q_T)\right)$ approximating u in the sense that

$$\lim_{m\to\infty} ||u_m - u||_{L_4(\Xi, L_\infty(0,T,L_2(\Omega)))} = 0,$$

$$\lim_{m\to\infty} \left|\left|\frac{\partial(u_m - u)}{\partial x}\right|\right|_{L_2(\Xi, L_2(0,T,L_2(\Omega)))} = 0.$$

For simplicity we denote from now on a solution $u^{(s_j)}$ of the problem (13)–(15) as $u^{(s)}$. We look for $u^{(s)}$ in the form

$$u^{(s)}(t, x) = u(t, x) - H_{1s}(t, x) - H_{2s}(t, x) + R_s(t, x), \tag{34}$$

where

$$H_{1s}(\omega,t,x)=\sum_{i\in I_s'} v_i^{(s)}(x)\,u(\omega,t,x)\,\psi_i^{(s)}(x),$$

$$H_{2s}(\omega,t,x)=\sum_{i\in I_s''} v_i^{(s)}(x)\,u(\omega,t,x)\,\psi_i^{(s)}(x),$$

and $R_s(\omega,t,x)$ is the remainder term.

A surprising fact is that the remainder turns out to be expressed in terms of stochastic integrals as

$$R_s(\omega,t,x)=\sum_{k=1}^{d}\int_0^t G_{sk}(\omega,\tau,x)\,dw_\tau^{(s)k}, \tag{35}$$

where $G_{sk}(\omega,t,x)$ are some appropriate G_t-measurable functions such that

$$E\int_0^t |G_{sk}(\omega,\tau,x)|^2\,d\tau<\infty,\ \text{ for } t\in[0,T].$$

We have the following corrector result.

Theorem 10 *As $s\to\infty$, under the conditions of Theorem 8 we have*

$$H_{1s}\to 0 \text{ strongly in } L_2\left(\Xi, L_2\left(0,T,H_0^1(\Omega)\right)\right), \tag{36}$$

$$H_{2s}\to 0 \text{ strongly in } L_p\left(\Xi, L_p\left(0,T,\overset{o}{W}{}_p^1(\Omega)\right)\right) \text{ with } p\in(1,2), \tag{37}$$

$$H_{2s}\to 0 \text{ weakly in } L_2\left(\Xi, L_2\left(0,T,H_0^1(\Omega)\right)\right), \tag{38}$$

$$R_s\to 0 \text{ strongly in } L_2\left(\Xi, L_2\left(0,T,H_0^1(\Omega)\right)\right). \tag{39}$$

Derivation of the Limit Problem

The construction of the limit problem (28)–(30) satisfied by the function $u(\omega,t,x)$ is carried through by substituting in the integral identity (6) the function η with the sequence of functions h_s given by

$$h_s(x)=h(x)-\sum_{i=1}^{I(s)} h(x)\,v_i^{(s)}(x)\,\psi_i^{(s)}(x),\ s=1,2,\ldots \tag{40}$$

where $h(x)$ is an arbitrary function in $C_o^{\infty}(\Omega)$, $\psi_i^{(s)}(x)$ are the test functions introduced in the previous subsection and the functions $v_i^{(s)}(x)$ are solutions of the problem (20). It is clear that $h_s(x) \in H_0^1(\Omega^{(s)})$. The convergences (16)–(19), combined with suitable asymptotic properties of the second term in the right side of (40) conclude the derivation of the homogenized problem (28)–(30).

3 Homogenization of Stochastic Heat Equation in Perforated Domains: The Cioranescu-Murat Approach

We maintain the same notations as in the previous section. Namely the set $F^{(s)} = \cup_{i=1}^{I(s)} F_i^{(s)}$ is the union of closed subsets from Ω and $\Omega^{(s)} = \Omega \backslash F^{(s)}$. Cioranescu and Murat in [11] developed an approach in which the assumptions on the perforated domain $F^{(s)}$ are formulated in a more general way using analogs of the solutions $v_i^{(s)}$ of the cell problems (20). The advantage of their approach lies in the simplification of the derivation of the homogenized problem. Now let us formulate their assumptions and deal with the homogenization of a stochastic heat equation in that framework.

We assume that there exists a sequence of functions $\left(v^{(s)}\right)_{s\in\mathbb{N}}$ and a distribution μ such that

A1 $v^{(s)} \in H^1(\Omega) \cap L_{\infty}(\Omega)$ and $v^{(s)}$ is uniformly bounded in $L_{\infty}(\Omega)$,
A2 $v^{(s)} = 0$ on $F^{(s)}$,
A3 $v^{(s)}$ converges to 1weakly in $H^1(\Omega)$ and almost everywhere in Ω,
A4 There exist some sequences of distributions $\mu^{(s)}$ and $\gamma^{(s)}$ in $H^{-1}(\Omega)$ such that

$$-\Delta v^{(s)} = \mu^{(s)} - \gamma^{(s)} \text{ in } \Omega,$$

and $\mu^{(s)}$ converges strongly to μ in $H^{-1}(\Omega)$, while $\langle \gamma^{(s)}, v^{(s)} \rangle = 0$, for any $v^{(s)} \in H_0^1(\Omega)$ such that $v^{(s)} = 0$ on $F^{(s)}$; $\langle \cdot, \cdot \rangle$ denotes the duality pairing of $H^{-1}(\Omega)$ and $H_0^1(\Omega)$. The current formulation of the conditions on the perforated domain was borrowed from [10].

It can be shown that the assumptions A1–A4 lead to the previous assumptions H1–H3 for some type of perforated domains; for instance when the sets $F_i^{(s)}$ are periodically distributed balls. This is evidenced by the example 1 in [22, p. 55] and example 2.1 [11, p. 54].

Now we consider the sequence of linear stochastic initial boundary value problems in perforated cylindrical domains $Q_T^{(s)} = (0, T) \times \Omega^{(s)}$

$$du^{(s)} = \left[\Delta u^{(s)} + f^{(s)}(t, x)\right] dt + \sum_{k=1}^{d} g_k^{(s)}(t, x)\, dw_t^k, \text{ in } Q_T^{(s)}, \tag{41}$$

$$u^{(s)}(t,x) = 0 \text{ on } \partial Q_T^{(s)}, \tag{42}$$

$$u^{(s)}(0,x) = u_0^{(s)}(x) \text{ in } \Omega^{(s)}. \tag{43}$$

The Eq. (41) is driven by a d-dimensional standard Wiener process $W(t) = \left(w_t^1, \ldots, w_t^d\right)$. The problem is considered on the filtered probability space $\left(\mathcal{S}, \mathcal{F}, \{\mathcal{F}_t\}_{0 \leq t \leq T}, \mathcal{P}\right)$, where the triple $(\mathcal{S}, \mathcal{F}, \mathcal{P})$ is a fixed probability space and $\{\mathcal{F}_t\}_{0 \leq t \leq T}$ the filtration generated by W. Our assumptions on the data are as follows.

A5 The sequences $f^{(s)}, g_k^{(s)} \in L_2\left(Q_T^{(s)}\right)$ and there exist some functions some extensions of $f^{(s)}, g_k^{(s)}$ by zero outside $Q_T^{(s)}$ denoted by the same symbols and some functions f, $g_k \in L_2(Q_T)$ such that the extended functions $f^{(s)}$ and $g_k^{(s)}$ converge weakly to f and g_k in $L_2(Q_T)$, respectively.

A6 The sequence $u_0^{(s)} \in L_2\left(\Omega^{(s)}\right)$ and has an extension by zero outside $\Omega^{(s)}$, denoted by the same symbol, such that the extended sequence converges weakly in $L_2(\Omega)$ to a function u_0.

Under the conditions A5 and A6, for each $s \in \mathbb{N}$, the problem (41)–(43) has a unique strong (probabilistic) solution $u^{(s)}$ such that

$$u^{(s)} \in L_q\left(\mathcal{S}, L_\infty\left(0, T, L_2\left(\Omega^{(s)}\right)\right)\right) \cap L_2\left(\mathcal{S}, L_2\left(0, T, H_0^1\left(\Omega^{(s)}\right)\right)\right), \; q \geq 1$$

and (41)–(43) is satisfied in the sense of distribution. Furthermore, we have that modulo an extension by zero outside $Q_T^{(s)}$, the duly extended sequence $u^{(s)}$ lives in the space $\tilde{\mathcal{X}}_{p,\theta}$ defined in the previous section and satisfies the uniform bound

$$\left\|u^{(s)}\right\|_{\tilde{\mathcal{X}}_{p,\theta}} \leq C \tag{44}$$

with the constant independent of s.

The implementation of Cioranescu-Murat approach to our homogenization problem for (41)–(43) requires the use of test functions of the type $\psi(t)\phi(x)v^{(s)}(x)$ in the weak formulation of (41)–(43); here $\psi \in C^1([0,T])$, $\phi \in C_0^\infty\left(\Omega^{(s)}\right)$ and $v^{(s)}$ are our basic test functions in the assumptions A1–A4. In the process of passage to the limit in the integral identity resulting from testing (41)–(43) with $\psi(t)\phi(x)v^{(s)}(x)$, there arises the need for the strong convergence of $u^{(s)}$ (modulo the extraction of a subsequence) in $L_2(0, T, L_2(\Omega))$ to the limit process u, pathwise. This is possible only through Prokhorov and Skorokhod compactness results used in the previous section where we dealt with the case of non Lipschitz forcings. The new feature of the homogenized problem for (41)–(43) in the present framework is the appearance of a new term involving the measure μ in assumption A4. An important fact involving that measure is that under the assumptions A1–A4,

μ is characterized by the following relation

$$\langle \mu, \varphi \rangle = \lim_{s \to \infty} \int_{\Omega} \varphi \left| \nabla v^{(s)} \right|^2 dx, \text{ for all } \varphi \in C_0^{\infty}(\Omega);$$

as a consequence μ is a positive Radon measure, lying in $H^{-1}(\Omega)$ and $\mu(\Omega) < \infty$.

We can now formulate our main result in this section.

Theorem 11 *Let the conditions A1–A6 hold. Then there exist a probability space* $\left(\Xi, \{G_t\}_{0 \le t \le T}, G, P\right)$ *and stochastic processes* $\left(W^{(s_j)}, u^{(s_j)}\right)$, *(*$W^{(s_j)} = \left(w_t^{1(s_j)}, \ldots, w_t^{d(s_j)}\right)$ *is a d-dimensional Wiener process) on this probability space such that* $\left(W^{(s_j)}, u^{(s_j)}\right)$ *has the same distribution as* $\left(W, u^{(s)}\right)$ *and converges to* (W, u) *strongly in* $C\left(0, T, \mathbb{R}^d\right) \times L_2(0, T, L_2(\Omega))$, *pathwise. Furthermore u satisfies almost surely in the sense of distributions the stochastic initial boundary value problem*

$$du = [\Delta u - \mu u + f(t, x)]\, dt + \sum_{k=1}^{d} g_k(t, x)\, dw_t^k \text{ in } Q_T \tag{45}$$

$$u(t, x) = 0 \text{ on } \partial Q_T \tag{46}$$

$$u(0, x) = u_0(x) \text{ in } \Omega. \tag{47}$$

In addition $u^{(s_j)}$ *strongly converges to u in* $L_p\left(\Xi, L_p\left(0, T, \overset{o}{W}{}_p^1(\Omega)\right)\right)$ *for all* $p \in (1, 2)$.

The Eq. (45) with the presence of the term μu seems to be new in the context to stochastic partial differential equations. The existence and uniqueness of a solution to the above problem can be established using standard techniques such as Galerkin method. The solution lives in the space $L_2(\Xi, L_\infty(0, T, L_2(\Omega))) \cap L_2\left(\Xi, L_2\left(0, T, L_{2\mu}(\Omega)\right)\right) \cap L_2\left(\Xi, L_2\left(0, T, H_0^1(\Omega)\right)\right)$, where $L_{2\mu}(\Omega)$ is the space endowed with the norm

$$||v||_{L_{2\mu}(\Omega)} = \left[\int_{\Omega} |v(x)|^2 \, d\mu(x)\right]^{1/2}.$$

4 Homogenization of Stochastic Heat Equation in Varying Domains

The most general approach to the homogenization of partial differential equations in varying domains (which include the classes of perforated domains considered in the two previous sections) was introduced in a series of papers by Dal Maso and its coworkers and inspired by some ideas from shape optimization and potential theory used in [6] and [13], respectively. The most definite exposition of that approach is in the fundamental work of Dal Maso and Murat [14]; It is based on the observation that instead of working directly with the problem (41)–(43), one may instead consider a similar problem in which (41) is replaced by the equation

$$du^{(s)} = \left[\Delta u^{(s)} - \mu^{(s)} u^{(s)} + f^{(s)}(t,x)\right] dt + \sum_{k=1}^{d} g_k^{(s)}(t,x)\, dw_t^k, \tag{48}$$

featuring a sequence of suitably constructed measures $\mu^{(s)}$. The advantage of the new problem (48), (42), (43) (called the relaxed version of (41)–(43) is that it has the same form as the expected homogenized problem (45)–(47), Besides such a consideration leads to an even further weakening of Cioranescu-Murat hypotheses. In particular the sequence of domains $\Omega^{(s)}$ needs no longer to be perforated. We omit any further elaboration on the issue.

Both Cioranescu-Murat and Dal Maso-Murat approaches to the homogenization of stochastic parabolic and hyperbolic equations will be the objects of forthcoming works of the authors.

Acknowledgements The research of the authors is supported by the National Research Foundation of South Africa under the grant CGRR 93459. The support of DFG and AIMS for the participation of Mamadou Sango to the AIMS-DFG workshop in Mbour, Senegal is gratefully appreciated.

References

1. Allaire, Grégoire Homogenization and two-scale convergence. SIAM J. Math. Anal. 23 (1992), no. 6, 1482–1518.
2. Bensoussan, A.: Homogenization of a class of stochastic partial differential equations, In Composite Media and Homogenization Theory, Birkhauser, Basel and Boston, MA, 1991, pp. 47–65.
3. Bensoussan, A.; Lions, J.L.; Papanicolaou, G.C.: Asymptotic analysis for periodic structures, North-Holland, Amsterdam, 1978.
4. Bensoussan A.: Some existence results for stochastic partial differential equations, in Stochastic Partial Differential Equations and Applications (Trento, 1990), Pitman Res. Notes Math. Ser. 268, Longman Scientific and Technical, Harlow, UK, 1992, pp. 37–53.
5. Bourgeat, A.; Mikelić, A.; Wright, S.: Stochastic two-scale convergence in the mean and applications. J. Reine Angew. Math. 456 (1994), 19–51.

6. Buttazzo, Giuseppe; Dal Maso, Gianni Shape optimization for Dirichlet problems: relaxed formulation and optimality conditions. Appl. Math. Optim. 23 (1991), no. 1, 17–49.
7. Casado Diaz, J., Gayte, I.: The two-scale convergence method applied to generalized Besicovitch spaces. Proc. R. Soc. Lond. A 458, 2925–2946 (2002).
8. Cioranescu, D.; Damlamian, A.; Griso, G. The periodic unfolding method in homogenization. In "Assyr, Banasiak, Damlamian, Sango: Multiple scales problems in biomathematics, mechanics, physics and numerics", 1–35, GAKUTO Internat. Ser. Math. Sci. Appl., 31, Gakkōtosho, Tokyo, 2009.
9. Cioranescu, Doina; Donato, Patrizia An introduction to homogenization. Oxford Lecture Series in Mathematics and its Applications, 17. The Clarendon Press, Oxford University Press, New York, 1999.
10. Cioranescu, D; Donato, P.; Murat, F.; Zuazua, E.: Homogenization and corrector for the wave equation in domains with small holes, Ann. Scuola Norm. Sup. Pisa 18 (1991), 251–293.
11. Cioranescu, Doina; Murat, François A strange term coming from nowhere [MR0652509; MR0670272]. Topics in the mathematical modelling of composite materials, 45–93, Progr. Nonlinear Differential Equations Appl., 31, Birkhäuser Boston, Boston, MA, 1997.
12. Da Prato G., Zabczyk J.: Stochastic equations in infinite dimensions. Encyclopedia of Mathematics and its Applications, 44. Cambridge University Press, Cambridge, 1992.
13. Dal Maso, Gianni; Mosco, Umberto Wiener's criterion and Γ-convergence. Appl. Math. Optim. 15 (1987), no. 1, 15–63
14. Dal Maso, G.; Murat, F.: Asymptotic behaviour and correctors for Dirichlet problems in perforated domains with homogeneous monotone operators. Ann. Scuola Norm. Sup. Pisa Cl. Sci. (4) 24 (1997), no. 2, 239–290.
15. Dal Maso, Gianni An introduction to Γ-convergence. Progress in Nonlinear Differential Equations and their Applications, 8. Birkhäuser Boston, Inc., Boston, MA, 1993. xiv+340 pp.
16. L.C. Evans, Partial Differential Equations : Graduate text in Maths, AMS, Providence, 1998.
17. L.C. Evans, R.F. Gariepy : Measure Theory and Fine Properties of Functions, CRS Press, 1992.
18. Gilbarg, David; Trudinger, Neil S.: Elliptic partial differential equations of second order. Reprint of the 1998 edition. Classics in Mathematics. Springer-Verlag, Berlin, 2001.
19. Ichihara, Naoyuki.: Homogenization for stochastic partial differential equations derived from nonlinear filterings with feedback. J. Math. Soc. Japan 57 (2005), no. 2, 593–603.
20. Ichihara, Naoyuki.: Homogenization problem for stochastic partial differential equations of Zakai type. Stoch. Stoch. Rep. 76 (2004), no. 3, 243–266.
21. Kozlov, S. M. The averaging of random operators. (Russian) Mat. Sb. (N.S.) 109(151) (1979), no. 2, 188–202, 327.
22. Marchenko, V.A.; Khruslov, E.Ya.: Boundary value problems in regions with fine grained boundaries, Naukova Dumka, Kiev, 1974.
23. Marchenko, V. A.; Khruslov, E. Ya.: Homogenization of partial differential equations. Progress in Mathematical Physics, 46. Birkhäuser, Boston, 2006.
24. Mohammed, M.; Sango, M., Homogenization of linear hyperbolic stochastic partial differential equation with rapidly oscillating coefficients: the two scale convergence method. Asymptot. Anal. 91 (2015), no. 3–4, 341–371.
25. Mohammed, M.; Sango, M.,Homogenization of Neumann problem for hyperbolic stochastic partial differential equations in perforated domains. Asymptot. Anal. 97 (2016), no. 3–4, 301–327.
26. Mohammed, M,; Homogenization of nonlinear hyperbolic stochastic equation via Tartar's method. J. Hyperbolic Differ. Equ. 14 (2017), no. 2, 323–340.
27. Nagase, N.: Remarks on nonlinear stochastic partial differential equations : an application of the splitting-up method, SIAM J. Control & Optim., 33(1995), 1716–1730.
28. Nguetseng, G.: A general convergence result for a functional related to the theory of homogenization. SIAM J. Math. Anal. 20 (1989), no. 3, 608–623.
29. Nguetseng, G., Homogenization structures and applications I, Z. Anal. Anwend 22 (2003) 73–107.

30. Nguetseng, Gabriel; Sango, Mamadou; Woukeng, Jean Louis Reiterated ergodic algebras and applications. Comm. Math. Phys. 300 (2010), no. 3, 835–876.
31. Papanicolaou, G.C.; Varadhan, S.R.S: Diffusions in regions with many small holes, Stochastic Differential Systems : Filtering and Control, Lecture Notes in Control and Information Sciences, Vol. 25, Springer Verlag, Berlin, . . . , 1980, pp190–206.
32. Papanicolaou, G.C.; Varadhan S.R.S.: Boundary value problems with rapidly oscillating random coefficients, Proceedings of a colloquium on random fields : rigorous results in Statistical Mechanics and Quantum Field Theory, Colloq. Math. Soc. J. Bolyai, North-Holland, Amsterdam, 1979.
33. Prokhorov, Yu. V.: Convergence of random processes and limit theorems in probability theory. Teor. Veroyatnost. i Primenen. 1 (1956), 177–238.
34. Razafimandimby, Paul André; Woukeng, Jean Louis Homogenization of nonlinear stochastic partial differential equations in a general ergodic environment. Stoch. Anal. Appl. 31 (2013), no. 5, 755–784.
35. Razafimandimby, Paul André; Sango, Mamadou; Woukeng, Jean Louis Homogenization of a stochastic nonlinear reaction-diffusion equation with a large reaction term: the almost periodic framework. J. Math. Anal. Appl. 394 (2012), no. 1, 186–212.
36. Sango, M.: Asymptotic behavior of a stochastic evolution problem in a varying domain. Stochastic Anal. Appl. 20 (2002), no. 6, 1331–1358.
37. Sango, M.: Homogenization of the Dirichlet problem for a system of quasilinear elliptic equations in a domain with fine-grained boundary. Ann. Inst. H. Poincaré Anal. Non Linéaire 20 (2003), no. 2, 183–212.
38. Sango, M.: Pointwise a priori estimates for solution of a system of quasilinear elliptic equation. Appl. Anal. 80 (2001), no. 3–4, 367–378.
39. Sango, M.: Homogenization of stochastic semilinear parabolic equations with non-Lipschitz forcings in domains with fine grained boundaries. Commun. Math. Sci. 12 (2014), no. 2, 345–382.
40. Sango, M.; Woukeng, J-L.: Stochastic sigma-convergence and applications, Dynamics of Partial Differential Equations, 8 (2011), 4, 261–310.
41. Simon, J.: Compact sets in the space $L^p(0, T; B)$, Annali Mat. Pura Appl., 1987, 146, IV, 65–96.
42. Skorokhod, A. V.: Limit theorems for stochastic processes. Teor. Veroyatnost. i Primenen. 1 (1956), 289–319.
43. Skrypnik I.V.: Methods for analysis of nonlinear elliptic boundary value problems, Nauka, Moscow, 1990. English translation in : Translations of Mathematical Monographs, 139, AMS, Providence, 1994.
44. Skrypnik, I. V.: Averaging of quasilinear parabolic problems in domains with fine-grained boundaries. (Russian) Differentsial'nye Uravneniya 31 (1995), no. 2, 350–363, 368; translation in Differential Equations 31 (1995), no. 2, 327–339.
45. Wang, W.; Cao, D.; Duan, J.: Effective macroscopic dynamics of stochastic partial differential equations in perforated domains. SIAM J. Math. Anal. 38 (2006/07), no. 5, 1508–1527
46. Wang, W.; Duan, J.: Homogenized dynamics of stochastic partial differential equations with dynamical boundary conditions. Comm. Math. Phys. 275 (2007), no. 1, 163–186.
47. Zhikov, V. V.; Kozlov, S.M.; Oleinik, O. A.: Homogenization of differential operators and integral functionals, Springer-Verlag, Berlin, 1994.
48. Zhikov, V.V., Krivenko, E.V.: Homogenization of singularly perturbed elliptic operators. Matem. Zametki. 33, 571–582 (1983)

Two-Scale Convergence: Obviousness of the Choice of Test Functions? Not Always

Hubert Nnang

Abstract The asymptotic behaviour of a bounded sequence of solutions for some physical problems via the two-scale convergence may not be a direct consequence. In the present example, that is, the homogenization of the weakly damped wave equation (1.1) with initial conditions (1.2), it is shown that the choice of test functions will be more complicated than for the classical homogenization problems.

Keywords Two-scale convergence · Periodic homogenization · Test functions

1991 Mathematics Subject Classification. 35B27, 35B40, 35M20

1 Introduction

The choice of classical test functions (I mean, $\phi_\varepsilon(x) = \psi_0(x) + \varepsilon\psi_1\left(x, \frac{x}{\varepsilon}\right)$) does not always work in the framework of two-scale convergence. Let us recall a technical homogenization problem, that is, the motion of a mixture of solid rigid particles and an incompressible viscous fluid, see [7], (*joint work with: G. Nguetseng, and (the late) N. Svanstedt* [6]), the following weakly damped wave equation, where it is very fastidious in building such test functions:

$$\int_\Omega \rho^\varepsilon \frac{\partial^2 u_\varepsilon}{\partial t^2}(t)\cdot v dx + \varepsilon^2 b^\varepsilon\left(\frac{\partial u_\varepsilon}{\partial t}(t), v\right) + a^\varepsilon(u_\varepsilon(t), v) = \int_\Omega f(t)\cdot v dx$$
$$\text{for all } v = (v^1, v^2, v^3) \in V_\varepsilon \ (0 < t < T), \tag{1.1}$$

H. Nnang (✉)
University of Yaounde I, Higher Teachers' Training College, Yaounde, Cameroon
e-mail: hnnang@uy1.uninet.cm

V. Schulz, D. Seck (eds.), *Shape Optimization, Homogenization and Optimal Control*, International Series of Numerical Mathematics 169,
https://doi.org/10.1007/978-3-319-90469-6_3

with initial conditions

$$u_\varepsilon(0) = \frac{\partial u_\varepsilon}{\partial t}(0) = \omega \text{ (the origin in } \mathbb{R}^3) \tag{1.2}$$

where Ω is a smooth bounded open set in the numerical space $\mathbb{R}^3_x$ of variables $x = (x_1, x_2, x_3)$, T is a fixed positive real, the dot denotes the usual Euclidean inner product in $\mathbb{R}^3$, $f = (f^1, f^2, f^3)$ is a given source in $L^2(0, T; \mathbb{L}^2(\Omega; \mathbb{R}))$ (with $\mathbb{L}^2(\Omega; \mathbb{R}) = L^2(\Omega; \mathbb{R})^3$), dx denotes the Lebesgue measure on $\mathbb{R}^3_x$,

$$V_\varepsilon = \bigcap_{i,j=1}^{3} V_\varepsilon^{ij} \text{ with } V_\varepsilon^{ij} = \{v \in \mathbb{H}^1_0(\Omega; \mathbb{R}) : E_{ij}(v) = 0 \text{ in } \Omega_s^\varepsilon\}, \tag{1.3}$$

and $E_{ij}(v) = \frac{1}{2}\left(\frac{\partial v^i}{\partial x_j} + \frac{\partial v^j}{\partial x_i}\right)$, the function ρ^ε and the set Ω_s^ε are defined in what follows, and a^ε and b^ε are two continuous bilinear forms on $\mathbb{H}^1(\Omega; \mathbb{R}) \times \mathbb{H}^1(\Omega; \mathbb{R})$ defined by

$$b^\varepsilon(u, v) = \int_{\Omega_f^\varepsilon} b_{ijkh} \frac{\partial u^k}{\partial x_h} \frac{\partial v^i}{\partial x_j} dx \text{ and } a^\varepsilon(u, v) = \vartheta \int_{\Omega_f^\varepsilon} \operatorname{div} u \operatorname{div} v dx,$$

where $\operatorname{div} u = \frac{\partial u^1}{\partial x_1} + \frac{\partial u^2}{\partial x_2} + \frac{\partial u^3}{\partial x_3}$ for $u = (u^i)$, and where $\alpha, \lambda, \mu, \rho_s, \rho_f, \vartheta, c_0, b_{ijkh}$ are real coefficients with

$$\begin{aligned} &b_{ijkh} = \lambda\delta_{ij}\delta_{kh} + \mu(\delta_{ik}\delta_{jh} + \delta_{ih}\delta_{jk}) \text{ for } i, j, k, h = 1, 2, 3, \\ &0 < \alpha < 1, \mu > 0, \rho_s > 0, \rho_f > 0, \tfrac{\lambda}{\mu} \geq \left(-\tfrac{2}{3}\right)\alpha, \vartheta = c_0^2\rho_f \end{aligned}$$

(δ_{ij} being the Kronecker symbol), so that symmetry and ellipticity conditions below are verifying:

$$\begin{aligned} &b_{ijkh} = b_{jikh} = b_{ijhk} = b_{khij}, \\ &b_{ijkh}\xi_{kh}\xi_{ij} \geq \delta\xi_{ij}\xi_{ij} \text{ (with } \delta = 2\mu(1-\alpha)) \text{ for all symmetric} \\ &\text{matrices of real numbers } (\xi_{ij}). \end{aligned}$$

For the definition of the function ρ^ε and the description of domains Ω_s^ε and Ω_f^ε, let Y_s be a smooth open set in $\mathbb{R}^3_y$ such that

$$\overline{Y}_s \subset Y = (0, 1)^3 \;\; (\overline{Y}_s \text{ the closure of } Y_s),$$

and let Y_f be the open set $Y \backslash \overline{Y}_s$. Let $\rho \in L^\infty(\mathbb{R}^3_y)$ be given by

$$\rho(y) = \rho_s \text{ in } \widetilde{Y}_s \text{ and } \rho_f \text{ in } \widetilde{Y}_f \tag{1.4}$$

where, denoting by $\mathbb{Z}$ the set of integers, we put

$$\widetilde{Y}_s = \bigcup_{k\in\mathbb{Z}^3} (k + Y_s) \text{ and } \widetilde{Y}_f = \mathbb{R}^3_y \backslash \overline{\widetilde{Y}}_s \ (\overline{\widetilde{Y}}_s \text{ the closure of } \widetilde{Y}_s \text{ in } \mathbb{R}^3_y).$$

For $\varepsilon > 0$, ρ^ε is the $L^\infty(\mathbb{R}^3_x)$-function defined by

$$\rho^\varepsilon(x) = \rho\left(\frac{x}{\varepsilon}\right) \quad (x \in \mathbb{R}^3).$$

We endow $\mathbb{H}^1(\Omega; \mathbb{R}) = H^1(\Omega; \mathbb{R})^3$ with the norm

$$|||v||| = \left(\int_\Omega v^i v^i dx + \int_\Omega E_{ij}(v)E_{ij}(v)dx\right)^{\frac{1}{2}} \quad (v = (v^i) \in \mathbb{H}^1(\Omega; \mathbb{R})).$$

Thanks to Korn's inequality (see [7]), the norm $|||\cdot|||$ is equivalent to the usual $\mathbb{H}^1(\Omega)$-norm; and the seminorm

$$|||v|||_0 = \left(\int_\Omega E_{ij}(v)E_{ij}(v)dx\right)^{\frac{1}{2}} \quad (v \in \mathbb{H}^1(\Omega; \mathbb{R}))$$

is a norm in $\mathbb{H}^1_0(\Omega; \mathbb{R})$ equivalent to the $\mathbb{H}^1(\Omega)$-norm. We set

$$\Omega^\varepsilon_s = \Omega \cap \varepsilon\widetilde{Y}_s \text{ and } \Omega^\varepsilon_f = \Omega \cap \varepsilon\widetilde{Y}_f,$$

which defines two open subsets of Ω; and the space V_ε in (1.3) is then clearly a closed vector subspace of $\mathbb{H}^1_0(\Omega; \mathbb{R})$, and the seminorm

$$|||v|||_\varepsilon = \left(\int_{\Omega^\varepsilon_f} E_{ij}(v)E_{ij}(v)dx\right)^{\frac{1}{2}} \quad (v \in \mathbb{H}^1_0(\Omega; \mathbb{R}))$$

is a norm on V_ε equivalent to $|||\cdot|||_0$.

In the framework of homogenization via the two-scale convergence method, after proving the existence result, the following consists of passing to the limit (as $\varepsilon \to 0$) using test functions of the form $\phi_\varepsilon(x) = \psi_0(x) + \varepsilon\psi_1\left(x, \frac{x}{\varepsilon}\right)$, where $\psi_0 \in \mathcal{D}(\Omega)$ and $\psi_1 \in \mathcal{D}(\Omega) \otimes \mathcal{C}^1_{per}(Y)$. But in the present context there is no obvious way (in our knowledge) in building such test functions belonging to V_ε.

1. In fact, preliminary results (Lemma 3.2 below) suggest to take test functions of the form $\boldsymbol{\phi}_\varepsilon(x) = \boldsymbol{\psi}_0(x) + \varepsilon\boldsymbol{\psi}_1\left(x, \frac{x}{\varepsilon}\right)$, $\boldsymbol{\psi}_0 = \left(\psi^i_0\right) \in [\mathcal{D}(\Omega)]^3$ and $\boldsymbol{\psi}_1 = \left(\psi^i_1\right) \in \mathcal{D}(\Omega) \otimes \left[\mathcal{C}^1_{per}(Y)\right]^3$ with $e_{ij}(\boldsymbol{\psi}_1) = \frac{1}{2}\left(\frac{\partial\psi^i_1}{\partial y_j} + \frac{\partial\psi^j_1}{\partial y_i}\right) = 0$ in Y_s for all

$x \in \Omega$. Unfortunately with this choice, one has

$$E_{ij}\left(\boldsymbol{\phi}_{\varepsilon}(x)\right) = E_{ij}\left(\boldsymbol{\psi}_0\right)(x) + \varepsilon E_{ij}\left(\boldsymbol{\psi}_1\right)\left(x, \frac{x}{\varepsilon}\right) \text{ in } \Omega_s^{\varepsilon};$$

which is different to zero, that is, $E_{ij}\left(\boldsymbol{\phi}_{\varepsilon}\right) \neq 0$ in Ω_s^{ε}, so that $\boldsymbol{\psi}_{\varepsilon}$ does not belong to V_{ε}.

2. Besides, we can also take $E_{ij}\left(\boldsymbol{\psi}_0\right) = E_{ij}\left(\boldsymbol{\psi}_1\right) = 0$ in Ω_s^{ε} $(i, j = 1, 2, 3)$, but this choice is not correct according to the instability of the domain Ω_s^{ε} due to the variation of the parameter ε.

We present here what we builded in the co-authored paper [6] published in 2010; we are convinced that this way can be used or can be adapted in some appropriate research works.

By some methods (see, e.g., [2]), for each given $\varepsilon > 0$, there exists one and only one function u_{ε} from $[0, T]$ into V_{ε} defined by (1.1)–(1.2) and satisfying $u_{\varepsilon} \in \mathcal{C}([0, T]; V_{\varepsilon})$, $\frac{\partial u_{\varepsilon}}{\partial t} \in \mathcal{C}([0, T]; \mathbb{L}^2(\Omega; \mathbb{R})) \cap L^2(0, T; V_{\varepsilon})$, $\frac{\partial^2 u_{\varepsilon}}{\partial t^2} \in L^2(0, T; \mathbb{L}^2(\Omega; \mathbb{R}))$. The highlights of the periodic homogenization of (1.1)–(1.2) are to point out some difficulties we get in [6] when proving the convergence of the corresponding homogenization process.

The work is organized as follows. In Sect. 2, we recall briefly some fundamentals of homogenization. Section 3 is devoted to the building of appropriate test functions. In Sect. 4 we study the homogenization problem for (1.1)–(1.2) in the periodic setting when $f \in \mathcal{C}^1([0, T]; \mathbb{L}^2(\Omega; \mathbb{R}))$ with $f(0) = \omega$ (the origin in $\mathbb{R}^3$).

2 Fundamentals of Homogenization

Using the two-scale convergence method of Nguetseng [4] in studying the homogenization of (1.1)–(1.2), we need to recall the notion of weak two-scale convergence in $L^2(Q)$; that is, given $(v_{\varepsilon})_{\varepsilon>0} \subset L^2(Q)$ and $v_0 \in L^2(Q; L^2_{per}(Z))$, where

$$Q = \Omega \times (0, T), \; Z = Y \times \Theta \; \text{ with } \Theta = (0, 1),$$

we say that $v_{\varepsilon} \to v_0$ in $L^2(Q)$-weak two-scale if, as $\varepsilon \to 0$,

$$\int_Q v_{\varepsilon} \Phi^{\varepsilon} dx dt \to \iint_{Q \times Z} v_0 \Phi dx dt dy d\tau \tag{2.1}$$

for all $\Phi \in L^2(Q; \mathcal{C}_{per}(Z))$, where $\Phi^{\varepsilon}(x, t) = \Phi(x, t, \frac{x}{\varepsilon}, \frac{t}{\varepsilon})$ in the sense of [5, Proposition 4.1], $(x, t) \in Q$, $\varepsilon > 0$. In particular (2.1) makes sense for $\Phi \in L^2(Q; \mathcal{C}_{per}(Y)) \otimes \mathcal{C}_{per}(\Theta)$, and for $\Phi \in \mathcal{C}(\overline{Q}; L^{\infty}_{per}(Z))$ [5, Proposition 4.5].

Let us also recall the following fundamental results used in [6]:

Lemma 2.1 ([3]) *Let E be a fundamental sequence, i.e., E is an ordinary sequence of positive real numbers tending to 0; and suppose the sequence* $(v_\varepsilon)_{\varepsilon\in E} \subset L^2(Q)$ *is bounded in* $L^2(Q)$. *Then, a subsequence* E' *can be extracted from E such that the sequence* $(v_\varepsilon)_{\varepsilon\in E'}$ *is weakly two-scale convergent in* $L^2(Q)$.

Lemma 2.2 ([1]) *Let* $(v_\varepsilon)_{\varepsilon\in E}$ *be a bounded sequence in* $L^2(Q)$. *Suppose that* $(\varepsilon D v_\varepsilon)_{\varepsilon\in E}$ *is bounded in* $\mathbb{L}^2(Q)$ *(where* $Dv_\varepsilon = \left(\frac{\partial v_\varepsilon}{\partial x_1}, \frac{\partial v_\varepsilon}{\partial x_2}, \frac{\partial v_\varepsilon}{\partial x_3}\right)$*). Then, there exist a subsequence* $E' \subset E$ *and a function* $v_0 \in L^2\left(Q; L^2_{per}(Z)\right)$ *with* $\frac{\partial v_0}{\partial y_i} \in L^2\left(Q; L^2_{per}(Z)\right)$ *such that, as* $E' \ni \varepsilon \to 0$,

$$\begin{aligned}&\int_Q v_\varepsilon \Phi^\varepsilon dxdt \to \iint_{Q\times Z} v_0 \Phi dxdtdyd\tau, \text{ and}\\ &\int_Q \varepsilon \frac{\partial v_\varepsilon}{\partial x_i}\Phi^\varepsilon dxdt \to \iint_{Q\times Z} \frac{\partial v_0}{\partial y_i}\Phi dxdtdyd\tau,\ i=1,2,3\end{aligned}$$

for all $\Phi \in L^2\left(Q; \mathcal{C}_{per}(Z)\right)$.

Lemma 2.3 ([6, Lemma 2.3]) *Let* $(v_\varepsilon)_{\varepsilon\in E}$ *be a bounded sequence in* $L^2(Q)$. *Suppose that* $(\frac{\partial v_\varepsilon}{\partial t})_{\varepsilon\in E}$ *is bounded in* $L^2(Q)$. *Then, there exist a subsequence* $E' \subset E$ *and two functions* $v_0 \in H^1(0,T; L^2(\Omega; L^2_{per}(Y)))$ *and* $v_1 \in L^2\left(Q; L^2_{per}\left(Y; H^1_\#(\Theta)\right)\right)$ *with* $H^1_\#(\Theta) = \{u \in H^1_{per}(\Theta) : \int_\Theta u d\tau = 0\}$, *such that as* $E' \ni \varepsilon \to 0$,

$$\begin{aligned}&\int_Q v_\varepsilon \Phi^\varepsilon dxdt \to \iint_{Q\times Z} v_0 \Phi dxdtdyd\tau, \text{ and}\\ &\int_Q \frac{\partial v_\varepsilon}{\partial t}\Phi^\varepsilon dxdt \to \iint_{Q\times Z} \left(\frac{\partial v_0}{\partial t} + \frac{\partial v_1}{\partial \tau}\right)\Phi dxdtdyd\tau\end{aligned}$$

for all $\Phi \in L^2\left(Q; \mathcal{C}_{per}(Z)\right)$.

3 An Appropriate Variational Formulation and Test Functions

3.1 Preliminaries

By using a Gronwall lemma, we get the following estimates.

Lemma 3.1 ([6]) *Let $(u_\varepsilon)_{\varepsilon>0}$ be the sequence of solutions of (1.1)–(1.2). Then,*

$\sup_{\varepsilon>0}\left(\|u'_\varepsilon\|_{L^\infty(0,T;\mathbb{L}^2(\Omega))}+\|\varepsilon u'_\varepsilon\|_{L^2(0,T;\mathbb{H}^1_0(\Omega))}+\|divu_\varepsilon\|_{L^\infty(0,T;L^2(\Omega))}\right)<+\infty$
and $\sup_{\varepsilon>0}\left(\|u_\varepsilon\|_{L^\infty(0,T;\mathbb{L}^2(\Omega))}+\|\varepsilon u_\varepsilon\|_{L^\infty(0,T;\mathbb{H}^1_0(\Omega))}\right)<+\infty.$

Furthermore, since $f\in\mathcal{C}^1([0,T];\mathbb{L}^2(\Omega))$ with $f(0)=\omega$, then

$$\sup_{\varepsilon>0}\left(\|u''_\varepsilon\|_{L^\infty(0,T;\mathbb{L}^2(\Omega))}+\|\varepsilon u''_\varepsilon\|_{L^2(0,T;\mathbb{H}^1_0(\Omega))}+\|divu'_\varepsilon\|_{L^\infty(0,T;L^2(\Omega))}\right)<+\infty.$$

The followings have been also proved in [6].

Lemma 3.2 *There exist a sequence $E'\subset E$ and two functions $p_0\in L^2(Q;L^2_{per}(Y))$ with $\frac{\partial p_0}{\partial t}\in L^2(Q;L^2_{per}(Y))$, and $u_0=(u_0^1,u_0^2,u_0^3)\in L^2(Q;\mathbb{H}^1_{per}(Y))$ with $\frac{\partial u_0}{\partial t}\in L^2(Q;\mathbb{H}^1_{per}(Y))$, $\frac{\partial^2 u_0}{\partial t^2}\in L^2(Q;\mathbb{L}^2_{per}(Y))$, such that*

$$e_{ij}(u_0)=e_{ij}\left(\frac{\partial u_0}{\partial t}\right)=p_0=0 \text{ in } Q\times Y_s, \tag{3.1}$$

$$\text{div}_y u_0=\text{div}_y\left(\frac{\partial u_0}{\partial t}\right)=0 \text{ in } Q\times Y_f, \tag{3.2}$$

$$\int_{Y_f}p_0dy+\vartheta\text{div}\left(\int_Y u_0dy\right)=0 \text{ in } Q.$$

Furthermore, as $E'\ni\varepsilon\to 0$, we have

$$u^i_\varepsilon\to u^i_0 \text{ and } p_\varepsilon\to p_0 \text{ in } L^2(Q)\text{-weak two-scale}$$

where $p_\varepsilon=-\vartheta\text{div}u_\varepsilon$, $e_{ij}(u_0)=\frac{1}{2}\left(\frac{\partial u_0^i}{\partial y_j}+\frac{\partial u_0^j}{\partial y_i}\right)$, $i,j=1,2,3$, and $\text{div}_y u_0=e_{ii}(u_0)=\frac{\partial u_0^1}{\partial y_1}+\frac{\partial u_0^2}{\partial y_2}+\frac{\partial u_0^3}{\partial y_3}$.

Lemma 3.3 $u_0,\frac{\partial u_0}{\partial t}\in\mathcal{C}([0,T];L^2(\Omega;\mathbb{L}^2_{per}(Y)))$ *with* $u_0(0)=\frac{\partial u_0}{\partial t}(0)=\omega$.

Lemma 3.4 *By letting*

$$p(x,t)=-\frac{\vartheta}{|Y_f|}\text{div}\left(\int_Y u_0(x,t,y)dy\right)\quad ((x,t)\in Q), \tag{3.3}$$

where $|Y_f|$ is the Lebesgue measure of Y_f, for almost all $(x,t)\in Q$ we have $p_0(x,t,\cdot)=0$ in Y_s and $p(x,t)$ in Y_f, with $\frac{\partial p}{\partial x_i}\in L^2(0,T;L^2(\Omega))$, $i=1,2,3$.

3.2 A Suitable Variational Equation

The following general result is the turning point of this work.

Lemma 3.5 *Let Ω_0 and Ω_1 be two smooth open bounded sets in $\mathbb{R}^N$ ($N \geq 1$) with the property*

$$\overline{\Omega}_0 \subset \Omega_1 \ (\overline{\Omega}_0 \text{ being the closure of } \Omega_0 \text{ in } \mathbb{R}^N), \tag{3.4}$$

and let $\gamma_\circ \in \mathcal{L}(H^1(\Omega_0); H^{\frac{1}{2}}(\partial\Omega_0))$, the usual trace operator on Ω_0. There exist two continuous linear mappings $\Gamma_\circ \in \mathcal{L}(H_0^1(\Omega_1); H^{\frac{1}{2}}(\partial\Omega_0))$ and $R_\circ \in \mathcal{L}(H^{\frac{1}{2}}(\partial\Omega_0); H_0^1(\Omega_1))$ ($H_0^1(\Omega_1)$ endowed with the $H^1(\Omega_1)$-norm) such that

$$(\Gamma_\circ \circ R_\circ)g = g \text{ for all } g \in H^{\frac{1}{2}}(\partial\Omega_0), \text{ and } (R_\circ \circ \Gamma_\circ)v = v \text{ for all } v \in (\ker \Gamma_\circ)^\perp,$$

where $(\ker \Gamma_\circ)^\perp$ is the orthogonal in $H_0^1(\Omega_1)$ of the kernel of $\Gamma_\circ$ and $\circ$ denotes usual composition.

Proof See [6]. □

Lemma 3.5 immediately leads us to the following

Lemma 3.6 *Let Ω_0 and Ω_1 be two smooth open bounded sets in $\mathbb{R}^N$ ($N \geq 1$) with the property (3.4), let $\Phi_\circ$ be a linear form on $H_0^1(\Omega_1)$, and let V be a fixed vector subspace of $H_0^1(\Omega_1)$. If $\Phi_\circ = 0$ in $V \cap \ker \Gamma_\circ$, then*

$$\Phi_\circ = \Phi_\circ \circ R_\circ \circ \Gamma_\circ \text{ in } V.$$

Proof See [6]. □

Applying Lemma 3.6, we get a suitable variational problem linked to (1.1)–(1.2). Precisely we have:

Proposition 3.1 *There is $0 < \varepsilon_0 \leq 1$ such that for $0 < \varepsilon \leq \varepsilon_0$, there exists $z_\varepsilon \in L^\infty(0, T; \mathbb{H}^1(\Omega_s^\varepsilon; \mathbb{R}))$ verifying*

$$\begin{aligned} &\int_Q \rho^\varepsilon \frac{\partial^2 u_\varepsilon}{\partial t^2} \cdot v dx dt + \varepsilon^2 \int_Q b_{ijkh} E_{kh}\left(\frac{\partial u_\varepsilon}{\partial t}\right) E_{ij}(v) dx dt \\ &+\vartheta \int_Q \mathrm{div} u_\varepsilon \mathrm{div} v dx dt = \int_{Q_s^\varepsilon} E_{ij}(z_\varepsilon) E_{ij}(v) dx dt + \int_Q f \cdot v dx dt \end{aligned} \tag{3.5}$$

for all $v \in L^2(0, T; \mathbb{H}_0^1(\Omega; \mathbb{R}))$, u_ε being the solution of (1.1)–(1.2). Furthermore,

$$\sup_{0<\varepsilon\leq\varepsilon_0} \left(\| E_{ij}(z_\varepsilon) \|_{L^2(0,T:L^2(\Omega_s^\varepsilon))} \right) < \infty \ (1 \leq i, j \leq 3). \tag{3.6}$$

Proof See [6]. □

3.3 Building of Appropriate Test Functions

We end this section by proving a crucial lemma for choice of test functions $\Psi_0\left(x, \frac{x}{\varepsilon}\right) + \varepsilon\Psi_1\left(x, \frac{x}{\varepsilon}\right)$ such that $\Psi_0(x, \cdot)$ satisfies at least (3.1) and (3.2) of Lemma 3.2. We denote by :

R^s_{per} the space of $w \in \mathbb{H}^1_{per}(Y_s;\mathbb{R})$, $e_{ij}(w) = 0$, $i, j = 1, 2, 3$,

$\mathbb{H}^1_{per}(Y_s;\mathbb{R})/R^s_{per}$ the space of $u \in \mathbb{H}^1_{per}(Y_s;\mathbb{R})$, $\int_{Y_s} u \cdot w dy = 0$ for all $w \in R^s_{per}$.

We have $\mathbb{H}^1_{per}(Y_s;\mathbb{R}) = R^s_{per} \oplus \left(\mathbb{H}^1_{per}(Y_s;\mathbb{R})/R^s_{per}\right)$, and the seminorm

$$u \to e_s(u) = \left(\int_{Y_s} e_{ij}(u)e_{ij}(u)dy\right)^{\frac{1}{2}}$$

is a norm on $\mathbb{H}^1_{per}(Y_s;\mathbb{R})/R^s_{per}$ equivalent to the usual $\mathbb{H}^1(Y_s)$-norm. We endow $L^2(Q;\mathbb{H}^1_{per}(Y_s;\mathbb{R})/R^s_{per})$ with the hilbertian norm

$$\|u\|_{Q,Y_s} = \left(\int_Q [e_s(u(x))]^2\, dx\right)^{\frac{1}{2}} \quad (u \in L^2(Q;\mathbb{H}^1_{per}(Y_s;\mathbb{R})/R^s_{per})).$$

Lemma 3.7 *Given* $\Psi_0 \in H^1\left(Q;\mathbb{H}^1_{per}(Y;\mathbb{R})\right)$ *with* $e_{ij}(\Psi_0) = 0$ *in* $Q \times Y_s$ *for all* $i, j = 1, 2, 3$, *there exists* $\Psi_1 \in L^2\left(Q;\mathbb{H}^1_{per}(Y_s;\mathbb{R})\right)$ *such that*

$$\iint_{Q\times Y_s} e_{ij}(\Psi_1)\, e_{ij}(v)\, dxdtdy = -\iint_{Q\times Y_s} E_{ij}(\Psi_0)\, e_{ij}(v)\, dxdtdy \tag{3.7}$$

for all $v \in L^2\left(Q;\mathbb{H}^1_{per}(Y_s;\mathbb{R})\right)$.

Proof There exists exactly one function $u_1 \in L^2\left(Q;\mathbb{H}^1(Y_s;\mathbb{R})/R^s\right)$ such that

$$\iint_{Q\times Y_s} e_{ij}(u_1)\, e_{ij}(v)\, dxdtdy = -\iint_{Q\times Y_s} E_{ij}(\Psi_0)\, e_{ij}(v)\, dxdtdy$$

for all $v \in L^2\left(Q;\mathbb{H}^1(Y_s;\mathbb{R})/R^s\right)$, where $R^s = \bigcap_{i,j=1}^{3} R^s_{ij}$ with $R^s_{ij} = \{w \in \mathbb{H}^1(Y_s;\mathbb{R}) : e_{ij}(w) = 0\}$. Extending by periodicity, we obtain a unique function, still denoted u_1, in $L^2\left(Q;\mathbb{H}^1_{per}(Y_s;\mathbb{R})/R^s_{per}\right)$ such that

$$\iint_{Q\times Y_s} e_{ij}(u_1)\, e_{ij}(v)\, dxdtdy = -\iint_{Q\times Y_s} E_{ij}(\Psi_0)\, e_{ij}(v)\, dxdtdy$$

for all $v \in L^2\left(Q; \mathbb{H}^1_{per}(Y_s; \mathbb{R}) / R^s_{per}\right)$. Hence, by extension, there exists a function $\Psi_1 \in L^2\left(Q; \mathbb{H}^1_{per}(Y_s; \mathbb{R})\right)$ such that

$$\iint_{Q\times Y_s} e_{ij}(\Psi_1)\, e_{ij}(v)\, dxdtdy = -\iint_{Q\times Y_s} E_{ij}(\Psi_0)\, e_{ij}(v)\, dxdtdy$$

for all $v \in L^2\left(Q; \mathbb{H}^1_{per}(Y_s; \mathbb{R})\right)$. □

Since $\mathcal{D}(0,T) \otimes \mathcal{D}(\Omega) \otimes \mathcal{C}^1_{per}(Y_s)^3$ is dense in $L^2\left(Q; \mathbb{H}^1_{per}(Y_s)\right)$, we will choose in (3.5) test functions of the form $v = \Phi_\varepsilon \theta^\varepsilon$ with

$$\Phi_\varepsilon = \Psi_0^\varepsilon + \varepsilon \Psi_{1n}^\varepsilon,\ n \in \mathbb{N}, \text{ and } \theta = 1 \text{ in } \mathbb{R}, \tag{3.8}$$

where

$$\begin{cases} \Psi_0 \in \mathcal{D}(0,T) \otimes \mathcal{D}(\Omega) \otimes \mathcal{W} \\ \Psi_{1n} \in \mathcal{D}(0,T) \otimes \mathcal{D}(\Omega) \otimes \mathcal{C}^1_{per}(Y)^3,\ n \in \mathbb{N}, \\ \text{the restriction } \Psi_{1n}\Big|_{Q\times \widetilde{Y}_s} \text{ of } \Psi_{1n} \text{ on } Q \times \widetilde{Y}_s \text{ satisfying :} \\ \Psi_{1n}\Big|_{Q\times \widetilde{Y}_s} \to \Psi_1 \text{ in } L^2\left(Q; \mathbb{H}^1_{per}(Y_s)\right) \text{ as } n \to \infty \\ \text{such that } \Psi_1 \text{ is the solution of (3.7) for } \Psi_0 \text{ above} \end{cases} \tag{3.9}$$

with $\mathcal{W} = \bigcap_{i,j=1}^{3} \mathcal{W}_{ij}$ where

$$\mathcal{W}_{ij} = \{w \in \mathcal{C}^1_{per}(Y; \mathbb{R})^3 : \mathrm{div}_y\, w = 0 \text{ in } Y_f \text{ and } e_{ij}(w) = 0 \text{ in } Y_s\},$$

and

$$\Psi_0^\varepsilon(x,t) = \Psi_0\left(x, t, \frac{x}{\varepsilon}\right),\ \Psi_{1n}^\varepsilon(x,t) = \Psi_{1n}\left(x, t, \frac{x}{\varepsilon}\right) \quad (x,t) \in Q.$$

4 The Homogenization Result for $f \in \mathcal{C}^1([0,T]; \mathbb{L}^2(\Omega; \mathbb{R}))$, $f(0) = \omega$

According to Lemma 3.2, we need to recall some vector spaces. It is well-known that $E(\Omega) = \{u = (u^i) \in \mathbb{L}^2(\Omega) : \mathrm{div}\, u \in L^2(\Omega)\}$ is a Hilbert space for the norm $\|u\|_{E(\Omega)} = \left(\|u\|^2_{\mathbb{L}^2(\Omega)} + \|\mathrm{div}\, u\|^2_{L^2(\Omega)}\right)^{\frac{1}{2}}$, $u \in E(\Omega)$ and $E_0(\Omega) = \{u \in E(\Omega; \mathbb{R}) : \gamma_n u = 0\}$ is one of its closed vector subspace, where γ_n is the continuous linear operator from $E(\Omega)$ into $H^{-\frac{1}{2}}(\partial\Omega)$ such that $\gamma_n u(x) = u^i(x) n^i(x)$ for all $x \in \partial\Omega$ and all $u = (u^i) \in \mathcal{D}(\overline{\Omega})^3$, $n = (n^i)$ being the outer unit normal to $\partial\Omega$.

Now, the set $W = \bigcap_{i,j=1}^{3} W_{ij}$ with

$$W_{ij} = \{w \in \mathbb{H}^1_{per}(Y; \mathbb{R}) : \text{div}_y\, w = 0 \text{ in } Y_f,\, e_{ij}(w) = 0 \text{ in } Y_s\},$$

is a closed vector subspace of $\mathbb{H}^1_{per}(Y; \mathbb{R})$, and the seminorm

$$w \to \|w\|_W = \left(\|w\|^2_{\mathbb{L}^2(Y)} + \int_{Y_f} e_{ij}(w)e_{ij}(w)dy \right)^{\frac{1}{2}}$$

is a norm in W equivalent to the usual norm of $\mathbb{H}^1_{per}(Y)$. Furthermore, the space $\mathcal{W}$ (at (3.9)) is dense in W. Let

$$E_0(\Omega; W) = \{u = (u^i) \in L^2(\Omega; W) : \widetilde{u} \in E_0(\Omega)\},$$

where $\widetilde{u} = (\widetilde{u}^i)$, $\widetilde{u}^i$ being the function on Ω defined by $\widetilde{u}^i(x) = \int_Y u^i(x, y)dy$ ($x \in \Omega$). Clearly, $E_0(\Omega; W)$ is a vector subspace of $L^2(\Omega; W)$. Furthermore, $E_0(\Omega; W)$ is a Hilbert space for the norm

$$\|u\|_{E_0(\Omega;W)} = \left(\|u\|^2_{L^2(\Omega;W)} + \|\text{div}\widetilde{u}\|^2_{L^2(\Omega)} \right)^{\frac{1}{2}} \quad (u \in E_0(\Omega; W)).$$

Theorem 4.1 *Let $(u_\varepsilon)_{\varepsilon>0}$ be the sequence of solutions of (1.1)–(1.2). Then, as $\varepsilon \to 0$, we have*

$$u_\varepsilon \to \widetilde{u}_0 \text{ in } L^\infty(0, T; E_0(\Omega))\text{-weak } *, \tag{4.1}$$

where u_0 is the unique function from $[0, T]$ into $L^2(\Omega; W)$ satisfying

$$u_0 \in L^2(0, T; E_0(\Omega; W)), \tag{4.2}$$

$$\frac{\partial u_0}{\partial t} \in L^2(0, T; E_0(\Omega; W)) \cap L^\infty(0, T; L^2(\Omega; \mathbb{H}^1_{per}(Y))), \tag{4.3}$$

$$\frac{\partial^2 u_0}{\partial t^2} \in L^2(0, T; L^2(\Omega; \mathbb{L}^2_{per}(Y))), \tag{4.4}$$

and the variational wave equation

$$\begin{aligned} &\iint_{\Omega\times Y} \rho \frac{\partial^2 u_0^i}{\partial t^2}(t) v_0^i dxdy + \iint_{\Omega\times Y_f} \mu_{ijkh} e_{kh}\left(\tfrac{\partial u_0}{\partial t}(t)\right) e_{ij}(v_0)dxdy \\ &+ \frac{\vartheta}{|Y_f|}\int_\Omega div\widetilde{u}_0(t) div\widetilde{v}_0 dx = \int_\Omega f^i(t)\widetilde{v}_0^i dx \quad \textit{for all } v_0 \in E_0(\Omega; W), \end{aligned} \tag{4.5}$$

$0 < t < T$, with initial conditions

$$u_0(0) = \frac{\partial u_0}{\partial t}(0) = \omega \tag{4.6}$$

where $\mu_{ijkh} = \mu(\delta_{ik}\delta_{jh} + \delta_{ih}\delta_{jk})$.

Proof According to (3.6), we can extend each z_ε such that

$$z_\varepsilon \in L^2(0,T;\widetilde{\mathbb{H}^1_0(\Omega)}),\ 0 < \varepsilon \le \varepsilon_0, \text{ and } \sup_{0<\varepsilon\le\varepsilon_0} \left(\|z_\varepsilon\|_{L^2(0,T;\mathbb{H}^1_0(\Omega))}\right) < \infty.$$

Then, there exists a fundamental sequence extracted from $E' \cap (0, \varepsilon_0]$, still denoted E' for simplicity, such that we have [4, Theorem 3], as $E' \ni \varepsilon \to 0$,

$$\int_Q E_{ij}(z_\varepsilon)\Phi^\varepsilon dxdt \to \iint_{Q\times Z} \left(E_{ij}(r) + e_{ij}(r_1)\right)\Phi dxdtdyd\tau \tag{4.7}$$

for all $\Phi \in L^2(Q; \mathcal{C}_{per}(Z))$ and for $i, j = 1, 2, 3$, where $r \in L^2(0,T; L^2_{per}(\Theta; \mathbb{H}^1(\Omega)))$, $r_1 \in L^2(Q; L^2_{per}(\Theta; \mathbb{H}^1_\#(Y)))$.

Thanks to Lemmas 3.1–3.4, it follows that $u_\varepsilon \in L^\infty(0,T; E_0(\Omega))$ with (4.1) as $E' \ni \varepsilon \to 0$, and properties (4.2), (4.3), (4.4), (4.6).

Now, we take in (3.5) test functions of the form $v = \left(\Psi_0^\varepsilon + \varepsilon\Psi_{1n}^\varepsilon\right)\theta^\varepsilon$, $n \in \mathbb{N}$, where Ψ_0 and Ψ_{1n} satisfy (3.9), $\theta \in \mathcal{C}_{per}(\Theta)$, $\theta = 1$. We obtain

$$\begin{aligned}
&\int_Q \rho^\varepsilon \frac{\partial^2 u_\varepsilon}{\partial t^2}\cdot\Psi_0^\varepsilon\theta^\varepsilon dxdt + \int_{Q_f^\varepsilon} \varepsilon b_{ijkh}E_{kh}(\frac{\partial u_\varepsilon}{\partial t})(e_{ij}(\Psi_0))^\varepsilon\theta^\varepsilon dxdt \\
&-\int_{Q_f^\varepsilon} p_\varepsilon(\mathrm{div}\Psi_0)^\varepsilon\theta^\varepsilon dxdt - \int_{Q_f^\varepsilon} p_\varepsilon(\mathrm{div}_y\Psi_{1n})^\varepsilon\theta^\varepsilon dxdt - \int_Q f\cdot\Psi_0^\varepsilon\theta^\varepsilon dxdt \\
&-\int_Q E_{ij}(z_\varepsilon)\chi_s^\varepsilon(E_{ij}(\Psi_0))^\varepsilon\theta^\varepsilon dxdt - \int_Q E_{ij}(z_\varepsilon)\chi_s^\varepsilon(e_{ij}(\Psi_{1n}))^\varepsilon\theta^\varepsilon dxdt \\
&= \varepsilon\int_Q f\cdot\Psi_{1n}^\varepsilon\theta^\varepsilon dxdt + \varepsilon\int_{Q_f^\varepsilon} \varepsilon b_{ijkh}E_{kh}(\frac{\partial u_\varepsilon}{\partial t})(E_{ij}(\Psi_0))^\varepsilon\theta^\varepsilon dxdt \\
&+\varepsilon\int_{Q_f^\varepsilon} p_\varepsilon(\mathrm{div}\Psi_{1n})^\varepsilon\theta^\varepsilon dxdt + \varepsilon\int_{Q_s^\varepsilon} E_{ij}(z_\varepsilon)(E_{ij}(\Psi_{1n}))^\varepsilon\theta^\varepsilon dxdt \\
&-\varepsilon\int_Q \rho^\varepsilon\frac{\partial^2 u_\varepsilon}{\partial t^2}\cdot\Psi_{1n}^\varepsilon\theta^\varepsilon dxdt + \varepsilon\int_{Q_f^\varepsilon} \varepsilon b_{ijkh}E_{kh}(\frac{\partial u_\varepsilon}{\partial t})(e_{ij}(\Psi_{1n}))^\varepsilon\theta^\varepsilon dxdt \\
&-\varepsilon^2\int_{Q_f^\varepsilon} \varepsilon b_{ijkh}E_{kh}(\frac{\partial u_\varepsilon}{\partial t})(E_{ij}(\Psi_{1n}))^\varepsilon\theta^\varepsilon dxdt + \frac{1}{\varepsilon}\int_{Q_f^\varepsilon} p_\varepsilon(\mathrm{div}_y\Psi_0)^\varepsilon\theta^\varepsilon dxdt \\
&+\tfrac{1}{\varepsilon}\int_{Q_s^\varepsilon} E_{ij}(z_\varepsilon)(e_{ij}(\Psi_0))^\varepsilon\theta^\varepsilon dxdt,
\end{aligned}$$

where χ_s is the characteristic function from $\widetilde{Y}_s$ into $\mathbb{R}^3_y$. But now, since $\mathrm{div}_y\Psi_0 = 0$ in $Q \times Y_f$ and $e_{ij}(\Psi_0) = 0$ in $Q \times Y_s$ $(i, j = 1, 2, 3)$, the last two terms at the right hand side are equal to zero. Next, we observe that $\chi_s E_{ij}(\Psi_0)\theta$ and $\chi_s e_{ij}(\Psi_{1n})\theta$ belong to $\mathcal{C}(\overline{Q}; L^\infty_{per}(Z))$. Thus, by using (4.7), Lemmas 3.2 and 3.4, at the limit (when $E' \ni \varepsilon \to 0$), we conclude that the right hand side converges to zero, and the left hand side converges to

$$\begin{aligned}
&\iint_{Q\times Y} \rho \frac{\partial^2 u_0^i}{\partial t^2}\Psi_0^i dxdtdy + \iint_{Q\times Y_f} \mu_{ijkh} e_{kh}\left(\tfrac{\partial u_0}{\partial t}\right) e_{ij}(\Psi_0)dxdtdy \\
&-\int_Q p\left(\int_{Y_f} \mathrm{div}_y\Psi_{1n}dy\right)dxdt - \iint_{Q\times Y_f} p\mathrm{div}\Psi_0 dxdtdy \\
&-\iiint_{Q\times Y_s\times\Theta} \left(E_{ij}(r) + e_{ij}(r_1)\right)\left(E_{ij}(\Psi_0) + e_{ij}(\Psi_{1n})\right) dxdtdyd\tau \\
&-\iint_{Q\times Y} f^i\Psi_0^i dxdtdy \qquad \text{for all } n \in \mathbb{N}.
\end{aligned}$$

Letting $z = \int_\Theta r d\tau$ and $z_1 = \int_\Theta r_1 d\tau$, we have $z \in L^2\left(0, T; \mathbb{H}^1(\Omega)\right)$, $z_1 \in L^2\left(Q;\ \mathbb{H}^1_{per}(Y)\right)$ with $E_{ij}(z) = \int_\Theta E_{ij}(r)\,d\tau$ and $e_{ij}(z_1) = \int_\Theta e_{ij}(r_1)\,d\tau$, and it follows

$$\begin{aligned}
&\iint_{Q\times Y} \rho \frac{\partial^2 u_0^i}{\partial t^2}\Psi_0^i dxdtdy + \iint_{Q\times Y_f} \mu_{ijkh} e_{kh}\left(\tfrac{\partial u_0}{\partial t}\right) e_{ij}(\Psi_0)dxdtdy \\
&-\int_Q p\left(\int_{Y_f} \mathrm{div}_y\Psi_{1n}dy\right)dxdt - \iint_{Q\times Y_f} p\mathrm{div}\Psi_0 dxdtdy = \iint_{Q\times Y} f^i\Psi_0^i dxdtdy \\
&+\iint_{Q\times Y_s} \left(E_{ij}(z) + e_{ij}(z_1)\right)\left(E_{ij}(\Psi_0) + e_{ij}(\Psi_{1n})\right) dxdtdy \text{ for all } n \in \mathbb{N}.
\end{aligned}$$

Let us consider the function $\xi = (\xi^i)$ from $\widetilde{Y}_s$ into $\mathbb{R}^3$ defined by : for $k = (k_i) \in \mathbb{Z}^3$ and $y \in Y_s$, $\xi^i(y + k) = k_i$, $i = 1, 2, 3$. Then the function $y \to y - \xi(y)$ from $\widetilde{Y}_s$ into $\mathbb{R}^3$ belongs to $\mathbb{H}^1_{per}(Y_s)$.

Since $-\int_{Y_f}\mathrm{div}_y\Psi_{1n}dy = \int_{Y_s}\mathrm{div}_y\Psi_{1n}dy$ in Q, we have

$$\begin{aligned}
-\iint_{Q\times Y_f} p\mathrm{div}_y\Psi_{1n}dxdtdy &= \iint_{Q\times Y_s} p\mathrm{div}_y\Psi_{1n}dxdtdy \\
&= \iint_{Q\times Y_s} e_{ij}(\Psi_{1n})e_{ij}\left(p(y - \xi(y))\right) dxdtdy.
\end{aligned}$$

According to (3.7) and (3.9) it follows that, as $n \to \infty$,

$$\begin{aligned}
-\iint_{Q\times Y_f} p\mathrm{div}_y\Psi_{1n}dxdtdy &\to -\iint_{Q\times Y_s} E_{ij}(\Psi_0)e_{ij}\left(p(y - \xi(y))\right) dxdtdy \\
&= -\iint_{Q\times Y_s} p\mathrm{div}\Psi_0 dxdtdy.
\end{aligned}$$

Otherwise, according to the convergence in (3.9), it follows that, up to a subsequence, still denoted $(\Psi_{1n})_n$,

$$\iint_{Q\times Y_s} E_{ij}(z))\left(E_{ij}(\Psi_0)+e_{ij}(\Psi_{1n})\right)dxdtdy \to \iint_{Q\times Y_s} E_{ij}(z)\left(E_{ij}(\Psi_0)+e_{ij}(\Psi_1)\right)dxdtdy$$

as $n\to\infty$. Therefore

$$\iint_{Q\times Y_s}\left(E_{ij}(z)+e_{ij}(z_1)\right)\left(E_{ij}(\Psi_0)+e_{ij}(\Psi_{1n})\right)dxdtdy \to \iint_{Q\times Y_s}\left(E_{ij}(z)+e_{ij}(z_1)\right)\left(E_{ij}(\Psi_0)+e_{ij}(\Psi_1)\right)dxdtdy.$$

But

$$\begin{aligned}&\iint_{Q\times Y_s} E_{ij}(z)\left(E_{ij}(\Psi_0)+e_{ij}(\Psi_1)\right)dxdtdy\\ &=\int_Q E_{ij}(z(x,t))\left(\int_{Y_s}\left(E_{ij}(\Psi_0(x,t))+e_{ij}(\Psi_1(x,t))\right)dy\right)dxdt=0\end{aligned}$$

and

$$\iint_{Q\times Y_s} e_{ij}(z_1)\left(E_{ij}(\Psi_0)+e_{ij}(\Psi_1)\right)dxdtdy=0 \text{ (Lemma 3.7)}.$$

Consequently

$$\begin{aligned}&\iint_{Q\times Y}\rho\frac{\partial^2 u_0^i}{\partial t^2}\Psi_0^i dxdtdy+\iint_{Q\times Y_f}\mu_{ijkh}e_{kh}\left(\frac{\partial u_0}{\partial t}\right)e_{ij}(\Psi_0)dxdtdy\\ &\quad-\iint_{Q\times Y}\frac{\partial p}{\partial x_i}\Psi_0^i dxdtdy=\iint_{Q\times Y} f^i\Psi_0^i dxdtdy.\end{aligned}$$

In particular (3.3) implies

$$\begin{aligned}&\iint_{Q\times Y}\rho\frac{\partial^2 u_0^i}{\partial t^2}\Psi_0^i dxdtdy+\iint_{Q\times Y_f}\mu_{ijkh}e_{kh}\left(\frac{\partial u_0}{\partial t}\right)e_{ij}(\Psi_0)dxdtdy\\ &\quad+\frac{\vartheta}{|Y_f|}\int_Q \operatorname{div}\widetilde{u}_0\operatorname{div}\left(\int_Y\Psi_0 dy\right)dxdt=\int_Q f^i\left(\int_Y\Psi_0^i dy\right)dxdt,\end{aligned}$$

for all $\Psi_0\in\mathcal{D}(\Omega)\otimes\mathcal{W}$. This yields (4.5) thanks to the density of $\mathcal{D}(\Omega)\otimes\mathcal{W}$ into $E_0(\Omega;W)$, where

$$\widetilde{v}_0(x,t)=\int_Y v_0(x,t,y)dy=\left(\int_Y v_0^i(x,t,y)dy\right).$$

It remains to prove that u_0 is the unique function satisfying (4.1)–(4.6). Let us assume that $f = \omega$ and $v_0 = \frac{\partial u_0}{\partial t}(t)$ in (4.5). Then, we have

$$\frac{d}{ds}\left(\iint_{\Omega\times Y} \rho \frac{\partial u_0^i}{\partial t}(s)\frac{\partial u_0^i}{\partial t}(s)dxdy + \frac{\vartheta}{|Y_f|}\int_\Omega \operatorname{div}\widetilde{u}_0(s)\operatorname{div}\widetilde{u}_0(s)dx\right)$$
$$+2\iint_{\Omega\times Y_f} \mu_{ijkh}e_{kh}(\tfrac{\partial u_0}{\partial t}(s))e_{ij}(\tfrac{\partial u_0}{\partial t}(s))dxdy = 0 \qquad (0 < s < t < T).$$

Integrating from 0 to t, and using initial conditions (see Lemma 3.3), we obtain

$$\iint_{\Omega\times Y} \rho \frac{\partial u_0^i}{\partial t}(t)\frac{\partial u_0^i}{\partial t}(t)dxdy + \int_\Omega \operatorname{div}\widetilde{u}_0(t)\operatorname{div}\widetilde{u}_0(t)dx$$
$$+\int_0^t\iint_{\Omega\times Y_f} \mu_{ijkh}e_{kh}\left(\frac{\partial u_0}{\partial t}(s)\right)e_{ij}(\frac{\partial u_0}{\partial t}(s))dxdyds \le 0 \quad \text{for } 0 < t < T.$$

Therefore $\|u_0\|_{L^\infty(0,T;E_0(\Omega;W))} = 0$, and u_0 of Lemma 3.2 is the unique function from $[0, T]$ into $E_0(\Omega; W)$ satisfying (4.1)–(4.6). □

For a physical description of the associate local problem, let u_0 be the function defined in Theorem 4.1. There exists a negligible set $\mathcal{N}_0 \subset \Omega$ such that for $x \in \Omega\backslash\mathcal{N}_0$ the function $t \to u_0(x, t, \cdot)$ from $[0, T]$ into $\mathbb{L}^2_{per}(Y)$, denoted by $u_0(x)$, belongs to $L^2(0, T; \mathbb{L}^2_{per}(Y))$. Thanks to some isometric isomorphism, we get $u_0(x)$, $\frac{\partial}{\partial t}u_0(x) \in L^2(0, T; \mathbb{H}^1_{per}(Y))$ and $\frac{\partial^2}{\partial t^2}u_0(x) \in L^2(0, T; \mathbb{L}^2_{per}(Y))$ with $\frac{\partial}{\partial t}u_0(x) = \frac{\partial u_0}{\partial t}(x, \cdot, \cdot)$ and $\frac{\partial^2}{\partial t^2}u_0(x) = \frac{\partial^2 u_0}{\partial t^2}(x, \cdot, \cdot)$ for all $x \in \Omega\backslash\mathcal{N}$ ($\mathcal{N}$ being a negligible set such that $\mathcal{N}_0 \subset \mathcal{N} \subset \Omega$).

Theorem 4.2 *Let u_0 be the function in Theorem 4.1. For almost all $x \in \Omega$, $u_0(x)$ is uniquely determined by $u_0(x) \in \mathcal{C}([0, T]; W)$, $\frac{\partial}{\partial t}u_0(x) \in L^2(0, T; W) \cap \mathcal{C}([0, T]; \mathbb{L}^2_{per}(Y))$, $\frac{\partial^2}{\partial t^2}u_0(x) \in L^2(0, T; \mathbb{L}^2_{per}(Y))$ and the variational problem*

$$\begin{aligned}&\int_Y \rho\frac{\partial^2}{\partial t^2}u_0^i(x)(t)w^i dy + \int_{Y_f}\mu_{ijkh}e_{kh}\left(\frac{\partial}{\partial t}u_0(x)(t)\right)e_{ij}(w)dy\\ &= \left(f^i(x,t) - \tfrac{\partial p}{\partial x_i}(x,t)\right)\int_Y w^i dy \qquad \textit{for all } w \in W,\ 0 < t < T\end{aligned} \tag{4.8}$$

with initial conditions

$$u_0(x)(0) = \frac{\partial}{\partial t}u_0(x)(0) = \omega. \tag{4.9}$$

Proof See [6]. □

The local problem (4.8)–(4.9) describes the motion of an incompressible viscous fluid with solid rigid particles in suspension; and the partial differential equations which govern this motion are (for almost all $x \in \Omega$) :

$$\begin{aligned}
&\rho_f \frac{\partial^2}{\partial t^2} u_0(x) = \operatorname{div}_y \sigma^0(x) + f(x) - Dp(x) \text{ in } Y_f \times (0,T) \\
&\operatorname{div}_y u_0(x) = 0 \text{ in } Y_f \times (0,T) \\
&e_{ij}(u_0(x)) = 0 \text{ in } Y_s \times (0,T) \quad (i,j = 1,2,3) \\
&u_0(x) \text{ is } Y\text{-periodic}
\end{aligned}$$

with formally

$$\begin{aligned}
&\int_{Y_s} \left(\rho_s \frac{\partial^2}{\partial t^2} u_0^i(x) - f^i(x) + \frac{\partial p}{\partial x_i}(x) \right) dy + \textstyle\int_{\partial Y_s} \sigma_{ij}^0(x) n^j d\sigma = 0 \\
&\int_{Y_s} y \wedge \left(\rho_s \frac{\partial^2}{\partial t^2} u_0(x) - f(x) + Dp(x) \right) dy + \textstyle\int_{\partial Y_s} y \wedge \sigma^0(x) n d\sigma = 0
\end{aligned}$$

(for details, see [6]).

References

1. G. Allaire, Homogenization and two-scale convergence, SIAM J. Math. Anal., **23** (1992), 1482–1518.
2. A.C. Biazutti, On a nonlinear evolution equation and its applications, Nonlinear Analysis TMA, **24** (1995), 1221–1234.
3. D. Lukkassen, G. Nguetseng and P. Wall, Two-scale convergence, Int. J. Pure and Appl. Math., **1** (2002), 35–86.
4. G. Nguetseng, A general convergence result for a functional related to the theory of homogenization, SIAM J. Math. Anal., **20** (1989), 608–623.
5. G. Nguetseng, Homogenization structures and Applications I, Z. Anal. Andwend., **22** (2003), 203–221.
6. G. Nguetseng, H. Nnang, N. Svanstedt, Asymptotic analysis for a weakly damped wave equation with application to a problem arising in elasticity, J. Funct. Spaces and Appl., **8** (2010), 17–54.
7. E. Sanchez-Palencia, *Nonhomogeneous media and vibration theory*, Lect. Notes in Physics **127**, Springer-verlag, Berlin, New-York, 1980.

Corrector Problem and Homogenization of Nonlinear Elliptic Monotone PDE

Jean Louis Woukeng

Abstract In the deterministic homogenization of nonlinear monotone elliptic PDEs, we prove the existence of a distributional corrector and we find an approximation scheme for the homogenized coefficient. The obtained results represent an important step towards the numerical implementation of the results from the deterministic homogenization theory beyond the periodic setting.

Keywords Corrector problem · Deterministic homogenization · Approximation

Mathematics Subject Classification (2010). Primary 35B40, 46J10

1 Introduction

We aim at establishing the existence of a solution, in the usual sense of distributions, to the corrector problem for a family of nonlinear second order elliptic monotone equations in divergence form with rapidly oscillating coefficients. Before we state our results, let us however, make precise the framework in which we will be working in.

Let A be an algebra with mean value on $\mathbb{R}^d$, that is, a closed subalgebra of the C^*-algebra of bounded uniformly continuous real-valued functions on $\mathbb{R}^d$, $\mathrm{BUC}(\mathbb{R}^d)$ (the vector space of bounded uniformly continuous functions defined on $\mathbb{R}^d$), which

This work was completed when the author was visiting the Interdisciplinary Center for Scientific Computing (IWR) of the University of Heidelberg in Germany under the sponsorship of the Alexander von Humboldt Foundation.

J. L. Woukeng (✉)
Department of Mathematics and Computer Science, University of Dschang, Dschang, Cameroon
Current address: Interdisciplinary Center for Scientific Computing (IWR), University of Heidelberg, Heidelberg, Germany

V. Schulz, D. Seck (eds.), *Shape Optimization, Homogenization and Optimal Control*, International Series of Numerical Mathematics 169,
https://doi.org/10.1007/978-3-319-90469-6_4

contains the constants, is translation invariant and is such that any of its elements possesses a mean value in the following sense: for every $u \in A$, the sequence $(u^\varepsilon)_{\varepsilon>0}$ $(u^\varepsilon(x) = u(x/\varepsilon))$ weakly$*$-converges in $L^\infty(\mathbb{R}^d)$ to some real number $M(u)$ (called the mean value of u) as $\varepsilon \to 0$. The mean value expresses as

$$M(u) = \lim_{R\to\infty} \fint_{B_R} u(y)dy \text{ for } u \in A \tag{1.1}$$

where we have set

$$\fint_{B_R} = \frac{1}{|B_R|}\int_{B_R}.$$

For $1 \leq p < \infty$, we define the Marcinkiewicz space $\mathfrak{M}^p(\mathbb{R}^d)$ to be the set of functions $u \in L^p_{loc}(\mathbb{R}^d)$ such that

$$\limsup_{R\to\infty} \fint_{B_R} |u(y)|^p \, dy < \infty.$$

Endowed with the seminorm

$$\|u\|_p = \left(\limsup_{R\to\infty} \fint_{B_R} |u(y)|^p \, dy\right)^{1/p},$$

$\mathfrak{M}^p(\mathbb{R}^d)$ is a complete seminormed space. Next, we define the *generalized Besicovitch space* $B^p_A(\mathbb{R}^d)$ $(1 \leq p < \infty)$ as the closure of the algebra with mean value A in $\mathfrak{M}^p(\mathbb{R}^d)$. Then for any $u \in B^p_A(\mathbb{R}^d)$ we have that

$$\|u\|_p = \left(\lim_{R\to\infty} \fint_{B_R} |u(y)|^p \, dy\right)^{\frac{1}{p}} = (M(|u|^p))^{\frac{1}{p}}. \tag{1.2}$$

In this regard, we consider the space

$$B^{1,p}_A(\mathbb{R}^d) = \{u \in B^p_A(\mathbb{R}^d) : \nabla u \in (B^p_A(\mathbb{R}^d))^d\}$$

endowed with the seminorm

$$\|u\|_{1,p} = \left(\|u\|^p_p + \|\nabla u\|^p_p\right)^{\frac{1}{p}},$$

which is a complete seminormed space. The Banach counterpart of the previous spaces are defined as follows. We set $\mathcal{B}^p_A(\mathbb{R}^d) = B^p_A(\mathbb{R}^d)/\mathcal{N}$ where $\mathcal{N} = \{u \in B^p_A(\mathbb{R}^d) : \|u\|_p = 0\}$. We define $\mathcal{B}^{1,p}_A(\mathbb{R}^d)$ mutatis mutandis: replace $B^p_A(\mathbb{R}^d)$ by

$\mathcal{B}_A^p(\mathbb{R}^d)$ and $\partial/\partial y_i$ by $\overline{\partial}/\partial y_i$, where $\overline{\partial}/\partial y_i$ is defined by

$$\frac{\overline{\partial}}{\partial y_i}(u+\mathcal{N}) := \frac{\partial u}{\partial y_i} + \mathcal{N} \text{ for } u \in B_A^{1,p}(\mathbb{R}^d). \tag{1.3}$$

It is important to note that $\overline{\partial}/\partial y_i$ is also defined as the infinitesimal generator in the ith direction coordinate of the strongly continuous group $\mathcal{T}(y) : \mathcal{B}_A^p(\mathbb{R}^d) \to \mathcal{B}_A^p(\mathbb{R}^d)$; $\mathcal{T}(y)(u+\mathcal{N}) = u(\cdot + y) + \mathcal{N}$. Let us denote by $\varrho : B_A^p(\mathbb{R}^d) \to \mathcal{B}_A^p(\mathbb{R}^d) = B_A^p(\mathbb{R}^d)/\mathcal{N}$, $\varrho(u) = u + \mathcal{N}$, the canonical surjection. We remark that if $u \in B_A^{1,p}(\mathbb{R}^d)$ then $\varrho(u) \in \mathcal{B}_A^{1,p}(\mathbb{R}^d)$ with further

$$\frac{\overline{\partial}\varrho(u)}{\partial y_i} = \varrho\left(\frac{\partial u}{\partial y_i}\right),$$

as seen above in (1.3).

We assume in the sequel that the algebra A is ergodic, that is, any $u \in \mathcal{B}_A^p(\mathbb{R}^d)$ that is invariant under $(\mathcal{T}(y))_{y\in\mathbb{R}^d}$ is a constant in $\mathcal{B}_A^p(\mathbb{R}^d)$, i.e., if $\mathcal{T}(y)u = u$ for every $y \in \mathbb{R}^d$, then $u = c$, c a constant. Let us also recall the following property [6, 7]:

- The mean value M viewed as defined on A, extends by continuity to a positive continuous linear form (still denoted by M) on $B_A^p(\mathbb{R}^d)$. For each $u \in B_A^p(\mathbb{R}^d)$ and all $a \in \mathbb{R}^d$, we have $M(u(\cdot + a)) = M(u)$, and $\|u\|_p = \left[M(|u|^p)\right]^{1/p}$.

  To the space $B_A^p(\mathbb{R}^d)$ we also attach the *corrector* space defined as follows:

$$B_{\#A}^{1,p}(\mathbb{R}^d) = \{u \in W_{loc}^{1,p}(\mathbb{R}^d) : \nabla u \in B_A^p(\mathbb{R}^d)^d \text{ and } M(\nabla u) = 0\}.$$

In $B_{\#A}^{1,p}(\mathbb{R}^d)$ we identify two elements by their gradients: $u = v$ in $B_{\#A}^{1,p}(\mathbb{R}^d)$ iff $\nabla(u-v) = 0$, i.e. $\|\nabla(u-v)\|_p = 0$. We may therefore equip $B_{\#A}^{1,p}(\mathbb{R}^d)$ with the gradient norm $\|u\|_{\#,p} = \|\nabla u\|_p$. This defines a Banach space [2, Theorem 3.12] containing $B_A^{1,p}(\mathbb{R}^d)$ as a subspace.

We are now able to define the Σ-convergence concept. A sequence $(u_\varepsilon)_{\varepsilon>0}$ in $L^p(\Omega)$ $(1 \le p < \infty)$ is said to:

(i) *weakly Σ-converge* in $L^p(\Omega)$ to $u_0 \in L^p(\Omega; \mathcal{B}_A^p(\mathbb{R}^d))$ if, as $\varepsilon \to 0$,

$$\int_\Omega u_\varepsilon(x) f\left(x, \frac{x}{\varepsilon}\right) dx \to \int_\Omega M(u_0(x,\cdot) f(x,\cdot)) dx \tag{1.4}$$

for any $f \in L^{p'}(\Omega; A)$ $(p' = p/(p-1))$;

(ii) *strongly Σ-converge* in $L^p(\Omega)$ to $u_0 \in L^p(\Omega; \mathcal{B}_A^p(\mathbb{R}^d))$ if (1.4) holds and further $\|u_\varepsilon\|_{L^p(\Omega)} \to \|u_0\|_{L^p(\Omega;\mathcal{B}_A^p(\mathbb{R}^d))}$.

We denote (i) by "$u_\varepsilon \to u_0$ in $L^p(\Omega)$-weak Σ", and (ii) by "$u_\varepsilon \to u_0$ in $L^p(\Omega)$-strong Σ". It is to be noted that this is a generalization of the well known concept of two-scale convergence (for which, for example, (1.4) holds true provided that $A = \mathcal{C}_{per}(Y)$).

The following are the main properties of the above concept.

- Any bounded ordinary sequence in $L^p(\Omega)$ $(1 < p < \infty)$ possesses a subsequence that weakly Σ-converges in $L^p(\Omega)$.
- If $(u_\varepsilon)_{\varepsilon\in E}$ is a bounded sequence (E an ordinary sequence of positive real numbers converging to zero) in $W^{1,p}(\Omega)$, then there exist a subsequence E' of E and a couple $(u_0, u_1) \in W^{1,p}(\Omega) \times L^p(\Omega; B^{1,p}_{\#A}(\mathbb{R}^d))$ such that

$$u_\varepsilon \to u_0 \text{ in } W^{1,p}(\Omega)\text{-weak}$$

$$\frac{\partial u_\varepsilon}{\partial x_j} \to \frac{\partial u_0}{\partial x_j} + \frac{\partial u_1}{\partial y_j} \text{ in } L^p(\Omega)\text{-weak } \Sigma \quad (1 \le j \le d)$$

- If $u_\varepsilon \to u_0$ in $L^p(\Omega)$-weak Σ and $v_\varepsilon \to v_0$ in $L^q(\Omega)$-strong Σ, then $u_\varepsilon v_\varepsilon \to u_0 v_0$ in $L^r(\Omega)$-weak Σ, where $1 \le p, q, r < \infty$ and $\frac{1}{p} + \frac{1}{q} = \frac{1}{r}$.

With this in mind, for fixed $\varepsilon > 0$, we consider the problem

$$-\nabla \cdot a\left(\frac{x}{\varepsilon}, \nabla u_\varepsilon\right) = f \text{ in } \Omega, u_\varepsilon \in W^{1,p}_0(\Omega) \tag{1.5}$$

where $1 < p < \infty$, Ω is an open bounded set of $\mathbb{R}^d$ (integer $d \ge 1$) with smooth boundary $\partial\Omega$, $f \in W^{-1,p'}(\Omega)$ $(p' = p/(p-1))$ and $(y, \lambda) \to a(y, \lambda)$ is a function from $\mathbb{R}^d \times \mathbb{R}^d$ to $\mathbb{R}^d$ with the properties:

For each fixed $\lambda \in \mathbb{R}^d$, the function $y \to a(y, \lambda)$ (denoted by $a(\cdot, \lambda)$) from $\mathbb{R}^d$ to $\mathbb{R}^d$ is measurable. (1.6)

$$a(y, 0) = 0 \text{ almost everywhere (a.e.) in } y \in \mathbb{R}^d. \tag{1.7}$$

There are four constants $c_1, c_2 > 0$, $0 < \alpha_1 \le \min(1, p-1)$ and $\alpha_2 \ge \max(p, 2)$ such that, a.e. in $y \in \mathbb{R}^d$ and for $\lambda, \mu \in \mathbb{R}^d$:
(i) $(a(y, \lambda) - a(y, \mu)) \cdot (\lambda - \mu) \ge c_1 (|\lambda| + |\mu|)^{p-\alpha_2} |\lambda - \mu|^{\alpha_2}$
(ii) $|a(y, \lambda) - a(y, \mu)| \le c_2 (1 + |\lambda| + |\mu|)^{p-1-\alpha_1} |\lambda - \mu|^{\alpha_1}$
where the dot denotes the usual Euclidean inner product in $\mathbb{R}^d$ and $|\cdot|$ the associated norm. (1.8)

$$a(\cdot, \lambda) \in (B^{p'}_A(\mathbb{R}^d))^d \text{ for all } \lambda \in \mathbb{R}^d. \tag{1.9}$$

It is well known that the above assumptions (see especially (1.6)–(1.8)) entail the existence and uniqueness of $u_\varepsilon \in W_0^{1,p}(\Omega)$ solving problem (1.5). Under the additional assumption (1.9), the following theorem is our first main result.

Theorem 1.1 *There exists $u_0 \in W_0^{1,p}(\Omega)$ such that $u_\varepsilon \to u_0$ weakly in $W_0^{1,p}(\Omega)$ and strongly in $L^p(\Omega)$ (as $\varepsilon \to 0$) and u_0 solves uniquely the problem*

$$-\nabla \cdot a^*(\nabla u_0) = f \text{ in } \Omega, \tag{1.10}$$

where a^* is the homogenized coefficient defined by

$$a^*(r) = M\left(a(\cdot, r + \nabla_y \chi_r)\right) \text{ for } r \in \mathbb{R}^d \tag{1.11}$$

with, $\chi_r \in B_{\#A}^{1,p}(\mathbb{R}^d)$ being the unique solution (up to an additive constant) of the problem

$$-\nabla \cdot a(\cdot, r + \nabla \chi_r) = 0 \text{ in } \mathbb{R}^d. \tag{1.12}$$

Remark 1.2 Problem (1.12) is the so-called *corrector problem*. Let us emphasize that (1.12) appears here as a PDE in the sense of distributions on $\mathbb{R}^d$ (it is worth noticing that it was not the case before in almost all work dealing with deterministic homogenization beyond the periodic framework), thereby alloying the study of the regularity theory in the general deterministic homogenization theory. This is the starting point of what will lead to the numerical implementation and numerical simulations of deterministic homogenization results beyond the periodic setting.

In almost all the previous works dealing with deterministic homogenization theory beyond the periodic setting, the corrector problem was posed either on an abstract space (the **spectrum** of an algebra with mean value, which is known to be a compact topological semigroup whose kernel is a compact topological group; this characterization has been obtained recently in [11]), or in $\mathbb{R}^d$ but however solved in the latter case not in the sense of distributions, but rather in the sense of a duality arising from the mean value extension defined on the Sobolev-Besicovitch spaces as follows: for $\mathbf{u} \in (B_A^{p'}(\mathbb{R}^d))^d$ and $v \in B_A^{1,p}(\mathbb{R}^d)$, $\langle \nabla \cdot \mathbf{u}, v \rangle = -M(\mathbf{u} \cdot \nabla v)$. In order to get quantitative results beyond the periodic framework, it is therefore an urgent matter to solve (1.12) in the distributional sense, since quantitative results rely on the regularity theory, which is well developed in the literature for distributional solutions of PDEs. It is worth recalling that apart from the periodic setting, the spectrum of an algebra with mean value is abstract and hence useless in practical problems of daily-life. In order to perform numerical simulations, we need to get rid of the use of the concept of spectrum and pose the corrector problem on numerical space such as $\mathbb{R}^d$ or its subsets. The reformulation (as seen in (1.12)) is readily understandable to Applied Scientists. So in order to prove Theorem , we will need to prove the following result, which is the second main result of this work.

Theorem 1.3 *Let $r \in \mathbb{R}^d$. There exists a unique (up to an additive constant) function $\chi_r \in W^{1,p}_{loc}(\mathbb{R}^d))$ such that $\nabla\chi_r \in B^p_A(\mathbb{R}^d)^d$, and which solves the equation*

$$\nabla \cdot a(\cdot, r + \nabla\chi_r) = 0 \text{ in } \mathbb{R}^d. \tag{1.13}$$

The proof of Theorem 1.3 will be obtained as a consequence of Lemma 2.1 in Sect. 2 below.

With this in mind, after having performed the homogenization process (that is, the proof of Theorem 1.1), we shall proceed to the next step which will consist in finding a suitable approximation scheme for the homogenized matrix a^* (see (1.11)). This is not an issue in the periodic setting since under the periodicity, the corrector problem is posed on a bounded domain (namely the periodic cell $Y = (0, 1)^d$), as in that case, χ_r is a periodic function. In contrast with the periodic setting, the corrector problem in the general deterministic framework, is posed on the whole space $\mathbb{R}^d$, and cannot be reduced (as in the periodic situation) to a problem posed on a bounded domain. Therefore, the solution of the corrector problem (1.12) (and hence the homogenized coefficient, which depends on this solution) can not be computed directly. As a result, truncations of (1.12) must be considered, particularly on large domains $Q_R = (-R, R)^d$ with appropriate boundary conditions, and the homogenized coefficients will therefore be captured in the asymptotic regime. This is done in Theorem 1.4 (see below). To be more precise, we consider the problem

$$-\nabla \cdot a(\cdot, r + \nabla\chi_{r,R}) = 0 \text{ in } Q_R, \quad \chi_{r,R} \in W^{1,p}_0(Q_R).$$

and define the approximate effective coefficient a^*_R as follows:

$$a^*_R(r) = \fint_{Q_R} a(y, r + \nabla\chi_{r,R}(y))dy.$$

Our third main result reads as follows.

Theorem 1.4 *For fixed $r \in \mathbb{R}^d$, let $a^*_R(r)$ and $a^*(r)$ be defined above. Then, as $R \to \infty$, we have that $a^*_R(r) \to a^*(r)$.*

Remark 1.5 Theorems 1.3 and 1.4 are new, even in the linear case obtained by taking $a(y, \lambda) = B(y)\lambda$ with B a symmetric matrix satisfying suitable properties ensuring the existence of the solution to (1.5). Lemma 2.1 below is also new and establishes the existence of a unique approximate corrector.

The rest of the work is organized as follows. In Sect. 2, we prove the result about the existence of the corrector, namely, Theorem 1.3. We also prove Theorem 1.1 therein. Section 3 deals with the approximation of the homogenized coefficient.

Unless otherwise specified, the vector spaces throughout are assumed to be real vector spaces, and the scalar functions are assumed to take real values. We shall always assume that the numerical space $\mathbb{R}^d$ (integer $d \geq 1$) and its open sets are each provided with the Lebesgue measure denoted by $dx = dx_1 \dots dx_d$.

2 Proofs of Theorems 1.1 and 1.3

The assumptions and notation are as in the previous section. Our aim is to solve the corrector problem (1.12) and to prove Theorem 1.1. We begin by proving the following general result.

Lemma 2.1 *Let $T \geq 1$ and $r \in \mathbb{R}^d$ be freely fixed. Then there exists at least a function $v_{r,T} \equiv v_T \in B_A^{1,p}(\mathbb{R}^d)$ solution of*

$$-\nabla \cdot a(\cdot, r + \nabla v_T) + T^{-p} |v_T|^{p-2} v_T = 0 \text{ in } \mathbb{R}^d \tag{2.1}$$

satisfying further

$$\sup_{z \in \mathbb{R}^d, R \geq T} \fint_{B_R(z)} \left(T^{-p} |v_T|^p + |\nabla v_T|^p\right) dy \leq C \tag{2.2}$$

where the constant C depends only on d, c_1, c_2 and r. The function v_T is unique up to an additive function $w_T \in B_A^{1,p}(\mathbb{R}^d)$ satisfying $M(|w_T|^p) = 0$. Moreover, if $a(\cdot, \lambda)$ is linear with respect to λ and $p = 2$, then the solution v_T is unique in the class of functions satisfying (2.2).

Proof For the proof, we assume, without lost of generality, that $p \geq 2$ (the case $1 < p < 2$ is very similar to this one). We proceed in two steps.

Step 1. **Existence**. Let $R > 0$ and let $v_R \equiv v_{T,R} \in W_0^{1,p}(B_R)$ be defined by

$$-\nabla \cdot a(\cdot, r + \nabla v_R) + T^{-p} |v_R|^{p-2} v_R = 0 \text{ in } B_R.$$

We extend v_R by 0 off B_R to obtain a sequence $(v_R)_R$ in $W_{loc}^{1,p}(\mathbb{R}^d)$. The next point is to show that the sequence $(v_R)_R$ is bounded in $W_{loc}^{1,p}(\mathbb{R}^d)$. We choose as test function, the function $\eta_z^p v_R$, where $\eta_z(y) = \exp(-c|y - z|)$ for a fixed $z \in \mathbb{R}^d$, the positive constant c to be chosen later. Then

$$\begin{aligned}
&\int_{B_R} \eta_z^p a(\cdot, r + \nabla v_R) \cdot (r + \nabla v_R) + T^{-p} \int_{B_R} \eta_z^p |v_R|^p \\
&= -p \int_{B_R} \eta_z^{p-1} v_R \nabla \eta_z \cdot a(\cdot, r + \nabla v_R) \\
&+ \int_{B_R} \eta_z^p a(\cdot, r + \nabla v_R) \cdot r.
\end{aligned}$$

Using the properties of the function a (see (1.6)–(1.8)) we get the following inequality

$$c_1 \int_{B_R} \eta_z^p |r + \nabla v_R|^p + T^{-p} \int_{B_R} \eta_z^p |v_R|^p \tag{2.3}$$

$$\leq -p \int_{B_R} \eta_z^{p-1} v_R \nabla \eta_z \cdot a(\cdot, r + \nabla v_R) + \int_{B_R} \eta_z^p a(\cdot, r + \nabla v_R) \cdot r$$

$$\leq pc_2 \int_{B_R} \eta_z^{p-1} |v_R| \left|\nabla \eta_z\right| (1 + |r + \nabla v_R|^{p-1})$$

$$+ c_2 \int_{B_R} \eta_z^p |r| (1 + |r + \nabla v_R|^{p-1}).$$

The right-hand side of the above inequality is equal to

$$pc_2 \int_{B_R} \eta_z^{p-1} |v_R| \left|\nabla \eta_z\right| + pc_2 \int_{B_R} \eta_z^{p-1} |v_R| \left|\nabla \eta_z\right| |r + \nabla v_R|^{p-1}$$

$$+ c_2 \int_{B_R} \eta_z^p |r| + c_2 \int_{B_R} \eta_z^p |r| |r + \nabla v_R|^{p-1}$$

$$= I_1 + I_2 + I_3 + I_4.$$

We consider each term above separately. We fix a constant $k > 0$ to be determined later.

For I_2, applying Young inequality, we have

$$I_2 = p \int_{B_R} \left[\left(\frac{c_1 c_2}{k} \right)^{\frac{1}{p}} T^{-1} |v_R| \left|\nabla \eta_z\right| \right] \left[c_2^{\frac{1}{p'}} T \left(\frac{k}{c_1} \right)^{\frac{1}{p}} \eta_z^{p-1} |r + \nabla v_R|^{p-1} \right]$$

$$\leq (p-1) c_2 \left(\frac{k}{c_1} \right)^{\frac{1}{p-1}} T^{p'} \int_{B_R} \eta_z^p |r + \nabla v_R|^p$$

$$+ \frac{c_1 c_2}{k} T^{-p} \int_{B_R} |v_R|^p \left|\nabla \eta_z\right|^p.$$

Concerning I_4, we use once again Young inequality to get

$$I_4 = \int_{B_R} \left([c_2 p]^{\frac{1}{p'}} T \left(\frac{k}{c_1} \right)^{\frac{1}{p}} \eta_z^{p-1} |r + \nabla v_R|^{p-1} \right) \left(\frac{c_2^{\frac{1}{p}}}{p^{\frac{1}{p'}}} |r| T^{-1} \eta_z \right)$$

$$\leq (p-1) c_2 \left(\frac{k}{c_1} \right)^{\frac{1}{p-1}} T^{p'} \int_{B_R} \eta_z^p |r + \nabla v_R|^p + \frac{c_2}{p^p} |r|^p T^{-p} \int_{B_R} \eta_z^p.$$

Next, for I_1 and I_3 we easily get

$$I_3 \leq c_2 |r| \int_{B_R} \eta_z^p$$

and

$$\begin{aligned} I_1 &= p \int_{B_R} \left[\left(\frac{c_1 c_2}{k} \right)^{\frac{1}{p}} T^{-1} |v_R| \left| \nabla \eta_z \right| \right] \left(c_2^{\frac{1}{p'}} \eta_z^{p-1} \right) \\ &\leq \frac{c_1 c_2}{k} T^{-p} \int_{B_R} |v_R|^p \left| \nabla \eta_z \right|^p + (p-1) c_2 \int_{B_R} \eta_z^p. \end{aligned}$$

Using the fact that $\left|\nabla \eta_z\right| = c\eta_z$, we get at once in (2.3)

$$\begin{aligned} & c_1 \int_{B_R} \eta_z^p \left(1 - 2(p-1)c_2 \left(\frac{k}{c_1} \right)^{\frac{1}{p-1}} T^{p'} \right) |r + \nabla v_R|^p \qquad (2.4) \\ & + T^{-p} \int_{B_R} \eta_z^p \left(1 - 2 \frac{c_1 c_2 c^p}{k} \right) |v_R|^p \\ & \leq \left(\frac{c_2}{p^p} |r|^p T^{-p} + (p-1)c_2 + |r| c_2 \right) \int_{B_R} \eta_z^p. \end{aligned}$$

Now, choosing k and c such that

$$1 - 2(p-1)c_2 \left(\frac{k}{c_1} \right)^{\frac{1}{p-1}} T^{p'} = \frac{1}{2} \text{ and } 1 - 2 \frac{c_1 c_2 c^p}{k} = \frac{1}{2},$$

that is,

$$k = \frac{c_1}{(4(p-1)c_2)^{p-1} T^p} \text{ and } c = \frac{1}{c_2 T \left[4((4(p-1))^{p-1} \right]^{\frac{1}{p}}},$$

we readily get the estimate

$$c_1 \int_{B_R} \eta_z^p |r + \nabla v_R|^p + T^{-p} \int_{B_R} \eta_z^p |v_R|^p \leq C \int_{\mathbb{R}^d} \eta_z^p \qquad (2.5)$$

where $C = \frac{c_2}{p^p} |r|^p + (p-1)c_2 + |r| c_2$.

It emerges that the sequence $(v_R)_R$ is bounded in $W_{loc}^{1,p}(\mathbb{R}^d)$ for if both $(v_R)_R$ and $(r + \nabla v_R)_R$ (and hence $(\nabla v_R)_R$) are bounded in $L_{loc}^p(\mathbb{R}^d)$. Hence there is a subsequence of $(v_R)_R$ (still denoted by $(v_R)_R$) and a function v in $W_{loc}^{1,p}(\mathbb{R}^d)$ such that $(v_R)_R$ weakly converges in $W^{1,p}_{loc}(\mathbb{R}^d)$ towards v, and it is easy to see that v solves, in the sense of distributions in $\mathbb{R}^d$, the equation

$$-\nabla \cdot a(\cdot, r + \nabla v) + T^{-p} |v|^{p-2} v = 0 \text{ in } \mathbb{R}^d.$$

We take the lim inf in (2.5) get

$$c_1 \int_{\mathbb{R}^d} \eta_z^p |r + \nabla v|^p + T^{-p} \int_{\mathbb{R}^d} \eta_z^p |v|^p \leq C \int_{\mathbb{R}^d} \eta_z^p. \tag{2.6}$$

Using the inequality

$$\int_{\mathbb{R}^d} \eta_z^p |\nabla v|^p \leq 2^{p-1} \left(\int_{\mathbb{R}^d} \eta_z^p |r + \nabla v|^p + \int_{\mathbb{R}^d} \eta_z^p |r|^p \right),$$

we infer from (2.6)

$$c_1 \int_{\mathbb{R}^d} \eta_z^p |\nabla v|^p + T^{-p} \int_{\mathbb{R}^d} \eta_z^p |v|^p \leq C \int_{\mathbb{R}^d} \eta_z^p. \tag{2.7}$$

The above inequality implies (2.2) for $R = T$ since the right-hand side does not depend on z. The case $R \geq T$ follows at once.

Let us now show that $v \in B^{1,p}_A(\mathbb{R}^d)$. It suffices to check that v solves the equation

$$M(a(\cdot, r + \nabla v) \cdot \nabla \phi + T^{-p} |v|^{p-2} v\phi) = 0, \text{ all } \phi \in B^{1,p}_A(\mathbb{R}^d). \tag{2.8}$$

To this end, let $\varphi \in \mathcal{C}_0^\infty(\mathbb{R}^d)$ and $\phi \in B^{1,p}_A(\mathbb{R}^d)$. Define (for fixed $\varepsilon > 0$), $\psi(y) = \varphi(\varepsilon y)\phi(y)$. Then $\psi \in W^{1,p}_{loc}(\mathbb{R}^d)$ with compact support. Taking ψ as a test function in the variational formulation of (2.1) yields

$$\int_{\mathbb{R}^d} \left[a(\cdot, r + \nabla v) \cdot (\varepsilon \phi \nabla \varphi(\varepsilon \cdot) + \varphi(\varepsilon \cdot) \nabla \phi) + T^{-p} |v|^{p-2} v \varphi(\varepsilon \cdot) \phi \right] dy = 0.$$

The change of variables $t = \varepsilon y$ leads (after simplification by ε^{-d}) to

$$\int_{\mathbb{R}^d} \left[a^\varepsilon(\cdot, r + (\nabla_y v)^\varepsilon) \cdot (\varepsilon \phi^\varepsilon \nabla \varphi + \varphi (\nabla_y \phi)^\varepsilon) + T^{-p} |v|^{p-2} v \varphi \phi^\varepsilon \right] dt = 0$$

where $w^\varepsilon(t) = w(t/\varepsilon)$. Letting $\varepsilon \to 0$ above yields

$$\int_{\mathbb{R}^d} M(a(\cdot, r + \nabla_y v) \cdot \nabla \phi + T^{-p} |v|^{p-2} v \phi) \varphi dt = 0 \; \forall \varphi \in \mathcal{C}_0^\infty(\mathbb{R}^d), \phi \in B^{1,p}_A(\mathbb{R}^d),$$

which amounts to (2.8). So, we have just shown that, if $v \in W^{1,p}_{loc}(\mathbb{R}^d)$ solves (2.1) in the sense of distributions in $\mathbb{R}^d$, then it solves (2.8).

In order to proceed to the next step, we need to check that (2.8) possesses a unique solution in $B^{1,p}_A(\mathbb{R}^d)$ up to an additive function $w \in B^{1,p}_A(\mathbb{R}^d)$ satisfying $M(|w|^p) = 0$. This amounts to show that the solution of (2.8) is unique in the space $\mathcal{B}^{1,p}_A(\mathbb{R}^d) = B^{1,p}_A(\mathbb{R}^d)/\mathcal{N}$ (where $\mathcal{N} = \{u \in B^{1,p}_A(\mathbb{R}^d) : \|u\|_{1,p} = 0\}$), which is a Banach space with equipped with the norm

$$\|u + \mathcal{N}\|_{1,p} = \left(\|u\|_p^p + \|\nabla u\|_p^p\right)^{\frac{1}{p}} \text{ for } u \in B^{1,p}_A(\mathbb{R}^d).$$

Let us remind that if $w \in \mathcal{N}$ then $M(w) = 0$, since $|M(w)| \leq M(|w|) \leq (M(|w|^p))^{1/p} = \|w\|_p = 0$, so that $\|u + \mathcal{N}\|_{1,p}$ is well defined. Now, coming back to (2.8), one can easily show that the operator

$$P(u + \mathcal{N}) = -\nabla \cdot a(\cdot, r + \nabla u) + T^{-p} |u|^{p-2} u$$

defined on $\mathcal{B}^{1,p}_A(\mathbb{R}^d)$ by duality

$$\langle P(u + \mathcal{N}), v + \mathcal{N}\rangle = M(a(\cdot, r + \nabla u) \cdot \nabla v + T^{-p} |u|^{p-2} uv)$$

satisfies the required assumptions of Browder-Minty theorem in $\mathcal{B}^{1,p}_A(\mathbb{R}^d)$, so that there exists $u \in B^{1,p}_A(\mathbb{R}^d)$ such that $u + \mathcal{N} \in \mathcal{B}^{1,p}_A(\mathbb{R}^d)$ solves uniquely the equation

$$\langle P(u + \mathcal{N}), v + \mathcal{N}\rangle = 0 \text{ for all } v + \mathcal{N} \in \mathcal{B}^{1,p}_A(\mathbb{R}^d). \tag{2.9}$$

We deduce from (2.9) the existence of a unique solution (up to an additive function $w \in B^{1,p}_A(\mathbb{R}^d)$ satisfying $M(|w|^p) = 0$) to (2.8). This yields at once $v \in B^{1,p}_A(\mathbb{R}^d)$.

Step 2. **Uniqueness**. We have shown that for any $T \geq 1$, the problem (2.1) possess a distributional solution $v_T \in B^{1,p}_A(\mathbb{R}^d)$ that is unique up to an additive function $w_T \in B^{1,p}_A(\mathbb{R}^d)$ satisfying $M(|w_T|^p) = 0$. In the linear setting the uniqueness is insured in the class of functions in $B^{1,2}_A(\mathbb{R}^d)$ satisfying (2.2); see e.g. [3, 4]. Since we are in the general nonlinear setting, we consider the uniqueness in the above sense. This concludes the proof. □

We are now able to prove Theorem 1.3.

Proof of Theorem 1.3 *Step 1.* **Existence**. Let us denote by $(v_T)_{T\geq 1}$ a net constructed in Lemma 2.1. It satisfies (2.2), so that by the weak compactness, the sequence $(\nabla v_T)_{T\geq 1}$ weakly converges in $L^p_{loc}(\mathbb{R}^d)^d$ (up to extraction of a subsequence) to some $V \in L^p_{loc}(\mathbb{R}^d)^d$. From the equality $\partial^2 v_T/\partial y_i \partial y_j = \partial^2 v_T/\partial y_j \partial y_i$, a passage to the limit in the distributional sense yields $\partial V_i/\partial y_j = \partial V_j/\partial y_i$, where $V = (V_j)_{1\leq j\leq d}$. This implies $V = \nabla u$ for some $u \in W^{1,p}_{loc}(\mathbb{R}^d)$. Using the boundedness of $(T^{-1} v_T)_{T\geq 1}$ in $L^p_{loc}(\mathbb{R}^d)$, we pass to the limit in the variational formulation of (2.1) (as $T \to \infty$) to get that u solves (1.13). Repeating the proof of (2.8) (in Lemma 2.1), we are led to $V \in B_A^p(\mathbb{R}^d)^d$. Moreover, since $v_T \in B_A^{1,p}(\mathbb{R}^d)$, we have $M(\nabla v_T) = 0$, hence $M(\nabla u) = 0$.

Step 2. **Uniqueness**. As we saw above, repeating the proof of (2.8) we are able to show that $\nabla u \in B_A^p(\mathbb{R}^d)^d$ solves the equation

$$M(a(\cdot, r + \nabla u) \cdot \nabla\phi) = 0 \text{ for any } \phi \in B_{\#A}^{1,p}(\mathbb{R}^d).$$

Now, if u and v are two solutions of (1.13) then the following holds

$$M((a(\cdot, r + \nabla u) - a(\cdot, r + \nabla v)) \cdot \nabla\phi) = 0 \text{ for any } \phi \in B_{\#A}^{1,p}(\mathbb{R}^d). \tag{2.10}$$

Taking in (2.10) $\phi = u - v$, we get, using the monotonicity of a and the reverse Hölder inequality,

$$M(|\nabla u - \nabla v|^p) = 0, \text{ i.e., } \|u - v\|_{\#,p} = 0,$$

thereby proving the uniqueness of the solution in $B_{\#A}^{1,p}(\mathbb{R}^d)$. □

Remark 2.2 The uniqueness proved above entails that the homogenized coefficient is well defined for if u and v are two solutions of (1.13), and if we set $a^*(r) = M(a(\cdot, r + \nabla u))$ and $b^*(r) = M(a(\cdot, r + \nabla v))$, then

$$\left|a^*(r) - b^*(r)\right| \leq \|u - v\|_{\#,p} = 0,$$

so that $a^*(r) = b^*(r)$ for all $r \in \mathbb{R}^d$.

We are now able to prove the homogenization result. We recall that this is a classical result, and the main contribution here is that we have solved the corrector problem in a distributional sense, which has not been done before.

Proof of Theorem 1.1 Let $\Phi_\varepsilon(x) = \psi_0(x) + \varepsilon\psi_1(x, x/\varepsilon)$ where $\psi_0 \in \mathcal{C}_0^\infty(\Omega)$ and $\psi_1 \in \mathcal{C}_0^\infty(\Omega) \otimes A^\infty$, $A^\infty = \{u \in A : D^\alpha u \in A \text{ for all } \alpha \in \mathbb{N}^d\}$. Taking Φ_ε ($\in \mathcal{C}_0^\infty(\Omega)$) as a test function in the variational formulation of (1.5) yields

$$\int_\Omega a^\varepsilon(\cdot, \nabla u_\varepsilon) \cdot \nabla\Phi_\varepsilon dx = \langle f, \Phi_\varepsilon \rangle. \tag{2.11}$$

It is not difficult in seeing that the sequence $(u_\varepsilon)_{\varepsilon>0}$ is bounded in $W_0^{1,p}(\Omega)$, so that, considering an ordinary sequence $E \subset \mathbb{R}_+^*$, there exist a couple $(u_0, u_1) \in W_0^{1,p}(\Omega) \times L^p(\Omega; B_{\#A}^{1,p}(\mathbb{R}^d))$ and a subsequence E' of E such that, as $E' \ni \varepsilon \to 0$,

$$u_\varepsilon \to u_0 \text{ in } W_0^{1,p}(\Omega)\text{-weak and in } L^p(\Omega)\text{-strong} \tag{2.12}$$

$$\nabla u_\varepsilon \to \nabla u_0 + \nabla_y u_1 \text{ in } L^p(\Omega)^d\text{-weak } \Sigma. \tag{2.13}$$

On the other hand

$$\nabla \Phi_\varepsilon = \nabla \psi_0 + (\nabla_y \psi_1)^\varepsilon + \varepsilon (\nabla \psi_1)^\varepsilon \to \nabla \psi_0 + \nabla_y \psi_1 \text{ in } L^p(\Omega)^d\text{-strong } \Sigma. \tag{2.14}$$

Now, using the monotonicity of a we get

$$\int_\Omega \left(a^\varepsilon(\cdot, \nabla u_\varepsilon) - a^\varepsilon(\cdot, \nabla \Phi_\varepsilon)\right) \cdot (\nabla u_\varepsilon - \nabla \Phi_\varepsilon) dx \geq 0,$$

or equivalently,

$$\langle f, u_\varepsilon - \Phi_\varepsilon \rangle - \int_\Omega a^\varepsilon(\cdot, \nabla \Phi_\varepsilon) \cdot (\nabla u_\varepsilon - \nabla \Phi_\varepsilon) dx \geq 0. \tag{2.15}$$

A classical process (see e.g., [8–10]) associated to the convergence results (2.12)–(2.14) yields the following limit problem

$$\begin{cases} \int_\Omega M\left(a(\cdot, \nabla u_0 + \nabla_y u_1) \cdot (\nabla \psi_0 + \nabla_y \psi_1)\right) dx = \langle f, \psi_0 \rangle \\ \forall (\psi_0, \psi_1) \in \mathcal{C}_0^\infty(\Omega) \times (\mathcal{C}_0^\infty(\Omega) \otimes \Lambda^\infty). \end{cases} \tag{2.16}$$

Problem (2.16) above is equivalent to the system

$$\int_\Omega M\left(a(\cdot, \nabla u_0 + \nabla_y u_1) \cdot \nabla \psi_0\right) dx = \langle f, \psi_0 \rangle \;\; \forall \psi_0 \in \mathcal{C}_0^\infty(\Omega) \tag{2.17}$$

$$\int_\Omega M\left(a(\cdot, \nabla u_0 + \nabla_y u_1) \cdot \nabla_y \psi_1\right) dx = 0 \;\; \forall \psi_1 \in \mathcal{C}_0^\infty(\Omega) \otimes A^\infty. \tag{2.18}$$

Taking in (2.18) $\psi_1(x, y) = \varphi(x) v(y)$ with $\varphi \in \mathcal{C}_0^\infty(\Omega)$ and $v \in A^\infty$, we get that it is equivalent to

$$M\left(a(\cdot, \nabla u_0 + \nabla_y u_1) \cdot \nabla_y v\right) = 0 \;\; \forall v \in A^\infty, \tag{2.19}$$

which is, thanks to the density of A^∞ in $B_A^{1,p}(\mathbb{R}^d)$, the weak form of

$$\nabla_y \cdot a(\cdot, \nabla u_0 + \nabla_y u_1) = 0 \text{ in } \mathbb{R}^d, \tag{2.20}$$

of course with respect to the duality defined by (2.19). So fix $r \in \mathbb{R}^d$ and consider the problem

$$-\nabla_y \cdot a(\cdot, r + \nabla_y v_r) = 0 \text{ in } \mathbb{R}^d; \;\; v_r \in B_{\#A}^{1,p}(\mathbb{R}^d). \tag{2.21}$$

Appealing to Theorem 1.3, Eq. (2.21) possesses a unique solution $v_r \in B_{\#A}^{1,p}(\mathbb{R}^d)$. Choosing there $r = \nabla u_0(x)$, we infer from the uniqueness of the solution that

$u_1(x, y) = v_{\nabla u_0(x)}(y)$. Coming back to (2.17) and replacing there u_1 by the above function, we end up with

$$\int_{\Omega} (M(a(\cdot, \nabla u_0 + \nabla_y v_{\nabla u_0}) \cdot \nabla \psi_0 dx = \langle f, \psi_0 \rangle \;\; \forall \psi_0 \in \mathcal{C}_0^{\infty}(\Omega),$$

that is,

$$-\nabla \cdot a^*(\nabla u_0) = f \text{ in } \Omega. \qquad \square$$

3 Proof of Theorem 1.4

In the preceding section, we saw that the corrector problem is posed on the whole of $\mathbb{R}^d$. If besides the coefficient $a(y, \lambda)$ is periodic (that is, the function $y \mapsto a(y, \lambda)$ is Y-periodic ($Y = (0, 1)^d$), then the homogenized coefficient is defined on a bounded domain, say

$$a^*(r) = \int_Y a(y, r + \nabla \chi_r) dy,$$

χ_r being the solution of

$$-\nabla \cdot a(\cdot, r + \nabla \chi_r) = 0 \text{ in } Y, \chi_r \in W_{\#}^{1,p}(Y)$$

with

$$W_{\#}^{1,p}(Y) = \{u \in W_{per}^{1,p}(Y) : \int_Y u dy = 0\}.$$

Contrasting with the periodic setting, the corrector problem in the general deterministic setting cannot be reduced to a problem on a bounded domain. Therefore, truncations must be considered, particularly on large domains like B_R (or Q_R, the closed cube centered at the origin and of side length R) with appropriate boundary conditions. To this end, we proceed exactly as in the random setting (see [1]).

We make a truncation on the cube Q_R ($R > 0$) and impose homogeneous Dirichlet boundary condition on ∂Q_R:

$$-\nabla \cdot a(\cdot, r + \nabla \chi_{r,R}) = 0 \text{ in } Q_R, \;\; \chi_{r,R} \in W_0^{1,p}(Q_R). \tag{3.1}$$

Then problem (3.1) possesses a unique solution. In all what that follows, we use the following notation: for any $u \in L^1_{loc}(\mathbb{R}^d)$ that has a mean value $M(u)$, we set $M(u) = \langle u \rangle$.

The following holds true.

Lemma 3.1 *Problem* (3.1) *possesses a unique solution u that satisfies the estimate*

$$\left(\fint_{Q_R} |\nabla \chi_{r,R}|^p \, dy\right)^{\frac{1}{p}} \leq C \text{ for any } R \geq 1 \tag{3.2}$$

where C is a positive constant independent of R.

Proof The existence of $\chi_{r,R}$ is classical. Let us verify (3.2). Taking $w = \chi_{r,R}$ in the variational formulation of (3.1),

$$\frac{1}{|Q_R|}\int_{Q_R} a(y, r + \nabla \chi_{r,R}) \cdot \nabla_y \chi_{r,R} dy = 0.$$

But

$$a(y, r + \nabla \chi_{r,R}) \cdot \nabla \chi_{r,R} = a(y, r + \nabla \chi_{r,R}) \cdot (r + \nabla \chi_{r,R}) - a(y, r + \nabla \chi_{r,R}) \cdot r,$$

hence

$$\frac{1}{|Q_R|}\int_{Q_R} a(y, r + \nabla \chi_{r,R}) \cdot (r + \nabla \chi_{r,R}) dy = \frac{1}{|Q_R|}\int_{Q_R} a(y, r + \nabla \chi_{r,R}) \cdot r dy,$$

and, because of (1.8),

$$\begin{aligned}
\frac{c_1}{|Q_R|}\int_{Q_R} |r + \nabla \chi_{r,R}|^p \, dy &\leq \frac{1}{|Q_R|}\int_{Q_R} a(y, r + \nabla \chi_{r,R}) \cdot r dy \\
&\leq \frac{|r|}{|Q_R|}\int_{Q_R} |a(y, r + \nabla \chi_{r,R})| \, dy \\
&\leq \frac{c_2 |r|}{|Q_R|}\int_{Q_R} (1 + |r + \nabla \chi_{r,R}|^{p-1}) dy \\
&\leq c_2 |r| + c_2 |r| \left(\frac{1}{|Q_R|}\int_{Q_R} |r + \nabla \chi_{r,R}|^p \, dy\right)^{\frac{1}{p'}} \\
&= c_2 |r| \left(\frac{1}{|Q_R|}\int_{Q_R} |r + \nabla \chi_{r,R}|^p \, dy\right)^{\frac{1}{p'}} + c_2 |r|,
\end{aligned}$$

whence

$$\left(\frac{1}{|Q_R|}\int_{Q_R} |r + \nabla_y \chi_{r,R}|^p \, dy\right)^{\frac{1}{p}} \leq 2\frac{c_2}{c_1}|r| + \left(2\frac{c_2}{c_1}|r|\right)^{\frac{1}{p}}.$$

It follows at once that

$$\begin{aligned}\left(\frac{1}{|Q_R|}\int_{Q_R}|\nabla\chi_{r,R}|^p\,dy\right)^{\frac{1}{p}} &= \left(\frac{1}{|Q_R|}\int_{Q_R}|(r+\nabla\chi_{r,R})-r|^p\,dy\right)^{\frac{1}{p}}\\ &\leq \left(\frac{1}{|Q_R|}\int_{Q_R}|r+\nabla\chi_{r,R}|^p\,dy\right)^{\frac{1}{p}}+|r|\\ &\leq \left(2\frac{c_2}{c_1}+1\right)|r|+\left(2\frac{c_2}{c_1}|r|\right)^{\frac{1}{p}}.\end{aligned}$$

This completes the proof. □

Let $\chi_{r,R}$ be the solution of (3.1). We define the effective and approximate effective coefficients a^* and a^*_R respectively, as follows

$$\begin{cases} a^*(r)=\langle a(\cdot,r+\nabla\chi_r)\rangle \\ a^*_R(r)=⨍_{Q_R} a(y,r+\nabla\chi_{r,R}(y))dy. \end{cases} \tag{3.3}$$

We are able to prove Theorem 1.4.

Proof of Theorem 1.4 We set $w_r^R(y)=\frac{1}{R}\chi_{r,R}(Ry)$ for $y\in Q_1$ and consider the rescaled version of (3.1) whose w_r^R is solution. It reads as

$$-\nabla\cdot a(Ry,r+\nabla_y w_r^R)=0 \text{ in } Q_1,\ \ w_r^R=0 \text{ on } \partial Q_1. \tag{3.4}$$

Then (3.4) possesses a unique solution $w_r^R\in H_0^1(Q_1)$ satisfying the estimate

$$\left\|\nabla w_r^R\right\|_{L^p(Q_1)}\leq C \tag{3.5}$$

where $C>0$ is independent of $R>0$; see (3.2) in the Lemma 3.1. Based on (3.5) and for a fixed $r\in\mathbb{R}^d$, let $w_r\in W_0^{1,p}(Q_1)$ be the weak limit in $W_0^{1,p}(Q_1)$ of a weakly convergent subnet $(w_r^{R'})_{R'}$ of $(w_r^R)_R$. Then it is an easy exercise to see that w_r solves the equation

$$-\nabla\cdot a^*(r+\nabla w_r)=0 \text{ in } Q_1, \tag{3.6}$$

and further thanks to [5, Theorem 5.2], the convergence result (as $R'\to\infty$)

$$a(\cdot,r+\nabla_y w_r^{R'})\to a^*(r+\nabla w_r) \text{ in } L^p(Q_1)^d\text{-weak} \tag{3.7}$$

is satisfied. From the ellipticity property of $a^*(r)$ and the uniqueness of the solution to (3.6) in $W_0^{1,p}(Q_1)$, we deduce that $w_r=0$. We infer that the whole sequence

$(w_r^R)_R$ weakly converges towards 0 in $W_0^{1,p}(Q_1)$. Therefore, integrating (3.7) over Q_1, we get

$$a_R^*(r) = \fint_{Q_1} a(y, r + \nabla\chi_{r,R}(y))dy \to \fint_{Q_1} a^*(r + \nabla w_r)dy = a^*(r)$$

as $R \to \infty$. This completes the proof. □

References

1. A. Bourgeat, A.L. Piatnitski, Approximations of effective coefficients in stochastic homogenization, Ann. Inst. H. Poincaré Probab. Statist. **40** (2004) 153–165.
2. J. Casado Diaz, I. Gayte, The two-scale convergence method applied to generalized Besicovitch spaces. Proc. R. Soc. Lond. A **458** (2002) 2925–2946.
3. A. Gloria, F. Otto, Quantitative results on the corrector equation in stochastic homogenization, J. Eur. Math. Soc. (to appear), arXiv:1409.0801.
4. W. Jäger, J.L. Woukeng, Approximation of homogenized coefficients and convergence rates in deterministic homogenization, Preprint, 2017.
5. V.V. Jikov, S.M. Kozlov, O.A. Oleinik, Homogenization of differential operators and integral functionals, Springer-Verlag, Berlin, 1994.
6. G. Nguetseng, M. Sango, J.L. Woukeng, Reiterated ergodic algebras and applications, Commun. Math. Phys **300** (2010) 835–876.
7. M. Sango, N. Svanstedt, J.L. Woukeng, Generalized Besicovitch spaces and application to deterministic homogenization, Nonlin. Anal. TMA **74** (2011) 351–379.
8. J.L. Woukeng, Deterministic homogenization of non-linear non-monotone degenerate elliptic operators, Adv. Math. **219** (2008) 1608–1631.
9. J.L. Woukeng, Homogenization of nonlinear degenerate non-monotone elliptic operators in domains perforated with tiny holes, Acta Appl. Math. **112** (2010) 35–68.
10. J.L. Woukeng, Σ-convergence of nonlinear monotone operators in perforated domains with holes of small size, Appl. Math. **54** (2009) 465–489.
11. J.L. Woukeng, Introverted algebras with mean value and applications, Nonlinear Anal. TMA **99** (2014) 190–215.

Instantaneous Optimal Control of Friction Dominated Flow in a Gas-Network

Günter Leugering and Gisèle Mophou

Abstract We consider optimal control problems for the flow of gas in a pipe network. The equations of motions are taken to be represented by a nonlinear model derived from a semi-linear approximation of the fully nonlinear isothermal Euler gas equations. We formulate an optimal control problem on a given network and introduce a time discretization thereof. We then study the well-posedness of the corresponding time-discrete optimal control problem. In order to further reduce the complexity, we consider an instantaneous control strategy. This involves a p-Laplace-type problem on the graph with $p = \frac{3}{2}$. We prove well-posedness, existence of optimal controls and derive a first order optimality condition.

Keywords Optimal control · Gas networks · p-Laplace problem on a graph · Optimality system

Mathematics Subject Classification (2010). 35J70, 49J20, 49J45, 93C73

1 Introduction

1.1 Modeling of Gas Flow in a Single Pipe

The Euler equations are given by a system of nonlinear hyperbolic partial differential equations (PDEs) which represent the motion of a compressible non-viscous fluid or a gas. They consist of the continuity equation, the balance of moments and

G. Leugering (✉)
Lehrstuhl Angewandte Mathematik, Lehrstuhl II Universität Erlangen, Erlangen, Germany
e-mail: guenter.leugering@fau.de

G. Mophou
Humboldt Research Chair A.I.M.S. Cameroon, Limbé, Cameroon
e-mail: gisele.mophou@aims-cameroon.org

V. Schulz, D. Seck (eds.), *Shape Optimization, Homogenization and Optimal Control*, International Series of Numerical Mathematics 169,
https://doi.org/10.1007/978-3-319-90469-6_5

the energy equation. The full set of equations is given by (see [4, 9, 10, 12]) Let ρ denote the density, v the velocity of the gas and p the pressure. We further denote g the gravitational constant, λ the friction coefficient of the pipe, D the diameter, a the area of the cross section. The state variables of the system are ρ, the flux $q = \rho v$. We also denote c the speed of sound, i.e. $c^2 = \frac{\partial p}{\partial \rho}$ (for constant entropy). For natural gas we have 340 $\frac{\mathrm{m}}{\mathrm{s}}$. In particular, in the subsonic case ($|v| < c$), the one which we consider in the sequel, two boundary conditions have to be imposed on the left end and one at the right end of the pipe. We consider here the isothermal case only. Thus, for horizontal pipes

$$\begin{aligned} \frac{\partial \rho}{\partial t} + \frac{\partial}{\partial x}(\rho v) &= 0 \\ \frac{\partial}{\partial t}(\rho v) + \frac{\partial}{\partial x}(p + \rho v^2) &= -\frac{\lambda}{2D}\rho v\, |v|\,. \end{aligned} \tag{1.1}$$

In the particular case, where the we have a constant speed of sound $c = \sqrt{\frac{p}{\rho}}$, for small velocities $|v| \ll c$, we arrive at the semi-linear model

$$\begin{aligned} \frac{\partial \rho}{\partial t} + \frac{\partial}{\partial x}(\rho v) &= 0 \\ \frac{\partial}{\partial t}(\rho v) + \frac{\partial p}{\partial x} &= -\frac{\lambda}{2D}\rho v\, |v|\,. \end{aligned} \tag{1.2}$$

If we further neglect the inertia with respect to the flux and introduce $q = \rho v a$, we arrive at

$$\begin{aligned} \frac{\partial p}{\partial t} + \frac{c^2}{a}\frac{\partial}{\partial x} q &= 0 \\ \frac{\partial p^2}{\partial x} &= -\frac{\lambda c^2}{D a^2} q\, |q| =: -\gamma^2 q\, |q|\,. \end{aligned} \tag{1.3}$$

We now set $y := p^2$ and obtain from the second equation in (1.3)

$$q = -\frac{1}{\gamma}\frac{\frac{\partial y}{\partial x}}{\sqrt{\left|\frac{\partial y}{\partial x}\right|}}.$$

With $\alpha := \frac{\gamma a}{c}$ we obtain

$$\alpha \frac{\partial}{\partial t}\frac{y}{\sqrt{|y|}} - \frac{\partial}{\partial x}\frac{\frac{\partial y}{\partial x}}{\sqrt{\left|\frac{\partial y}{\partial x}\right|}} = 0. \tag{1.4}$$

We introduce the monotone function $\beta(s) := \frac{s}{\sqrt{|s|}}$. With this (1.4) reads as

$$\alpha \frac{\partial}{\partial t}\beta(y) - \frac{\partial}{\partial x}\beta(\frac{\partial y}{\partial x}) = 0. \tag{1.5}$$

It is also possible to write this down in the p-Laplace format: (1.4) reads as

$$\alpha \frac{\partial}{\partial t}\left(|y|^{p-2}y\right) - \frac{\partial}{\partial x}\left(|\frac{\partial y}{\partial x}|^{p-2}\frac{\partial y}{\partial x}\right) = 0, \tag{1.6}$$

where $p = \frac{3}{2}$. Equation (1.6) has come to be known as doubly nonlinear parabolic equation of p-Laplace type. See [11]. Notice that $p < 2$ and that the system is, therefore, singular for $\frac{\partial}{\partial x}y(x) = 0$. For $p > 2$ such equations exhibit instead degeneration. Equations similar to (1.5) have been considered in the literature, see e.g. [2, 3]. In this contribution, we aim at a discussion of such equations together with optimal control problems on networks. A more recent study of doubly nonlinear parabolic equations in the context of friction dominated flow has been provided in [1]. Equations of the type (1.5) are known to exhibit positive solutions, finite speed of propagation and satisfy a maximum principle. As a matter of fact, to the best knowledge of the authors, there are no studies on optimal control of such systems on general graphs available from the literature. Therefore, these notes can be seen as the first attempt in that direction. We refer to a forthcoming publication, where the additional properties mentioned, full proofs and numerical analysis are provided [8].

1.2 Network Modeling

Let $G = (V, E)$ denote the graph of the gas network with vertices (nodes) $V = \{n_1, n_2, \dots, n_{|V|}\} = \{n_j | j \in \mathcal{J}$ an edges $E = \{e_1, e_2, \dots, e_{|E|}\} = \{e_i | i \in \mathcal{I}\}$. For the sake of uniqueness, we associate to each edge a direction.

$$d_{ij} = \begin{cases} -1, & \text{if node } n_j \text{ if the edge } e_i \text{ starts at node } n_j e_i, \\ +1, & \text{if node } n_j \text{ if the edges } e_i \text{ end at node } n_j e_i, \\ 0, & \text{else.} \end{cases}$$

The pressure variables $y_i(n_j)$ coincide for all $i \in \mathcal{I}_j := \{i \in 1, \dots E | d_{ij} \neq 0\}$. We express the transmission conditions at the nodes in the following way. We introduce the edge degree $d_j := |\mathcal{I}_j|$. Then the continuity conditions read as follows

$$y_i(n_j, t) = y_k(n_j, t), \; \forall i, k \in \mathcal{I}_j, \; d_j > 1. \tag{1.7}$$

The nodal balance equation for the fluxes can be written as the classical Kirchhoff-type condition

$$\sum_{i \in \mathcal{I}_j} d_{ij}\beta(\partial_x y_i(n_j, t)) = 0, \; d_j > 1. \tag{1.8}$$

We use the d_j in order to decompose the index set for nodes $\mathcal{J}$ into $\mathcal{J} = \mathcal{J}^M \cup \mathcal{J}^S$, where $\mathcal{J}^M = \{j \in \mathcal{J} | d_j > 1\}$ represents the multiple nodes and $\mathcal{J}^S = \{j \in \mathcal{J} | d_j = 1\}$ the simple nodes. According to Dirichlet or Neumann boundary conditions a the simple nodes, we further decompose $\mathcal{J}^S = \mathcal{J}^S_D \cup \mathcal{J}^S_N\}$. We summarize the equations as follows:

$$\begin{aligned} \alpha_i \partial_t \beta(y_i(x,t)) + \partial_x \left(\beta(\partial_x y_i(x,t))\right) = 0,\ i \in \mathcal{I},\ x \in (0,\ell_i),\ t \in (0,T) \\ y_i(n_j,t) = y_k(n_j,t),\ \forall i,k \in \mathcal{I}_j,\ j \in \mathcal{J}^M,\ t \in (0,T) \\ \sum_{i \in \mathcal{I}_j} d_{ij}\beta(\partial_x y_i(n_j,t)) = 0,\ j \in \mathcal{J}^M,\ t \in (0,T) \\ y_i(n_j,t) = 0, i \in \mathcal{I}_j, j \in \mathcal{J}^S_D,\ t \in (0,T) \\ d_{ij}\beta(\partial_x y_i(n_j,t) = u_j(t),\ \ i \in \mathcal{I}_j, j \in \mathcal{J}^S_N,\ t \in (0,T) \\ p_i(x,0) = p_{i,0}(x),\ q_i(x,0) = q_{i0}(x),\ x \in (0,\ell_i) \end{aligned} \tag{1.9}$$

To the best knowledge of the authors, problem (1.9), no published result seems to be available.

1.2.1 Optimal Control Problems and Outline

We are now in the position to formulate optimal control problems on the level of the gas networks. We first describe the general format for an optimal control problem associated with the semi-linear model equations. This involves a cost function that assigns to admissible each pair (y,u) a 'cost' $I(y,u)$, which is represented on each individual edge by a contribution on the state $I_i(y)$ and the controls acting at simple nodes. The typical example, the one that we will use in the sequel is given by

$$I_i(y_i)(x) := \frac{1}{p}|y_i(x) - y_i^d(x)|^p,\ x \in (0,\ell_i), p \in \{\frac{3}{2}, 2\}.$$

$$\begin{aligned} \min_{(y,u)\in\Xi} I(y,u) := \sum_{i \in \mathcal{I}} \int_0^T \int_0^{\ell_i} I_i(y_i) dx dt + \frac{\nu}{2} \sum_{j \in \mathcal{J}^S_N} \int_0^T |u_j(t)|^2 dt \\ s.t. \\ (y,u) \text{ satisfies } (1.9), \end{aligned} \tag{1.10}$$

$$\Xi := \{(y,u) : \underline{y}_i \le y_i \le \overline{y}_i,\ i \in \mathcal{I}, \underline{u_j} \le u_j \le \overline{u_j}, j \in \mathcal{J}^S_N\}. \tag{1.11}$$

In (1.11), the quantities $\underline{y}_i, \overline{y}_i$ are given constants that determine the feasible pressures and flows in the pipe i, while $\underline{u}_i, \overline{u}_i$ describe control constraints. In the continuous-time case the inequalities are considered as being satisfied for all times and everywhere along the pipes. Due to limitations in space, we will not consider state- and control constraints here and refer instead to a forthcoming article [8]. Instead, we just penalize the control costs using $\nu > 0$. Moreover, we will restrict ourselves to time discretizations of (1.10) and, in fact, to the instantaneous control regime that has come to be known also as rolling horizon problem.

1.3 Time Discretization

We, therefore, consider the time discretization (1.9) such that $[0, T]$ is decomposed into break points $t_0 = 0 < t_1 < \cdots < t_N = T$ with widths $^{\Delta}t_n := t_{n+1} - t_n, n = 0, \ldots, N-1$ (we use $N+1$ as the number of break points which is not related to N as indicating Neumann conditions). Accordingly, we denote $p_i(x, t_n) := p_{i,n}(x), q_i(x, t_n) := q_{i,n}(x), n = 0, \ldots, N-1$. We consider a semi-implicit Euler scheme which takes p_i in the friction term in an explicit manner.

$$
\begin{aligned}
\frac{1}{\Delta t}\beta(y_{i,n+1})(x) - \partial_x\left(\beta(\partial_x y_{i,n+1}(x))\right) &= \frac{1}{\Delta t}\beta(y_{i,n})(x),\ i \in \mathcal{I},\ x \in (0, \ell_l) \\
y_{i,n+1}(n_j) &= y_{k,n+1}(n_j),\ \forall i, k \in \mathcal{I}_j,\ j \in \mathcal{J}^M \\
\sum_{i \in \mathcal{I}_j} d_{ij}\beta(\partial_x y_{i,n+1})(n_j) = 0,\ j \in \mathcal{J}^M & \\
\beta(\partial_x y_{i,n+1})(n_j) &= u_{j,n+1},\ d_j = 1,\ i \in \mathcal{I}_j,\ j \in \mathcal{J}_N^S \\
y_{i,n+1}(n_j) &= 0,\ i \in \mathcal{I}_j,\ j \in \mathcal{J}_D^S \\
y_{i,0}(x) &= y_{i,0}(x),\ i \in \mathcal{I},\ x \in (0, \ell_i),\ n = 1, \ldots, N-1.
\end{aligned}
\tag{1.12}
$$

We then obtain the optimal control problem on the time-discrete level:

$$
\begin{aligned}
&\min_{(y,u)} I(y, u) := \sum_{i \in \mathcal{I}} \sum_{n=1}^{N} \int_0^{\ell_i} I_i(y_{i,n})dx + \frac{\nu}{2} \sum_{j \in \mathcal{J}_N^S} \sum_{n=1}^{N} |u_j(n)|^2 \\
&\quad s.t. \\
&\quad (y, u) \text{ satisfies } (1.12)
\end{aligned}
\tag{1.13}
$$

It is clear that (1.13) involves all time steps in the cost functional. We would like to reduce the complexity of the problem even further. To this aim we consider the *instantaneous control* regime. This amount to reducing the sums in the cost function of (1.13) to the time-level t_{n+1}. This strategy has also come to be known as *rolling horizon* approach, the simplest case of the *moving horizon* paradigm. Thus, for each $n = 1, \dots, N-1$ and given $y_{i,n}$, we consider the problems

$$\min_{(y,u)} I(y,u) := \sum_{i\in\mathcal{I}} \int_0^{\ell_i} I_i(y_i)dx + \frac{\nu}{2} \sum_{j\in\mathcal{J}_N^S} |u_j|^2$$
$$s.t. \tag{1.14}$$
$$(y,u) \text{ satisfies (1.12) at time level } n+1.$$

It is now convenient to discard the actual time level $n+1$ and redefine the states at the former time as input data. To this end, we replace $\alpha_i := \frac{1}{\Delta t}$, $f_i^1 := \alpha_i \beta(y_{i,n})$, rewrite (1.12) as

$$\begin{aligned}
\alpha_i \beta(y_i)(x) - \partial_x \left(\beta(\partial_x y_i(x))\right) &= f_i^1(x),\ i \in \mathcal{I},\ x \in (0,\ell_i)\\
y_i(n_j) &= y_k(n_j),\ \forall i,k \in \mathcal{I}_j,\ j \in \mathcal{J}^M\\
\sum_{i\in\mathcal{I}_j} d_{ij}\beta(\partial_x y_i)(n_j) &= 0,\ j \in \mathcal{J}^M\\
\beta(\partial_x y_i)(n_j) &= u_j,\ d_j = 1,\ i \in \mathcal{I}_j,\ j \in \mathcal{J}_N^S\\
y_i(n_j) &= 0,\ i \in \mathcal{I}_j,\ j \in \mathcal{J}_D^S
\end{aligned} \tag{1.15}$$

Example We consider a star-graph with three edges (tripod). We take four numerical examples in order to illustrate the behavior of the system with respect to parameter changes. In Fig. 1 we take the parameters $\alpha_i = 10$, $\gamma_i = 1$, $f_i = 10$; right: $\alpha_i = 10$, $\gamma_i = 100$, $f_i = 10$, where γ puts a weight on the nonlinearity. In the second Fig. 2

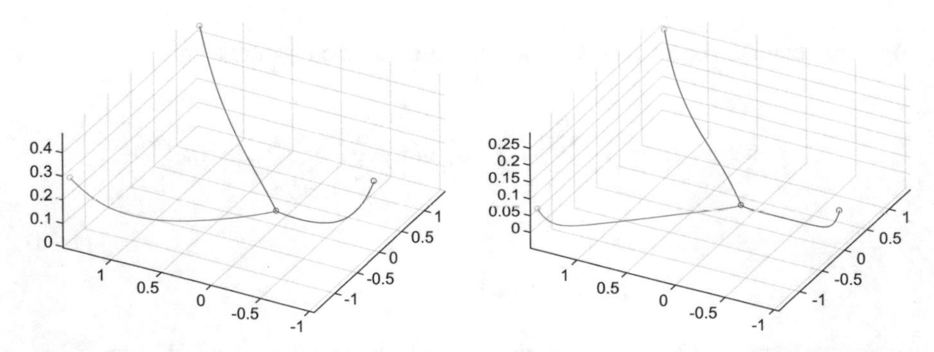

Fig. 1 Solutions $y_i, i = 1,2,3$ of (1.15) plotted as vertical displacement for different parameter settings; left: $\alpha = 10$, $\gamma = 1$, $f_i = 10$; right: $\alpha = 10$, $\gamma = 100$, $f_i = 10$

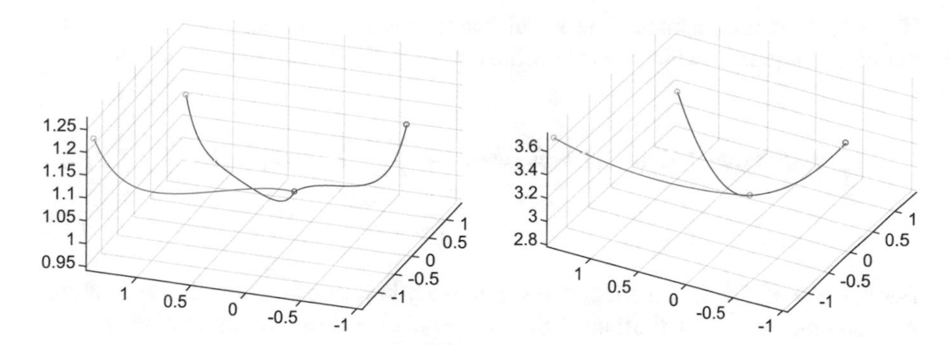

Fig. 2 Solutions $y_i, i = 1, 2, 3$ of (1.15) plotted as vertical displacement for different parameter settings; left: $\alpha = 100, \gamma = 1, f_i = 10$; right: $\alpha = 1, \gamma = 0.1, f_i = 10$

we take all parameters equal and set $\alpha = 100, \gamma = 1, f_i = 10$; right: $\alpha = 1, \gamma = 0.1, f_i = 10$. The calculations are done with the routine *bvp4c.m* from MATLAB (R2017a).

We now consider in the rest of the paper the following optimal control problem:

$$\min_{(y,u)} I(y, u) := \sum_{i \in \mathcal{I}} \int_0^{\ell_i} I_i(y_i)dx + \frac{\nu}{2} \sum_{j \in \mathcal{J}_N^S} |u_j|^2$$

$$s.t. \tag{1.16}$$

$$(y, u) \text{ satisfies (1.15).}$$

2 Well-Posedness

We begin our analysis of (1.15) with a Lagrange identity. To this end, we multiply the equation by a test function ϕ and integrate by parts.

$$0 = \sum_{i \in \mathcal{I}} \int_0^{\ell_i} (\alpha_i \beta(y_i) - \partial_x(\beta(\partial_x y_i)) - f_i) \phi_i dx \tag{2.1}$$

$$= \sum_{i \in \mathcal{I}} \int_0^{\ell_i} \left(\alpha_i \beta(y_i)\phi_i + (\beta(\partial_x y_i))\partial_x \phi_i - f_i \phi_i\right) \phi_i dx$$

$$- \sum_{j \in \mathcal{J}^M} \sum_{i \in \mathcal{I}_j} d_{ij} \beta(\partial_x y_i(n_j))\phi_i(n_j) - \sum_{j \in \mathcal{J}_N^S} \sum_{i \in \mathcal{I}_j} u_j \phi_i(n_j) \tag{2.2}$$

If the test function satisfies the nodal conditions of order zero, we arrive at the variational equation which serves for the proper definition of weak solutions

$$\sum_{i\in\mathcal{I}}\int_0^{\ell_i}\left(\alpha_i\beta(y_i)\phi_i+(\beta(\partial_x y_i))\partial_x\phi_i\right)\phi_i dx=\sum_{j\in\mathcal{J}_N^S}\sum_{i\in\mathcal{I}_j}u_j\phi_i(n_j)+\sum_{i\in\mathcal{I}}\int_0^{\ell_i}f_i\phi_i. \tag{2.3}$$

For the sake of simpler notation, we switch freely between the formulation involving $\beta(\cdot)$ and the explicit definition of $\beta(s)=\frac{s}{\sqrt{|s|}}$. Therefore, in our analysis $p=\frac{3}{2}$, even though, in many ways the results produced here are valid for general $p\le 2$. For the general case, we refer to [8]. It is clear then that the dual space of $L^p(0,\ell_i)$ is $L^q(0,\ell_i)=L^3(0,\ell_i)$. The mapping $\beta(\cdot)$ satisfies

$$\beta: L^p(0,\ell_i)\to L^q(0,\ell_i),\ \text{is bounded.}$$

Indeed, $\|\beta(\phi)\|_{L^q(0,\ell_i)}\le\|\phi\|_{L^p(0,\ell_i)}^{\frac{1}{2}}$. This applies of course also to $\beta(\partial_x\phi)$. We now define the strong version of the guiding operator $\mathcal{A}$:

$$\begin{aligned}(\mathcal{A}y)_{i\in\mathcal{I}}&:=(\alpha_i\beta(y_i)-\partial_x(\beta(\partial_x y_i)))_{i\in\mathcal{I}} \qquad (2.4)\\ D(\mathcal{A})&:=\Big\{(y_i)_{i\in\mathcal{I}}\in\Pi_{i\in\mathcal{I}}W^{1,p}(0,\ell_i)|\partial_x\beta(\partial_x y_i)\in L^q(0,\ell_i),\\ &\sum_{i\in\mathcal{I}_j}d_{ij}\beta(\partial_x y_i(n_j))\phi_i(n_j)=0,\\ &y_i(n_j)=y_k(n_j),i,k\in\mathcal{I}_j,j\in\mathcal{J}^M,\\ &y_i(n_j)=0,i\in\mathcal{I}_j,j\in\mathcal{J}_D^S,\beta(\partial_x y_i(n_j))=0,i\in\mathcal{I}_j,j\in\mathcal{J}_N^S\Big\}\end{aligned}$$

As for the weak formulation, we define the energy space

$$\begin{aligned}V:=\Big\{&(y_i)_{i\in\mathcal{I}}\in\Pi_{i\in\mathcal{I}}W^{1,p}(0,\ell_i)|y_i(n_j)=y_k(n_j),i,k\in\mathcal{I}_j,j\in\mathcal{J}^M,\\ &y_i(n_j)=0,i\in\mathcal{I}_j,j\in\mathcal{J}_D^S\Big\},\end{aligned} \tag{2.5}$$

where $\|\phi\|_V := \sum_{i\in\mathcal{I}} \|\phi_i\|_{W^{1,p}(0,\ell_i)}$. Thus, for $y \in D(\mathcal{A})$ we obtain

$$\langle \mathcal{A}(y), \phi\rangle := \sum_{i\in\mathcal{I}} \int_0^{\ell_i} (\alpha_i\beta(y_i) - \partial_x(\beta(\partial_x y_i))\,\phi_i dx \tag{2.6}$$

$$= \sum_{i\in\mathcal{I}_j} \int_0^{\ell_i} \left(\alpha_i\beta(y_i)\phi_i + (\beta(\partial_x y_i))\partial_x\phi_i\right)\phi_i dx$$

$$= \langle \mathcal{A}(y), \phi\rangle_{V^*,V},\ \forall \phi \in V \tag{2.7}$$

Using the boundedness of $\beta(\cdot)$, we infer that $\mathcal{A} : V \to V^*$ is bounded as a nonlinear operator. Indeed, for an open neighbourhood $U(0)$ of 0 in V,

$$\langle \mathcal{A}(y), \phi\rangle_{V^*,V}$$
$$\leq C\sum_{i\in\mathcal{I}} \|y_i\|_{W^{1,p}(0,\ell_i)}\|\phi\|_{W^{1,p}(0,\ell_i)} \leq C\|\phi\|_V, \forall y \in U(0) \subset V, \phi \in V.$$

We obtain the energy form

$$\langle \mathcal{A}(y), y\rangle = \sum_{i\in\mathcal{I}_j} \int_0^{\ell_l} \alpha_i|y_i|^p + |\partial_x y_i|^p dx \geq \underline{\alpha}\|y\|_V^p. \tag{2.8}$$

In order to further investigate the properties of $\mathcal{A}$, we recall [5, 7]

$$\left(|a|^p a - |b|^p b\right)(a-b) \geq (1 + |a| + |b|)^{p-2}|a-b|^2,\ 1 \leq p \leq 2.$$

From this it follows

$$\langle \mathcal{A}(y) - \mathcal{A}(z), y - z\rangle_{V^*,V} \geq c\|y - z\|_V^2 > 0,\ \forall y, z \in V, \tag{2.9}$$

for some $c > 0$. The inequality (2.9) implies that $\mathcal{A}$ is strictly monotone. We also easily verify

$$\frac{\langle \mathcal{A}(y), y\rangle_{V^*,V}}{\|y\|_V} \to \infty, \quad \text{as } \|y\|_V \to \infty,$$

i.e. the coercivity of $\mathcal{A}$. We have

$$\langle \mathcal{A}(y^k) - \mathcal{A}(y^0), \phi\rangle_{V^*,V}$$
$$= \sum_{i\in\mathcal{I}_j} \int_0^{\ell_i} \alpha_i(\beta(y_i^k) - \beta(y_i^0))\phi_i + (\beta(\partial_x y_i^k) - \beta(\partial_x y_i^0))\partial_x\phi_i dx \to 0,$$

as $y^k \to y^0$ in V for $k \to \infty$. This shows that $\mathcal{A}$ is demi-continuous. Summarizing, we have shown that $\mathcal{A}$ is demi-continuous, coercive and strictly, monotone. Applying the classical Brezis' result (see e.g. [11]), we infer

$$\forall f \in V^* \exists ! y \in V : \mathcal{A}(y) = f.$$

The sense of being a solution is weak, as $y \in V$. Clearly, more regular data imply strong solutions. If we consider the boundary controls as

$$f_u(\phi) := \sum_{j \in \mathcal{J}_N^S} \sum_{i \in \mathcal{I}_j} u_j \delta_{n_j}(\phi),$$

then $f_u \in V^*$ and we obtain the following

Theorem 2.1 *For $f \in \Pi_{i \in \mathcal{I}} L^3(0, \ell_i)$, $u \in \mathbb{R}^{|\mathcal{J}_N^S|}$, problem* (1.15) *admits a unique weak solution $y \in V$.*

Even though, the mapping $\beta(\cdot)$ is differentiable in $\mathbb{R} \setminus \{0\}$, the control-to-state-mapping $u \to y^u$ is not Gâteaux differentiable for $p < 2$. This has already been observed in [6]. However, the control-to-state-map is continuous. Indeed, let $(u^k)_k$ be a sequence of controls that converges to u^0 in $\mathbb{R}^{|\mathcal{J}_N^S|}$. For the corresponding solutions $y^k := y^{u^k}$, $y^0 := y^{u^0}$, we obtain

$$\sum_{i \in \mathcal{I}} \int_0^{\ell_i} |(|\beta(y_i^k) - \beta(y_i^0))(y_i^k - y_i^0)) + \beta(\partial_x y_i^k) - \beta(\partial_x y_i^0))(\partial_x y_i^k - \partial_x y_i^0) dx$$
$$= \sum_{j \in \mathcal{J}_N^S} \sum_{i \in \mathcal{I}_j} (u_j^k - u_i^0)(y_i^k - y_i^0)(n_j).$$

By the continuity of $\beta(\cdot)$ and the strong convergence of y^k to y^0 in V, we obtain

Theorem 2.2 *The mapping $u \to y^u$, where y^u solves* (1.15) *is continuous between $\mathbb{R}^{|\mathcal{J}_N^S|}$ and V.*

3 Optimal Control

We are now in the position to prove existence of optimal pairs (y, u) for the optimal control problem (1.16). Indeed, the control-to-state-map is continuous and the cost function is strictly convex and lower-semi-continuous with respect to the strong topologies.

Theorem 3.1 *The optimal control problem* (1.16) *admits a unique solution* $(\bar{y}, \bar{u}) \in V \times \mathbb{R}^{|\mathcal{J}_N^S|}$.

The proof follows classical arguments and is omitted here. It becomes now important to investigate optimality conditions. Clearly, this involves adjoint states, satisfying the adjoint problem. However, the adjoint problem, in turn, involves the linearization of the state equation along the optimal trajectory. As the nonlinearities of the state equation are governed by the mapping $\beta(\cdot)$ and the derivative of this mapping $\beta'(s) = \frac{1}{2\sqrt{|s|}}$ is unbounded inn the neighbourhood of $s = 0$, we need an argument that the optimal solution y_i together with its derivative $\partial_x y_i$ stays away from 0. We begin with a regularization of the problem, as proposed in [5, 6]. The regularization is made in order to bring us back into the standard $p = 2$ setting.

$$\begin{aligned}
&\epsilon y_i(x) + \alpha_i \beta(y_i)(x) - \partial_x (\epsilon \partial_x y_i(x) + \beta(\partial_x y_i(x))) = f_i^1(x),\ i \in \mathcal{I},\ x \in (0, \ell_i)\\
&y_i(n_j) = y_k(n_j),\ \forall i, k \in \mathcal{I}_j,\ j \in \mathcal{J}^M\\
&\sum_{i \in \mathcal{I}_j} d_{ij}(\epsilon \partial_x y_i(n_j) + \beta(\partial_x y_i)(n_j) = 0,\ j \in \mathcal{J}^M\\
&\epsilon \partial_x y_i(n_j) + \beta(\partial_x y_i)(n_j) = u_j,\ d_j = 1,\ i \in \mathcal{I}_j,\ j \in \mathcal{J}_N^S\\
&y_i(n_j) = 0,\ i \in \mathcal{I}_j,\ j \in \mathcal{J}_D^S.
\end{aligned} \tag{3.1}$$

We formulate the $\epsilon - perturbed$ optimal control problem as follows.

$$\begin{aligned}
&\min_{(y^\epsilon, u^\epsilon)} I_\epsilon(y^\epsilon, u^\epsilon) := \sum_{i \in \mathcal{I}} \int_0^{\ell_i} I_i(y_i^\epsilon) dx + \frac{\nu}{2} \sum_{j \in \mathcal{J}^S} |u_j^\epsilon|^2 + \frac{\nu}{2} \sum_{j \in \mathcal{J}^S} |u_j^\epsilon - \bar{u}_j|^2\\
&\qquad s.t.\\
&(y^\epsilon, u^\epsilon) \text{ satisfies (3.1)},
\end{aligned} \tag{3.2}$$

where $\bar{u}$ denotes the optimal control for the unperturbed problem. Problem (3.2) is a standard optimal control problem, where the control-to-state-map is now continuously differentiable. The following theorem, therefore, can be stated without dwelling on the proof.

Theorem 3.2 *For each* $\epsilon > 0$, *there exists an optimal pair* y^ϵ, u^ϵ *and an adjoint state* p^ϵ *such that* y^ϵ, p^ϵ *stay in*

$$W := \Big\{ y \in \Pi_{i \in \mathcal{I}} H^1(0, \ell_i) | y_i(n_j) = y_k(n_j), i, k \in \mathcal{I}_j, j \in \mathcal{J}^M, y_i(n_j) = 0\ ,\\ i \in \mathcal{I}_j, j \in \mathcal{J}_D^S. \Big\}$$

and satisfy the following optimality system.

$$
\begin{aligned}
&\epsilon y_i + \alpha_i \beta(y_i) - \partial_x\left(\epsilon \partial_x y_i + \beta(\partial_x y_i)\right) = f_i,\ i \in \mathcal{I},\ x \in (0,\ell_i)\\
&\epsilon p_i + \alpha_i \beta'(y_i) p_i - \partial_x\left(\epsilon \partial_x p_i + \beta'(\partial_x y_i \partial_x p_i)\right)\\
&\quad = \kappa(|y_i - y_i^d|^{p-2}(y_i - y_i^d),\ i \in \mathcal{I},\ x \in (0,\ell_i)\\
&\quad y_i(n_j) = y_k(n_j),\ \forall i,k \in \mathcal{I}_j,\ p_i(n_j) = p_k(n_j),\ \forall i,k \in \mathcal{I}_j,\ j \in \mathcal{J}^M\\
&\quad \sum_{i \in \mathcal{I}_j} d_{ij}(\epsilon \partial_x y_i(n_j) + \beta(\partial_x y_i)(n_j) = 0,\ j \in \mathcal{J}^M\\
&\quad \sum_{i \in \mathcal{I}_j} d_{ij}(\epsilon \partial_x p_i(n_j) + \beta'(\partial_x y_i)(n_j)\partial_x p_i(n_j) = 0,\ j \in \mathcal{J}^M\\
&\quad \epsilon \partial_x y_i(n_j) + \beta(\partial_x y_i)(n_j) = u_j,\ d_j = 1,\ i \in \mathcal{I}_j,\ j \in \mathcal{J}_N^S\\
&\quad \epsilon \partial_x p_i(n_j) + \beta'(\partial_x y_i)(n_j)\partial_x p_i(n_j) = 0,\ d_j = 1,\ i \in \mathcal{I}_j,\ j \in \mathcal{J}_N^S\\
&\quad y_i(n_j) = 0,\ p_i(n_j) = 0\ i \in \mathcal{I}_j,\ j \in \mathcal{J}_D^S\\
&\quad p_i^\epsilon(n_j) + \nu u_j^\epsilon + (u_j^\epsilon - \bar{u}_j) = 0,\ i \in \mathcal{I}_j, j \in \mathcal{J}_N^S.
\end{aligned}
\tag{3.3}
$$

In order to understand the limiting procedure as $^\epsilon \to 0$, we look at the adjoint equation and multiply it by $p\epsilon$. We obtain

$$
\int_0^{\ell_i} \epsilon(p_i^\epsilon)^2 + \alpha_i \beta'(y_i^\epsilon)(p_i^\epsilon)^2 + (\epsilon \partial_x p_i^\epsilon + \beta'(\partial_x y_i^\epsilon)\partial_x p_i^\epsilon)\partial_x p_i^\epsilon dx
$$

$$
= \int_0^{\ell_i} |y_i^\epsilon - y_i^d|^{p-2}(y_i^\epsilon - y_i^d) p_i^\epsilon dx.
$$

If now $y_i^\epsilon \to y_i$ strongly in W, then

$$
\epsilon + \alpha_i \beta'(y_i^\epsilon) \Longrightarrow \alpha_i \beta'(y_i^\epsilon), \quad \epsilon + \alpha_i \beta'(\partial_x y_i^\epsilon) \Longrightarrow \alpha_i \beta'(\partial_x y_i^\epsilon).
$$

Moreover, for bounded $(y_i^\epsilon)_\epsilon, (p_i^\epsilon)_\epsilon$, the terms in the equation stay bounded and if $p_i^\epsilon \to p_i$ (even weakly), the limiting equation is

$$
\int_0^{\ell_i} \alpha_i \beta'(y_i)(p_i)^2 + (\beta'(\partial_x y_i)\partial_x p_i)\partial_x p_i^\epsilon dx = \int_0^{\ell_i} |y_i - y_i^d|^{p-2}(y_i - y_i^d) p_i dx.
$$

We are now in the position to show the following limiting result.

Theorem 3.3 *There exists $\bar{p} \in V$ satisfying together with the optimal pair $(\bar{y}, \bar{u})$ the first order optimality condition.*

$$
\begin{aligned}
\alpha_i \beta(\bar{y}_i) - \partial_x \left(\beta(\partial_x \bar{y}_i)\right) &= f_i,\ i \in \mathcal{I},\ x \in (0, \ell_i) \\
\alpha_i \beta'(\bar{y}_i)\bar{p}_i - \partial_x \left(\beta'(\partial_x \bar{y}_i \partial_x \bar{p}_i)\right) &= \kappa(|\bar{y}_i - y_i^d|^{p-2}(\bar{y}_i - y_i^d),\ i \in \mathcal{I},\ x \in (0, \ell_i) \\
\bar{y}_i(n_j) = \bar{y}_k(n_j),\ \forall i, k \in \mathcal{I}_j,\ \bar{p}_i(n_j) &= \bar{p}_k(n_j),\ \forall i, k \in \mathcal{I}_j,\ j \in \mathcal{J}^M \\
\sum_{i \in \mathcal{I}_j} d_{ij}(+ \beta(\partial_x \bar{y}_i)(n_j) &= 0,\ j \in \mathcal{J}^M \\
\sum_{i \in \mathcal{I}_j} d_{ij}(+ \beta'(\partial_x \bar{y}_i)(n_j)\partial_x \bar{p}_i(n_j) &= 0,\ j \in \mathcal{J}^M \\
\beta(\partial_x \bar{y}_i)(n_j) &= \bar{u}_j,\ d_j = 1,\ i \in \mathcal{I}_j,\ j \in \mathcal{J}_N^S \\
\beta'(\partial_x \bar{y}_i)(n_j)\partial_x \bar{p}_i(n_j) &= 0,\ d_j = 1,\ i \in \mathcal{I}_j,\ j \in \mathcal{J}_N^S \\
\bar{y}_i(n_j) = 0,\ \bar{p}_i(n_j) &= 0\ i \in \mathcal{I}_j,\ j \in \mathcal{J}_D^S \\
\bar{p}_i^\epsilon(n_j) + \nu \bar{u}_j &= 0,\ i \in \mathcal{I}_j, j \in \mathcal{J}_N^S.
\end{aligned}
\tag{3.4}
$$

Example We consider the tripod above, where we use $\alpha_i = \alpha = 10$, $f_i = f = 10$ and demonstrate the effect of the second order nonlinear operator by choosing coefficients $\gamma_i = \gamma = 1$ or 100. The penalty for $\kappa_i = \kappa$ is chosen equal to 1000. For the calculations, we used a regularization, as in the text as $\epsilon = 0.01$. The numerical studies reveal that the system behaviour does not largely depend on ϵ as long this number is in order of magnitude smaller than the resolution tolerance which was set here to $1e - 4$. The optimality system is solved using the MATLAB routine *bvp4c*. See Figs. 3 and 4.

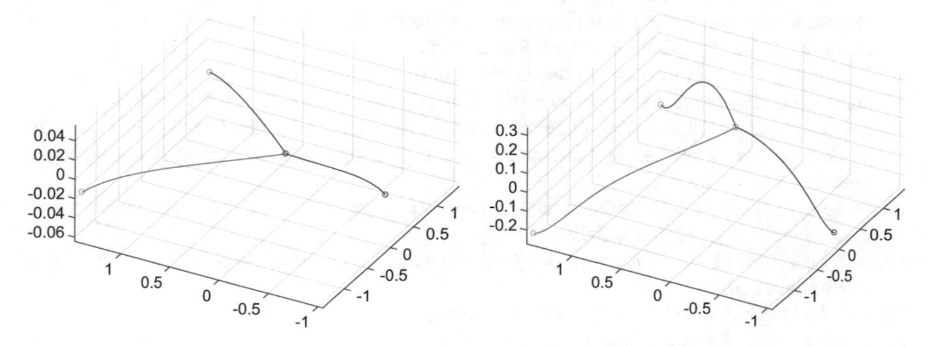

Fig. 3 Parameter setting: $\alpha = 10$, $\gamma = 1$, $f_i = 10$, $\kappa = 1000$, left: optimal states y_i, $i = 1, 2, 3$; right: adjoint states p_i, $i = 1, 2, 3$

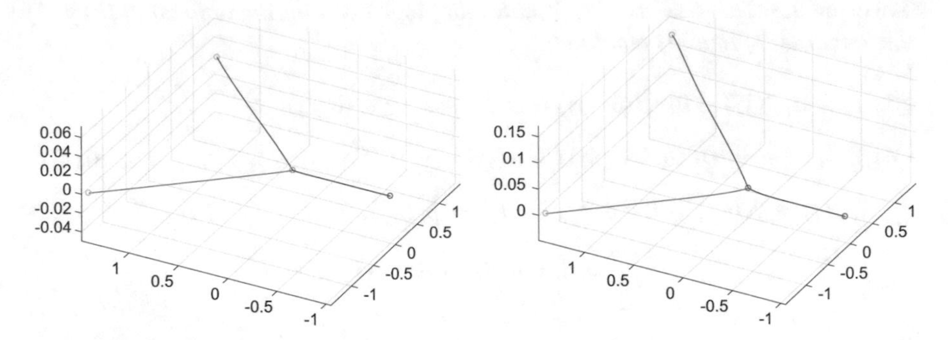

Fig. 4 Parameter setting $\alpha = 10, \gamma = 100, f_i = 10, \kappa = 1000$, left: optimal states $y_i, i = 1, 2, 3$; right: adjoint states $p_i, i = 1, 2, 3$

Acknowledgment The first author was supported by the DFG-TRR 154 "Modellierung Simulation und Optimierung am Beispiel von Gasnetzwerken" (TPA05), the second author was supported by the Alexander von Humboldt foundation, under the programme financed by the BMBF entitled "German research Chairs".

References

1. R. Alonso, M. Santillana, and C. Dawson. On the diffusive wave approximation of the shallow water equations. *European J. Appl. Math.*, 19(5):575–606, 2008.
2. A. Bamberger. étude d'une équation doublement non linéaire. *J. Functional Analysis*, 24(2):148–155, 1977.
3. A. Bamberger, M. Sorine, and J. P. Yvon. Analyse et contrôle d'un réseau de transport de gaz. In *Computing methods in applied sciences and engineering (Proc. Third Internat. Sympos., Versailles, 1977), II*, volume 91 of *Lecture Notes in Phys.*, pp. 347–359. Springer, Berlin/New York, 1979.
4. J. Brouwer, I. Gasser, and M. Herty. Gas pipeline models revisited: Model hierarchies, nonisothermal models, and simulations of networks. 9(2):601–623.
5. E. Casas and L. A. Fernández. Optimal control of quasilinear elliptic equations. In *Control of partial differential equations (Santiago de Compostela, 1987)*, volume 114 of *Lect. Notes Control Inf. Sci.*, pp. 92–99. Springer, Berlin, 1989.
6. E. Casas and L. A. Fernández. Optimal control of quasilinear elliptic equations with nondifferentiable coefficients at the origin. *Rev. Mat. Univ. Complut. Madrid*, 4(2–3):227–250, 1991.
7. P. I. Kogut and G. Leugering. *Optimal Control Problems for Partial Differential Equations on Reticulated Domains*. Systems and Control: Foundations and Applications. Springer, New York
8. G. Leugering and G. Mophou. Optimal control of a doubly nonlinear parabolic equation on a metric graph, in preparation, 2017.
9. R. J. Le Veque. *Finite Volume Methods for Hyperbolic Problems*. Cambridge University Press, Cambridge, 2002
10. R. J. Le Veque. *Numerical Methods for Conservation Laws*. Birkhäuser, Basel, 1992
11. T. Roubí ček. *Nonlinear partial differential equations with applications*, volume 153 of *International Series of Numerical Mathematics*. Birkhäuser/Springer, Basel, second edition, 2013.
12. Joel Smoller. *Shock Waves and Reaction-Diffusion Equations*, volume 258 of *Grundlehren der mathematischen Wissenschaften*. Springer, New York, 1983.

Multilevel Iterations for the Combined Moving Horizon Estimation and Nonlinear Model Predictive Control for PDE Models

Ekaterina Kostina and Gregor Kriwet

Abstract The application of model-based optimization methods has become indispensable in science and engineering. Since all models contain systematic errors, the result of optimization can be very sensitive to such errors and even useless. Therefore, application of mathematical optimization to real-life processes demands taking into account uncertainties. One of the remedies is Nonlinear Model Predictive Control (NMPC), which is based on two steps: a simultaneous on-line estimation of the system state and parameters (Moving Horizon Estimation, MHE) and re-optimization of the optimal control for the current parameter and state values (NMPC). The challenge is to solve these optimal control problems with high frequency in real time. During the last decade significant progress has been made in development of so-called multilevel real-time iterations for NMPC for ODE and DAE models to reduce the response time to a few milliseconds compared with standard methods. However, today's state-of-the-art is to perform MHE and NMPC separately. A next logical step is the development of a simultaneous MHE and NMPC in one step. Another challenge is to transfer recent advances of NMPC to a much more complex case of nonstationary PDE models.

In this paper we present combined multilevel real-time iterations based on coupling of the MHE and NMPC for nonstationary PDE models. These are the first steps towards real time feasibility of the NMPC technique for PDE models.

Keywords Optimal control · Feedback control · Nonstationary PDE · Moving horizon estimation · Nonlinear model predictive control · Multilevel iteration

Mathematics Subject Classification (2010). Primary 49J21, 49M30

E. Kostina (✉)
IWR and IAM, Heidelberg University, Heidelberg, Germany
e-mail: kostina@uni-heidelberg.de

G. Kriwet
P3 Group, Stuttgart, Germany
e-mail: Gregor.Kriwet@p3-group.com

V. Schulz, D. Seck (eds.), *Shape Optimization, Homogenization and Optimal Control*, International Series of Numerical Mathematics 169,
https://doi.org/10.1007/978-3-319-90469-6_6

1 Introduction

In industrial and manufacturing applications model-based optimal control of processes is a crucial task. Nevertheless, under uncertainties or perturbations the real process might behave different than predicted from the mathematical model. Remedy is feedback control techniques e.g. nonlinear model predictive control (NMPC) being the most prominent approach. NMPC is based on two steps: a simultaneous on-line estimation of the system states and parameters (Moving Horizon Estimation, MHE) and re-optimization of the optimal control for the current parameter and state values. The challenge is to solve these optimal control problems with high frequency in real time.

Recent years have seen tremendous advances in the theory of NMPC and the development of efficient numerical methods for ODE and DAE models that consider the different time scales, both in requirements of the control problem and in the computational efforts of the components of the numerical optimization procedures. In these new methods the optimization algorithm is no longer a lower level procedure driven by perturbations of the states, rather the progress of the real life process and of the iterative solution of the optimization process are properly nested. This nested approach has led to the development of the so-called "real time iteration" schemes for the DAE case, see e.g. [6, 7]. The main idea of the "real time iteration" approach is to embed perturbed values of states and parameters into linear but possibly infeasible constraint. This allows to divide the iteration into two phases: the feedback phase, where the current feedback control is computed fast, and the preparation phase during which all functions values, gradients, Jacobians, Hessian and the decomposition of quadratic subproblems are performed, before the current states and parameters are known. This so called real-time iteration reduces the delay in the control response to the feedback phase, which is only a small fraction of the cost of one SQP iteration. Control responses can be drastically accelerated even further if e.g. additional cheap inexact SQP iterations are performed during the preparation phase. Two cheap variants of the real-time iteration are feasibility (requires additional evaluation of the constraint residuals) and optimality (requires the solution of a dual problem to generate the Lagrange function gradient) improvement. A proper nesting of all the variants of this so-called "multilevel iteration" results in a drastic acceleration of the computation of feedback controls, such that even very fast processes can be properly controlled which require feedback response at high tact rate. The details of the "multilevel iteration" and results of numerical studies for DAE and ODE models can be found in [4, 8, 9].

In this paper we address two new issues. First, we would like to apply the above principle to nonstationary PDE models. To our knowledge, so far applications of NMPC to PDE were based on a reduction to the DAE case. However, in PDE and large-scale DAE case neither does measurement information yield accurate information of complete state and parameters nor is this necessary for effective, e.g. stabilizing, NMPC. Hence our second idea is to combine MHE and NMPC in "one step" and to generalize multilevel iteration for simultaneous MHE and NMPC.

This allows to exploit more structure in the problem taking into account that in practical applications changes in the system estimates should span a reduced space since they come from discrete measurements. The results of preliminary numerical experiments show that the MHE/NMPC multilevel iteration has a potential to make the NMPC technique real time feasible even for PDE applications.

The paper is organized as follows. The Sect. 2 describes the problem formulations and presents necessary theoretical background. Decompositions of NMPC and MHE problems are derived in Sects. 3 and 4. Section 5 presents the details of the MHE/NMPC multilevel iterations. Section 6 finishes the paper with the benchmark problem and the results of numerical experiments.

2 Problem Formulation: NMPC and MHE

We consider an optimal control problem for a process which is modeled by a nonlinear parabolic PDE. The objective to be minimized is given by the function

$$\|f[u,q]\|^2_{L^2([0,T]\times\Omega)} = \int_0^T \int_\Omega \|f[u(\tau,x),q(\tau,x)]\|^2\,dx\,d\tau.$$

Here the process u satisfies

$$\begin{aligned} \partial_t u + A(u,p,q) &= 0 \qquad \text{on } [0,T]\times\Omega, \\ u(0) &= u_0 \qquad \text{on } \quad \Omega. \end{aligned} \tag{2.1}$$

where the boundary conditions are already included in the elliptic operator $A(u,p,q)$.

The states of the process u live on the bounded spatial domain Ω and we are interested in a time evolution of the process over the time interval $[0,T]$. We assume that the dynamics of the process depends on the parameters $p \in \mathbb{R}^{n_p}$, and controls $q \in \mathbb{R}^{n_q}$. Under standard assumptions [12, 16] the nonlinear parabolic PDE has a unique solution in the vector space

$$u \in W([0,T]) := \left\{u \in L^2([0,T],H^1(\Omega)) \text{ and } \partial_t u \in L^2([0,T],H^{1^*}(\Omega))\right\}.$$

Since the embedding $H^1(\Omega) \overset{d}{\hookrightarrow} L^2(\Omega)$ is continuous and dense, according to [12, 16, 18] the Gelfand triple $H^1(\Omega) \overset{d}{\hookrightarrow} L^2(\Omega) \overset{d}{\hookrightarrow} H^1(\Omega)^*$ assures the embedding

$$W([0,T]) \hookrightarrow C([0,T],L^2(\Omega)).$$

Furthermore, for two arbitrary functions $u, v \in W([0, T])$ and time points $0 \leq s \leq t \leq T$ the integration by parts satisfies the expression

$$\langle u(t), v(t) \rangle_{L^2} - \langle u(s), v(s) \rangle_{L^2} = \int_s^t \left(\langle \partial_t u, v \rangle_{H^{1*} \times H^1} + \langle \partial_t v, u \rangle_{H^{1*} \times H^1} \right).$$

The notation $\langle \cdot, \cdot \rangle_{V \times V^*}$ is used for the dual pairing between the Banach space V and its dual space V^* with $< \cdot, \cdot >_H$ denoting the scalar product of a Hilbert space H. This embedding allows to impose initial values in $L^2(\Omega)$ and to perform evaluations of the states at specific time points.

For any real-world dynamic process the model (2.1) is in general only an approximation. In the presence of disturbances and modeling errors, the real process will never follow the "off-line" computed optimal solution. One of the possibilities to deal with such disturbances is to compute feedback controls by using e.g. NMPC, which is based on two steps: a simultaneous on-line estimation of the system state and parameters and re-optimization of the optimal control for the current parameter and state values, [1, 2, 5, 13, 14]. The second step is realized by solving an open-loop optimal control problem on a smaller moving prediction horizon $[t, t+k]$ subject to a PDE model, which depends on possibly perturbed initial values $\hat{s}_t$ and parameters $\hat{p}$ at a current time t:

$$\begin{aligned}
\min_{u \in W([t,t+k]), p \in \mathbb{R}^{n_p}, q \in C_q} \quad & \frac{1}{2} \| f[u, q] \|^2_{L^2(\Omega_t)} + \frac{1}{2} \| E[u(t+k) - \bar{u}_{t+k}] \|^2_{L^2(\Omega)} \\
\text{s.t.} \quad & \partial_t u + A(u, p, q) = 0, \\
& u\big|_t = \hat{s}_t, \; p = \hat{p}.
\end{aligned}$$

Here $\Omega_t = [t, t+k] \times \Omega$ and the second term in the cost functional is introduced for taking into account the remaining part of the whole time interval and for asymptotic convergence to a stabilized steady state. Obviously, using the first instant of the optimal open loop control $q(t; \hat{s}_t, \hat{p})$ we obtain a feedback control. Determination of the current values of states and parameters is left to an estimation procedure based on measurements of functions of the states. The most effective method appears to be a moving horizon estimation (MHE) approach in which the present states are estimated from measurements over a past estimation horizon.

In MHE we consider only measurements from the estimation horizon $[t-k, t]$. Measurements outside the estimation horizon are taken into account in the so called arrival costs. The parameters and states are estimated by minimizing the arrival costs and the norm of the deviation between the measurements η_i and the measurement function

$$\begin{aligned}
\mathcal{M} : L^2(\Omega) \times \mathbb{R}^{n_p} &\longrightarrow \mathbb{R} \\
(u(t_i), p) &\longmapsto \mathcal{M}(u(t_i), p).
\end{aligned}$$

Without loss of generality, we assume that we have only one measurement at each time point. Then the MHE problem with arrival costs can be expressed as follows:

$$\begin{aligned} \min_{u(t_i)\in L^2(\Omega),\, i=t-k,\dots,t,\, p} \quad & \sum_{i=t-k}^{t} \left(\mathcal{M}(u(t_i), p) - \eta_i\right)^2 + \| \left(p - \hat{p}\right) \|^2_{W_p} + \\ & \| \left(s_{t-k} - \hat{s}_{t-k}\right) \|^2_{W_s} \\ \text{s.t.} \quad & \partial_t u + A(u, p, q) = 0 \qquad \text{on } [t-k, t] \times \Omega. \end{aligned} \tag{2.2}$$

As we have only a finite number of measurements a meaningful feedback control can not be applied for distributed controls in $L^2(\Omega)$. We assume that the control is finite dimensional $q(t) \in \mathbb{R}^{n_{q(t)}}$, nevertheless the control can act on parts of the domain. Furthermore we approximate the control with piece-wise polynomials in time, thus we obtain a finite dimensional control denoted by $q \in C_q \subset \mathbb{R}^{n_q}$.

3 Decomposition of the MHE Problem

In order to estimate the current parameters and states the MHE problem (2.2) has to be solved over the estimation horizon. This MHE problem is parametrized with a direct multiple shooting approach as presented in [3, 10, 17]. We denote with $s_{t-i} = u(t-i) \in L^2(\Omega)$ the multiple shooting node values and with $S(s_{t-i}, p, q)$ the solution operator of the PDE on the multiple shooting interval $[t-i, t-i+1]$ with initial value s_{t-i}

$$\begin{aligned} \partial_t u + A(u, p, q) &= 0 \qquad && \text{on } [t-i, t-i+1] \times \Omega, \\ u\big|_{t-i} &= s_{t-i} && \text{on } \Omega. \end{aligned}$$

The continuity of the PDE solution is maintained by the matching conditions

$$r_{t-i} := S(t-i; s_{t-i-1}, p, q) - s_{t-i} = 0.$$

Then, the parametrized MHE problem to estimate the current parameter and states can be written as follows:

$$\begin{aligned} \min_{s_i \in L^2(\Omega),\, p \in \mathbb{R}^{n_p}} \quad & \left\| \begin{pmatrix} \mathcal{M}(s_{t-k}, p) - \eta_{t-k} \\ \vdots \\ \mathcal{M}(s_t, p) - \eta_t \end{pmatrix} \right\|^2 + \| \left(p - \hat{p}\right) \|^2_{W_p} \\ & + \| \left(s_{t-k} - \hat{s}_{t-k}\right) \|^2_{W_s}, \end{aligned}$$

$$s_{t-k+1} = S(s_{t-k}, p, q)(t-k+1),$$
$$\vdots$$
$$s_t = S(s_{t-1}, p, q)(t).$$

The MHE problem is solved with a Gauss Newton method. At each iteration of the method a linearized parameter estimation problem is solved for the current iterate $p, s_{t-k}, \ldots, s_t$. For notation purposes we combine the model responses in one vector F_1 and define the corresponding Jacobian as

$$F_1 = \begin{pmatrix} \mathcal{M}(s_{t-k}, p) - \eta_{t-k} \\ \vdots \\ \mathcal{M}(s_t, p) - \eta_t \end{pmatrix} \in \mathbb{R}^m, \qquad J = \begin{pmatrix} \partial_{s_{t-k}}\mathcal{M} & & \\ & \ddots & \\ & & \partial_{s_t}\mathcal{M} \end{pmatrix},$$

where $\partial_{s_i}\mathcal{M} := \partial_{s_i}\mathcal{M}(s_i, p)$ is the Fréchet derivative of the model response to the states. The linearized MHE problem can be written as follows

$$\min_{\Delta s_i,\, \Delta p} \left\| J \begin{pmatrix} \Delta s_{t-k} \\ \vdots \\ \Delta s_t \\ \Delta p \end{pmatrix} + F_1 \right\|^2 + \| (p - \hat{p}) \|_{W_p}^2 + \| (s_{t-k} - \hat{s}_{t-k}) \|_{W_s}^2,$$
$$\Delta s_{t-k+1} = \partial_s S \Delta s_{t-k} + \partial_p S \Delta p + r_{t-k+1},$$
$$\vdots$$
$$\Delta s_t = \partial_s S \Delta s_{t-1} + \partial_p S \Delta p + r_t.$$

With a condensing strategy similarly to [15] the matching constraints and the shooting node values may be eliminated from the linearized problem. Thereby we obtain an affine dependency for the increments of the shooting node states, depending on the parameter $p \in \mathbb{R}^{n_p}$ and the initial state values $\Delta s_{t-k} \in L^2(\Omega)$:

$$\begin{aligned} \Delta s_i &= P^i \Delta p + G^i \Delta s_{t-k} + b^i, \qquad i = t-k, \ldots, t, \\ P^i &: \mathbb{R}^{n_p} \longrightarrow L^2(\Omega), \\ G^i &: L^2(\Omega) \longrightarrow L^2(\Omega), \\ b^i &\in L^2(\Omega). \end{aligned}$$

Exploiting the boundary value problem structure of the constraints, the linear condensing operators P^i, G^i and the affine term b^i can be computed using the following recursion

$$\begin{aligned} P^{t-k} &= 0 \quad \text{and } P^{i+1} = \partial_s S(t_{i+1})(s_i, p) P^i + \partial_p S(t_{i+1})(s_i, p), \\ G^{t-k} &= \mathbb{I} \quad \text{and } G^{i+1} = \partial_s S(t_{i+1})(s_i, p) G^i, \\ b^{t-k} &= r_{t-k} \text{ and } b^{i+1} = \partial_s S(t_{i+1})(s_i, p) b^i + r_{i+1}. \end{aligned}$$

For the further notation we define the reduced Jacobians J_p, J_g and combine the residuals at the shooting nodes in a vector b

$$J_p = \begin{pmatrix} \partial_{s_{t-k}} \mathcal{M} P^{t-k} \\ \vdots \\ \partial_{s_t} \mathcal{M} P^t \end{pmatrix}, \qquad J_g = \begin{pmatrix} \partial_{s_{t-k}} \mathcal{M} G^{t-k} \\ \vdots \\ \partial_{s_t} \mathcal{M} G^t \end{pmatrix}, \qquad b = \begin{pmatrix} b^{t-k} \\ \vdots \\ b^t \end{pmatrix}.$$

Exploiting the structure imposed by multiple shooting we obtain the reduced linearized MHE problem

$$\min_{\Delta s_{t-k}, \Delta p} \left\| \begin{pmatrix} 0 & W_s^{1/2} \\ W_p^{1/2} & 0 \\ J_p & J_g \end{pmatrix} \begin{pmatrix} \Delta p \\ \Delta s_{t-k} \end{pmatrix} + \begin{pmatrix} W_s^{1/2}(s_{t-k} - \hat{s}_{t-k}) \\ W_p^{1/2}(p - \hat{p}) \\ F_1 + J\, b \end{pmatrix} \right\|^2_{L^2(\Omega) \times \mathbb{R}^{n_p + m}} \tag{3.1}$$

Now let us discuss how to compute the reduced Jacobians and the condensed operators P^i and G^i.

The condensed linear operators $P^i : \mathbb{R}^{n_p} \longrightarrow L^2(\Omega)$ map the n_p dimensional vector space $\mathbb{R}^{n_p}$ onto the function space of the multiple shooting node values $L^2(\Omega)$. Since the condensed operators P_i are linear, it is sufficient to evaluate the operator only for a basis $(e_1, \ldots, e_{n_p})$ of $\mathbb{R}^{n_p}$. The operator applied to the basis element e_j can be computed by solving the linearized PDE

$$\begin{aligned} \partial_t y_j + \partial_u A(u, p, q) y_j &= -\partial_{p_j} A(u, p, q) \\ y_j \big|_{t-i} &= P^{t-i} e_j, \end{aligned}$$

and the condensed operator $P^{t-i+1} e_j = y_j(t-i+1)$ can be evaluated as indicated in [11]. Thus in order to compute the operators P^i we have to solve n_p times the linearized PDEs with different right hand sides.

The operator $G^i : L^2(\Omega) \longrightarrow L^2(\Omega)$ maps the infinite dimensional vector space of the initial states into an infinite dimensional space and cannot be approximated efficiently. Fortunately, we need only products

$$\partial_{s_i} \mathcal{M}\, G^i : L^2(\Omega) \longrightarrow \mathbb{R}, \qquad i = t-k, \ldots, t.$$

Since we have only a finite number of measurements m, each row of the product can be interpreted as an element of the dual space. Hence, we can compute the product $\partial_{s_i}\mathcal{M}\, G^i$ by solving m adjoint PDEs for each row g_i, $i = 1, \ldots, m$, of the reduced Jacobian J_g. In the case of one measurement at each time point, the row g_i solves the system

$$
\begin{aligned}
&-\partial_t g_i + \partial_u A^*(u, p, q)\, g_i = 0\\
&\qquad (g_i(i), \lambda)_{L^2(\Omega)} = \partial_{s_{t-k-1+i}}\mathcal{M}[\lambda], \quad \lambda \in L^2(\Omega).
\end{aligned}
$$

Finally, the reduced Jacobian can be computed using the vectors g_i, $i = 1, \ldots, m$, and the corresponding scalar product

$$
J_g\, \Delta s_{t-k} = \begin{pmatrix} \langle g_{t-k}(t-k), \Delta s_{t-k}\rangle_{L^2(\Omega)} \\ \vdots \\ \langle g_t(t-k), \Delta s_{t-k}\rangle_{L^2(\Omega)} \end{pmatrix}.
$$

The following theorem shows, that the solution of the reduced MHE problem can be decomposed into two parts $\Delta s_{\mathcal{N}}$ and $\Delta s_{\mathcal{R}}$, where $\Delta s_{\mathcal{N}}$ does not depend on the measurements and $\Delta s_{\mathcal{R}}$ solves a finite dimensional least squares problem. In the formulation $\mathcal{N}(J_g)$ denotes the kernel of J_g, $\mathcal{R}(W_s^{-1} J_g^*)$ denotes the range space of $W_s^{-1} J_g^*$, $\mathcal{P}_{\mathcal{N},\mathcal{R}}$ denotes the projection along the range space $\mathcal{R}(W_s^{-1} J_g^*)$ onto the space $\mathcal{N}(J_g)$ and finally $\mathcal{P}_{\mathcal{R},\mathcal{N}}$ denotes the projection along $\mathcal{N}(J_g)$ onto $\mathcal{R}(W_s^{-1} J_g^*)$:

$$
\begin{aligned}
&\mathcal{P}_{\mathcal{N},\mathcal{R}}\, s = s, \quad \forall s \in \mathcal{N}(J_g), \qquad \text{and} \quad \mathcal{P}_{\mathcal{N},\mathcal{R}}\, s = 0, \quad \forall s \in \mathcal{R}(W_s^{-1} J_g^*),\\
&\mathcal{P}_{\mathcal{R},\mathcal{N}}\, s = s, \quad \forall s \in \mathcal{R}(W_s^{-1} J_g^*), \quad \text{and} \quad \mathcal{P}_{\mathcal{R},\mathcal{N}}\, s = 0, \quad \forall s \in \mathcal{N}(J_g).
\end{aligned}
$$

Theorem *The initial values Δs_{t-k} of the reduced linearized MHE problem (3.1) can be decomposed into*

$$
\Delta s_{t-k} = \Delta s_{\mathcal{N}} + \Delta s_{\mathcal{R}}, \quad \Delta s_{\mathcal{N}} \in \mathcal{N}(J_g), \quad \textit{and} \quad \Delta s_{\mathcal{R}} \in \mathcal{R}(W_s^{-1} J_g^*), \tag{3.2}
$$

which satisfy

$$
\Delta s_{\mathcal{N}} = \mathcal{P}_{\mathcal{N},\mathcal{R}}\, \Delta s_{t-k} \qquad \textit{and} \qquad \Delta s_{\mathcal{R}} = \mathcal{P}_{\mathcal{R},\mathcal{N}}\, \Delta s_{t-k}.
$$

Furthermore, $\Delta s_{\mathcal{N}}$ does not depend on the measurements and is the projection of $\hat{s} - s$ along the range space $\mathcal{R}(W_s^{-1} J_g^*)$ onto the space $\mathcal{N}(J_g)$

$$
\Delta s_{\mathcal{N}} = \mathcal{P}_{\mathcal{N},\mathcal{R}}\, (\hat{s} - s).
$$

The part $\Delta s_{\mathcal{R}}$ solves the finite dimensional least squares problem

$$\min_{\substack{p \in \mathbb{R}^{n_p} \\ \Delta s_{\mathcal{R}} \in \mathcal{R}(W_s^{-1} J_g^*)}} \left\| \begin{pmatrix} W_p^{\frac{1}{2}} & 0 \\ 0 & W_s^{\frac{1}{2}} \\ J_p & J_g \end{pmatrix} \begin{pmatrix} \Delta p \\ \Delta s_{\mathcal{R}} \end{pmatrix} + \begin{pmatrix} W_p^{\frac{1}{2}}(p - \hat{p}) \\ W_s^{\frac{1}{2}}(s_{t-k} - \hat{s}_{t-k}) \\ F_1 \end{pmatrix} \right\|^2 . \quad (3.3)$$

Proof For the purpose of verifying the decomposition (3.2), we have to show that the intersection of the two spaces contains only the null element. Let $x \in \mathcal{N}(J_g) \cap \mathcal{R}(W_s^{-1} J_g^*)$ be arbitrary. Since $x \in \mathcal{R}(W_s^{-1} J_g^*)$ there exist $y \in \mathbb{R}^m$ with $x = W_s^{-1} J_g^* y$.

In order to simplify the notation we drop the time index of the initial values. The parameters and the initial values solve the normal equation system corresponding to problem (3.1)

$$\begin{pmatrix} W_p + J_p^* J_p & J_p^* J_g \\ J_g^* J_p & W_s + J_g^* J_g \end{pmatrix} \begin{pmatrix} \Delta p \\ \Delta s \end{pmatrix} = - \begin{pmatrix} W_p(p - \hat{p}) + J_p^* F_1 \\ W_s(s - \hat{s}) + J_g^* F_1 \end{pmatrix}.$$

The last row of the normal equation system can be rewritten as

$$\begin{aligned} & J_g^* J_p \Delta p + (W_s + J_g^* J_g)(\mathcal{P}_{\mathcal{N},\mathcal{R}} + \mathcal{P}_{\mathcal{R},\mathcal{N}})\Delta s = -W_s(s - \hat{s}) + J_g^* F_1 \\ \Leftrightarrow \quad & s^* \left(J_g^* J_p \Delta p + (W_s + J_g^* J_g)(\mathcal{P}_{\mathcal{N},\mathcal{R}} + \mathcal{P}_{\mathcal{R},\mathcal{N}})\Delta s \right) = \\ & - s^* \left(W_s(s - \hat{s}) + J_g^* F_1 \right) \quad \forall s \in \mathbb{R}^{n_s} \end{aligned}$$

$$\Leftrightarrow \quad \begin{cases} s^* \left(J_g^* J_p \Delta p + (W_s + J_g^* J_g)(\mathcal{P}_{\mathcal{N},\mathcal{R}} + \mathcal{P}_{\mathcal{R},\mathcal{N}})\Delta s \right) = \\ \qquad -s^* \left(W_s(s - \hat{s}) + J_g^* F_1 \right) \quad \forall s \in \mathcal{R}(W_s^{-1} J_g^*) \\ \\ s^* \left(J_g^* J_p \Delta p + (W_s + J_g^* J_g)(\mathcal{P}_{\mathcal{N},\mathcal{R}} + \mathcal{P}_{\mathcal{R},\mathcal{N}})\Delta s \right) = \\ \qquad -s^* \left(W_s(s - \hat{s}) + J_g^* F_1 \right) \quad \forall s \in \mathcal{N}(J_g) \end{cases}$$

$$\Leftrightarrow \quad \begin{cases} s^* \mathcal{P}_{\mathcal{R},\mathcal{N}}^* \left(J_g^* J_p \Delta p + (W_s + J_g^* J_g)(\mathcal{P}_{\mathcal{N},\mathcal{R}} + \mathcal{P}_{\mathcal{R},\mathcal{N}})\Delta s \right) = \\ \qquad -s^* \mathcal{P}_{\mathcal{R},\mathcal{N}}^* \left(W_s(s - \hat{s}) + J_g^* F_1 \right) \quad \forall s \in \mathbb{R}^{n_s} \\ s^* \mathcal{P}_{\mathcal{N},\mathcal{R}}^* \left(J_g^* J_p \Delta p + (W_s + J_g^* J_g)(\mathcal{P}_{\mathcal{N},\mathcal{R}} + \mathcal{P}_{\mathcal{R},\mathcal{N}})\Delta s \right) = \\ \qquad -s^* \mathcal{P}_{\mathcal{N},\mathcal{R}}^* \left(W_s(s - \hat{s}) + J_g^* F_1 \right) \quad \forall s \in \mathbb{R}^{n_s}. \end{cases}$$

In order to simplify the normal equation system we use the properties of the subspaces $\mathcal{R}(W_s^{-1}J_g^*)$ and $\mathcal{N}(J_g)$

$$J_g \mathcal{P}_{\mathcal{N},\mathcal{R}} = 0 \qquad \text{and} \qquad \mathcal{P}_{\mathcal{N},\mathcal{R}}^* J_g^* = 0.$$

Let $y \in \mathcal{R}(W_s^{-1}J_g^*)$ and $x \in \mathcal{N}(J_g)$, then there exist $\tilde{y} \in \mathbb{R}^m$ with $W_s^{-1}J_g^*\tilde{y} = y$ and

$$y^* W_s x = (W_s^{-1}J_g^*\tilde{y})^* W_s x = \tilde{y}^* J_g W_s^{-1} W_s x = \tilde{y}^* \underbrace{J_g x}_{=0} = 0.$$

Consequently, we obtain $\mathcal{P}_{\mathcal{R},\mathcal{N}}^* W_s \mathcal{P}_{\mathcal{N},\mathcal{R}} = 0$ and $\mathcal{P}_{\mathcal{N},\mathcal{R}}^* W_s \mathcal{P}_{\mathcal{R},\mathcal{N}} = 0$.

With these properties the normal equation system can be simplified to

$$\begin{pmatrix} W_p + J_p^T J_p & J_p^T J_g \mathcal{P}_{\mathcal{R},\mathcal{N}} & 0 \\ \mathcal{P}_{\mathcal{R},\mathcal{N}}^* J_g^T J_p & \mathcal{P}_{\mathcal{R},\mathcal{N}}^* \left(W_s + J_g^T J_g\right) \mathcal{P}_{\mathcal{R},\mathcal{N}} & 0 \\ 0 & 0 & \mathcal{P}_{\mathcal{N},\mathcal{R}}^* W_s \mathcal{P}_{\mathcal{N},\mathcal{R}} \end{pmatrix} \begin{pmatrix} \Delta p \\ \Delta s_{\mathcal{R}} \\ \Delta s_{\mathcal{N}} \end{pmatrix}$$

$$= - \begin{pmatrix} W_p(p - \hat{p}) + J_p^T F_1 \\ \mathcal{P}_{\mathcal{R},\mathcal{N}}^* \left(W_s(s - \hat{s}) + J_g^T F_1\right) \\ \mathcal{P}_{\mathcal{N},\mathcal{R}}^* W_s(s - \hat{s}) \end{pmatrix}.$$

Obviously, there is no coupling between $(\Delta p, \Delta s_{\mathcal{R}})$ and $\Delta s_{\mathcal{N}}$, thus we can compute them separately. For the solution on the subspace $\mathcal{N}(J_g)$ we obtain

$$\begin{aligned} & \mathcal{P}_{\mathcal{N},\mathcal{R}}^* W_s \mathcal{P}_{\mathcal{N},\mathcal{R}} \Delta s_{\mathcal{N}} = -\mathcal{P}_{\mathcal{N},\mathcal{R}}^* W_s(s - \hat{s}) \\ \Leftrightarrow \quad & \mathcal{P}_{\mathcal{N},\mathcal{R}}^* W_s \mathcal{P}_{\mathcal{N},\mathcal{R}} \Delta s_{\mathcal{N}} = \mathcal{P}_{\mathcal{N},\mathcal{R}}^* W_s \underbrace{\left(\mathcal{P}_{\mathcal{N},\mathcal{R}} + \mathcal{P}_{\mathcal{R},\mathcal{N}}\right)}_{=\mathbb{I}}(\hat{s} - s) \\ \Leftrightarrow \quad & \mathcal{P}_{\mathcal{N},\mathcal{R}}^* W_s \mathcal{P}_{\mathcal{N},\mathcal{R}} \Delta s_{\mathcal{N}} = \mathcal{P}_{\mathcal{N},\mathcal{R}}^* W_s \mathcal{P}_{\mathcal{N},\mathcal{R}}(\hat{s} - s) + \underbrace{\mathcal{P}_{\mathcal{N},\mathcal{R}}^* W_s \mathcal{P}_{\mathcal{R},\mathcal{N}}}_{=0}(\hat{s} - s) \\ \Leftrightarrow \quad & \mathcal{P}_{\mathcal{N},\mathcal{R}}^* W_s \mathcal{P}_{\mathcal{N},\mathcal{R}} \Delta s_{\mathcal{N}} = \mathcal{P}_{\mathcal{N},\mathcal{R}}^* W_s \mathcal{P}_{\mathcal{N},\mathcal{R}} \mathcal{P}_{\mathcal{N},\mathcal{R}}(\hat{s} - s) \\ \Leftrightarrow \quad & \Delta s_{\mathcal{N}} = \mathcal{P}_{\mathcal{N},\mathcal{R}}(\hat{s} - s). \end{aligned}$$

The part of the normal equation system corresponding to $(\Delta p, \Delta s_{\mathcal{R}}) \in \mathbb{R}^{n_p} \times \mathcal{R}(W_s^{-1}J_g^*)$ is the normal equations of the following least squares problem

$$\min_{p, \Delta s_{\mathcal{R}}} \left\| \begin{pmatrix} W_p^{1/2} & 0 \\ 0 & W_s^{1/2}\mathcal{P}_{\mathcal{R},\mathcal{N}} \\ J_p & J_g \mathcal{P}_{\mathcal{R},\mathcal{N}} \end{pmatrix} \begin{pmatrix} \Delta p \\ \Delta s_{\mathcal{R}} \end{pmatrix} + \begin{pmatrix} W_p^{1/2}(p - \hat{p}) \\ W_s^{1/2}(s - \hat{s}) \\ F_1 \end{pmatrix} \right\|^2.$$

As the projection $\mathcal{P}_{\mathcal{R},\mathcal{N}}$ acts like the identity on the subspace $\mathcal{R}(W_s^{-1}J_g^*)$ we can omit it. This finishes the proof of the theorem. □

Using the solution of the reduced MHE (3.1), the update in the shooting node values can be expressed as

$$\Delta s_i = P^i \Delta p + G^i \Delta s_{\mathcal{R}} + G^i \mathcal{P}_{\mathcal{R},\mathcal{N}}(s - \hat{s}) + b^i \quad i = t-k, \ldots, t.$$

The theorem implies the subspace $\mathcal{N}$ does not depend on $\hat{s} - s$ but only on measurements. Since $\Delta s_{\mathcal{N}}$ does not depend on the actual measurements, we can compute the update of the shooting node values online without solving a PDE. Furthermore, the weights W_s of the arrival costs $\| \left(s - \hat{s}\right) \|_{W_s}^2$ determine only the orientation of the subspace $\mathcal{R}$ compared to $\mathcal{N}$. The subspace $\mathcal{R}$ is spanned by the vectors $\tilde{g}_v := W_s^{-1} g_v,\ v = t-k, \ldots, t$, thus the operators G^i on this subspace can be pre-computed. Hence in order to compute $G^i\, \tilde{g}_v,\ v = t-k, \ldots, t$ we need to solve $k+1$ PDEs forward in time.

4 Decomposition of the NMPC Problem

For feedback techniques like the NMPC it is crucial to have a special initialization strategy in order to get a feedback control fast enough. A technique for an optimal transition from one optimal control problem to the next one is the initial value embedding, see e.g. [7]. According to the initial value embedding technique in order to compute the new feedback control, an optimal control problem has to be solved over the prediction horizon $[t, t+k]$

$$\begin{aligned} \min_{u\in W([t,t+k]),\, p\in \mathbb{R}^{n_p},\, s_t\in L^2(\Omega),\, q\in C_q} \quad & \frac{1}{2}\, \| f[u,q] \|_{L^2(\Omega_t)}^2 + \\ & \frac{1}{2}\, \| E[u(t+k) - \bar{u}_{t+k}] \|_{L^2(\Omega)}^2 \\ s.t. \quad \partial_t u + A(u,p,q) = 0,& \\ u\big|_t = s_t, \quad s_t = \hat{s}_t, \quad p = \hat{p}.& \end{aligned}$$

This optimization problem will be solved using an SQP-type method. In each iteration of an SQP-type method a quadratic subproblem has to be solved, where the cost functional presents a quadratic approximation of the Lagrange function

$$\mathcal{L} : W([t,t+k]) \times \mathbb{R}^{n_p} \times L^2(\Omega) \times \mathbb{R}^{n_q} \times W([t,t+k]) \times \mathbb{R}^{n_p} \times L^2(\Omega) \to \mathbb{R}$$

$$\begin{aligned} (u, p, s_t, q, \lambda, \lambda_p, \lambda_s) \longmapsto & \frac{1}{2}\, \| f(u,q) \|_{L^2(\Omega_t)}^2 + \frac{1}{2}\, \| u(t+k) - \bar{u}_{t+k} \|_E^2 \\ & - \int_t^{t+k} \langle \partial_t u(t) + A(u(t),p,q), \lambda(t) \rangle_{H^1(\Omega)^* \times H^1(\Omega)}\, dt \end{aligned}$$

$$
\begin{aligned}
&- (u(t) - s_t, \lambda(t))_{L^2(\Omega)} \\
&- \left(s_t - \hat{s}_t, \lambda_s\right)_{L^2(\Omega)} \\
&- (p - \hat{p})^T \lambda_p,
\end{aligned}
$$

subject to a linear approximation of the constraints. Using the notations $\nabla^2 \mathcal{L}$ and ∇f for the Hessian of the Lagrange function and the gradient of the objective function respectively, the quadratic subproblem can be written as

$$
\min_{\Delta u,\, \Delta p,\, \Delta s,\, \Delta q} \quad \frac{1}{2} \begin{pmatrix} \Delta u \\ \Delta p \\ \Delta q \end{pmatrix}^* \nabla^2 \mathcal{L} \begin{pmatrix} \Delta u \\ \Delta p \\ \Delta q \end{pmatrix} + \nabla f^* \begin{pmatrix} \Delta u \\ \Delta p \\ \Delta q \end{pmatrix}
$$

$$
\begin{aligned}
(\partial_t + \partial_u A(u, p, q))\, \Delta u \;&= -\partial_q A(u, p, q)\Delta q - \partial_p A(u, p, q)\Delta p - \partial_t u - A(u, p, q), \\
\Delta u\big|_t = \Delta s, \quad &\Delta s = \hat{s}_t - s_t, \quad \Delta p = \hat{p} - p.
\end{aligned}
$$

Let us assume that the linearized PDE $\partial_t + \partial_u A(u, p, q)$ has a unique solution for arbitrary initial values and right hand sides. In this case, the increment in the states depends linearly on the increment of the controls, parameters, and initial values

$$
\begin{aligned}
\Delta u &= \partial_{s_t} u \Delta s + \partial_p u \Delta p + \partial_q u \Delta q + \delta u \\
\partial_{s_t} u &: L^2(\Omega) \longrightarrow W([t, t+k]), \\
\partial_p u &: \mathbb{R}^{n_p} \quad \longrightarrow W([t, t+k]), \\
\partial_q u &: \mathbb{R}^{n_q} \quad \longrightarrow W([t, t+k]), \\
\delta u &\in W([t, t+k]).
\end{aligned}
$$

Let us give few comments on the computation of necessary operators. The operator $\partial_q u : \mathbb{R}^{n_q} \longrightarrow W([t, t+k])$ maps the finite dimensional control space onto the infinite dimensional solution space of the parabolic PDE. As the control space $\mathbb{R}^{n_q}$ has the finite dimension n_q, we can precompute this operator by solving n_q linear PDEs

$$
\begin{aligned}
\partial_t z_i + \partial_u A(u, p, q)\, z_i \;&= -\partial_q A(u, p, q)\, q_i, \\
z_i\big|_t &= 0,
\end{aligned}
$$

and evaluating the operator $\partial_q u\, q_i = z_i$, $i = 1, \ldots, n_q$. In a similar way, also the operator $\partial_p u$ can be precomputed by solving n_p linear PDEs

$$
\begin{aligned}
\partial_t z_i + \partial_u A(u, p, q)\, z_i \;&= -\partial_{p_i} A(u, p, q), \\
z_i\big|_t &= 0.
\end{aligned}
$$

The part δu improves feasibility of the PDE constraints and its computation is equivalent to a Newton iteration on the nonlinear PDE

$$\begin{aligned} \partial_t \delta u + \partial_u A(u, p, q)\, \delta u &= -\partial_t u - A(u, p, q), \\ \delta u\big|_t &= s_t - u\big|_t. \end{aligned}$$

The operator $\partial_{s_t} u$ maps the infinite dimensional vector space $L^2(\Omega)$ onto $W([t, t+k])$. Hence, we can not efficiently precompute the operator. Nevertheless, we can evaluate the operator for certain Δs by solving

$$\begin{aligned} \partial_t z + \partial_u A(u, p, q)\, z &= 0, \\ z\big|_t &= \Delta s. \end{aligned}$$

Let us consider again the quadratic subproblem. By eliminating constraints we get a reduced quadratic problem which has n_q controls $q \in C_q \subset \mathbb{R}^{n_q}$ as optimization variables

$$\min_{q+\Delta q \in C_q} \quad \frac{1}{2}\Delta q^* \mathcal{H} \Delta q + (\nabla J)^* \, \Delta q. \tag{4.1}$$

Now let us discuss how to compute $\mathcal{H}$ and ∇J. The Hessian of the Lagrange function $\nabla^2 \mathcal{L}$, the reduced Hessian $\mathcal{H}$ and ∇J are defined as

$$\nabla^2 \mathcal{L} = \begin{pmatrix} \partial^2_{u,u}\mathcal{L} & \partial^2_{u,p}\mathcal{L} & 0 & \partial^2_{u,q}\mathcal{L} \\ \partial^2_{p,u}\mathcal{L} & \partial^2_{p,p}\mathcal{L} & 0 & \partial^2_{p,q}\mathcal{L} \\ 0 & 0 & 0 & 0 \\ \partial^2_{q,u}\mathcal{L} & \partial^2_{q,p}\mathcal{L} & 0 & \partial^2_{q,q}\mathcal{L} \end{pmatrix},$$

$$\mathcal{H} = \begin{pmatrix} \partial_q u \\ 0 \\ 0 \\ \mathbb{I} \end{pmatrix}^* \nabla^2 \mathcal{L} \begin{pmatrix} \partial_q u \\ 0 \\ 0 \\ \mathbb{I} \end{pmatrix} = \partial_q u^* \partial^2_{u,u}\mathcal{L}\, \partial_q u + \partial_q u^* \partial^2_{u,q}\mathcal{L} + \partial^2_{q,u}\mathcal{L} \partial_q u + \partial^2_{q,q}\mathcal{L},$$

$$\nabla J = \nabla f + H_p(\hat{p} - p) + H_s(\hat{s} - s) + H_u \delta u,$$

$$H_p = \begin{pmatrix} \partial_q u \\ 0 \\ 0 \\ \mathbb{I} \end{pmatrix}^* \nabla^2 \mathcal{L} \begin{pmatrix} \partial_p u \\ \mathbb{I} \\ 0 \\ 0 \end{pmatrix} = \left[\left(\partial_q u^* \partial^2_{u,u}\mathcal{L} + \partial^2_{q,u}\mathcal{L}\right) \partial_p u + \partial_q u^* \partial^2_{u,p}\mathcal{L} + \partial^2_{q,p}\mathcal{L}\right],$$

$$H_s = \begin{pmatrix} \partial_q u \\ 0 \\ 0 \\ \mathbb{I} \end{pmatrix}^* \nabla^2 \mathcal{L} \begin{pmatrix} \partial_{s_t} u \\ 0 \\ \mathbb{I} \\ 0 \end{pmatrix} = \left[\left(\partial_q u^* \partial^2_{u,u}\mathcal{L} + \partial^2_{q,u}\mathcal{L}\right) \partial_{s_t} u\right],$$

$$H_u = \begin{pmatrix} \partial_q u \\ 0 \\ 0 \\ \mathbb{I} \end{pmatrix}^* \nabla^2 \mathcal{L} \begin{pmatrix} \mathbb{I} \\ 0 \\ 0 \\ 0 \end{pmatrix} = \left[\left(\partial_q u^* \partial^2_{u,u} \mathcal{L} + \partial^2_{q,u} \mathcal{L} \right) \right],$$

$$\nabla f_q = \begin{pmatrix} \partial_q u \\ 0 \\ 0 \\ \mathbb{I} \end{pmatrix}^* \begin{pmatrix} \partial_u f \\ 0 \\ 0 \\ \partial_q f \end{pmatrix} = \left[\partial_q u^* \partial_u f + \partial_q f \right].$$

The Hessian of the Lagrange function depends only on the Lagrange multiplier for the PDE constraint which can be obtained by solving the adjoint PDE

$$\begin{aligned} -\partial_t \lambda + \partial_u A^*(u, p, q)\, \lambda &= (\partial_u f[.], f)_{L^2([t,t+k]\times\Omega)} \\ \lambda\big|_{t+k} &= E^* E \left(u\big|_{t+k} - \hat{u}_{t+k} \right). \end{aligned} \tag{4.2}$$

Using the operators $\partial_p u,\ \partial_q u,\ \partial_{s_t} u$, and δu, we can establish the reduced quadratic problem using a basis $(q_1, \ldots, q_{n_q})$ of the control space $C_q \subset \mathbb{R}^{n_q}$ by evaluating the necessary derivatives of the Lagrange function. We show exemplary the evaluation for

$$\begin{aligned} q_i^* \partial_q u^* \partial^2_{u,q} \mathcal{L} q_j = \quad & (\partial^2_{u,q} f[\partial_q u(q_i), q_j], f)_{L^2(\Omega_t)} \\ & + (\partial_u f[\partial_q u(q_i)], \partial_q f[q_i])_{L^2(\Omega_t)} \\ & - \int \int \partial^2_{u,q} A(u, p, q)[\partial_q u(q_i), q_j] \lambda \, dx \, dt. \end{aligned}$$

The other derivatives of the Lagrange function can be computed in a similar way. Since we are able to precompute the operators $\partial_p u$ and $\partial_q u$, we can also precompute the reduced Hessian $\mathcal{H}$ and the term H_p. For the purpose of updating the reduced quadratic problem (4.1) without solving online a PDE whenever new estimates $\hat{p}$ and $\hat{s}$ are available, we also need the operator H_s. Fortunately, we need the operator $\partial_{s_t} u$ only to evaluate the expression

$$q_i^* \left(\partial_q u^* \partial^2_{u,u} \mathcal{L} + \partial^2_{q,u} \mathcal{L} \right) \partial_{s_t} u, \qquad i = 1, \ldots, n_q,$$

which are elements of the dual space for fixed q_i and can be evaluated by solving n_q adjoint PDEs

$$\begin{aligned} -\partial_t h_i + \partial_u A^*(u, p, q)\, h_i &= q_i^* \partial_q u^* \partial^2_{u,u} \mathcal{L}(.) + q_i^* \partial^2_{q,u} \mathcal{L}(.) \\ h_i\big|_{t+k} &= E^* E \left(\partial_q u[q_i]\big|_{t+k} \right). \end{aligned}$$

For the adjoint PDEs we have to evaluate the adjoint right hand side at time points $\tau \in [t, t+k]$

$$
\begin{aligned}
q_i^* \partial_q u^* \partial^2_{u,u} \mathcal{L}\big|_\tau(.) + q_i^* \partial^2_{q,u} \mathcal{L}\big|_\tau(.) = \;& (\partial^2_{u,q} f(u(\tau), p, q)[\,.\,, q_j], f(u(\tau), p, q))_{L^2(\Omega)} \\
&+ (\partial_u f(u(\tau), p, q)[\,.\,], \partial_q f(u(\tau), p, q)[q_i])_{L^2(\Omega)} \\
&+ (\partial^2_{u,u} f(u(\tau), p, q)[\partial_q u(q_i), \,.\,], f(u(\tau), p, q))_{L^2(\Omega)} \\
&+ (\partial_u f(u(\tau), p, q)[\partial_q u(q_i)], \partial_u f(u(\tau), p, q)[\,.\,])_{L^2(\Omega)} \\
&+ (\partial_{u,u} A(u(\tau), p, q)[\partial_q u\, q_i, \,.\,], \lambda(\tau))_{L^2(\Omega)} \\
&+ (\partial_{q,u} A(u(\tau), p, q)[q_i, \,.\,], \lambda(\tau))_{L^2(\Omega)}.
\end{aligned}
$$

With the solutions h_i of the adjoint PDEs we can evaluate the operator

$$
H_s \Delta s = \begin{pmatrix} < h_1(t), \Delta s >_{L^2(\Omega)} \\ \vdots \\ < h_{n_q}(t), \Delta s >_{L^2(\Omega)} \end{pmatrix}.
$$

Let us note, that the controls Δq does not depend on a part of the initial values which is orthogonal to the functions h_i.

5 Multilevel Iterations

In the previous sections we have shown how to establish the reduced MHE and NMPC problems. Now we show how the established reduced problems are used in the multilevel iterations, in order to achieve real time feasibility for the response times.

5.1 Establishing the Reduced Problems

In order to combine the MHE and NMPC step we use the estimation of the initial values of the MHE problem s_{t-k} which together with the estimation of the parameters $\hat{p}$ determine the initial values for the NMPC problem. First we present the algorithm for establishing the reduced problems and the condensed operators.

Let us note that the first m adjoint PDEs can be solved in parallel. Afterwards, the n_p+m PDEs as well can be solved in parallel over $[t-k, t+k]$. The n_q+1 adjoint PDEs over $[t, t+k]$, and the n_q forward PDEs over $[t, t+k]$ have no coupling, thus they can also be solved in parallel. The same is true for the establishment of the reduced Hessian and gradients.

Algorithm 1 Establishing the reduced problems

1: **procedure** SOLVE m ADJOINT PDES OVER $[t-k,t]$
2: Solve for g_i or respectively J_g

$$-\partial_t g_i + \partial_u A^*(u,p,q)g_i = 0$$
$$\langle g_i(i),\lambda\rangle_{L^2(\Omega)} = \partial_{s_{t-k-1+i}}\mathcal{M}[\lambda], \qquad \forall\lambda\in L^2(\Omega)$$

3: Compute a basis of the range space $\mathcal{R}$ by solving

$$W_s\,\tilde{g}_i = g_i(t-k)$$

4: **procedure** SOLVE $m+n_p$ LINEARIZED PDES OVER $[t-k,t+k]$
5: Solve for $P^i\,e_j = y_j(t_i)$ and $\partial_p u\,e_j = y_j\big|_{[t,t+k]}$

$$\partial_t y_i + \partial_u A(u,p,q)y_i = -\partial_{p_i}A(u,p,q)$$
$$y_i\big|_{t-k} = 0$$

6: Establish $J_p = \begin{pmatrix}\partial_{s_{t-k}}\mathcal{M}P^{t-k}\\ \vdots \\ \partial_{s_t}\mathcal{M}P^t\end{pmatrix}$
7: Solve for $G^i\,\tilde{g}_j = z(t_i)$ and $\partial_{s_t}u\,(\tilde{g}_j) = z_j\big|_{[t,t+k]}$

$$\partial_t z_j + \partial_u A(u,p,q)z_j = 0$$
$$z_j\big|_{t-k} = \tilde{g}_j$$

8: **procedure** SOLVE n_q+1 ADJOINT PDES OVER $[t,t+k]$
9: Solve for Lagrange multiplier λ

$$-\partial_t\lambda + \partial_u A^*(u,p,q)\,\lambda = <\partial_u f[.], f>_{L^2([t,t+k]\times\Omega)}$$
$$\lambda\big|_{t+k} = E^*E\left(u\big|_{t+k} - \hat{u}_{t+k}\right)$$

10: Solve for ∇f_s or respectively h_i

$$-\partial_t h_i + \partial_u A^*(u,p,q)\,h_i = q_i^*\partial_q u^*\partial^2_{u,u}\mathcal{L}(.) + q_i^*\partial^2_{q,u}\mathcal{L}(.)$$
$$h_i\big|_{t+k} = E^*E\left(\partial_q u[q_i]\big|_{t+k}\right)$$

11: **procedure** SOLVE n_q LINEARIZED PDES OVER $[t,t+k]$
12: Solve for $\partial_q u\,(q_j) = z_j$

$$\partial_t z_i + \partial_u A(u,p,q)\,z_i = -\partial_q A(u,p,q)\,q_i,$$
$$z_i\big|_t = 0$$

13: **procedure** ESTABLISH THE REDUCED PROBLEMS
14: Compute QR decomposition of

$$J := \begin{pmatrix} 0 & W_s^{\frac{1}{2}}\tilde{g}_{t-k} & \cdots & W_s^{\frac{1}{2}}\tilde{g}_t \\ W_p^{\frac{1}{2}} & 0 & & \\ J_p & J_g\tilde{g}_{t-k} & \cdots & J_g\tilde{g}_t \end{pmatrix}$$

15: Evaluate $\mathcal{H}$ and H_p

5.2 *Iteration on the Level A*

The iteration on the level A consists of a small number of matrix vector multiplications and the solution of the reduced parameter estimation problems (3.3) as soon as we get new measurements and the reduced quadratic problem (4.1) as soon as we get the new estimation of the states and parameters. On this level we ignore non-linearities, and the mismatch of the shooting node values. The iteration on the level A uses the precomputed operators on the space $\mathcal{R}$ and is presented in the Algorithm 2.

Algorithm 2 Iteration on the level A

1: **procedure** INITIALIZATION

2: $$F_1 = \begin{pmatrix} \frac{\mathcal{M}(t-k, s_{t-k}, p) - \eta_{t-k}}{\sigma} \\ \vdots \\ \frac{\mathcal{M}(t, s_t, p) - \eta_t}{\sigma} \end{pmatrix}$$

3: **procedure** MOVING HORIZON ESTIMATION

4: Solve the MHE problem for updates of parameters $\Delta p \in \mathbb{R}^{n_p}$ and the part of the initial state values $\Delta \bar{s} \in \mathbb{R}^m$

$$\min_{\Delta p \in \mathbb{R}^{n_p}, \Delta \bar{s} \in \mathbb{R}^m} \left\| J \begin{pmatrix} \Delta p \\ \Delta \bar{s} \end{pmatrix} + \begin{pmatrix} W_s^{\frac{1}{2}} (s_{t-k} - \hat{s}_{t-k}) \\ W_p^{\frac{1}{2}} (p - \hat{p}) \\ F_1 \end{pmatrix} \right\|^2$$

5: $$\Delta s_i = P_i \Delta p + \sum_{j=1}^{m} G_i \tilde{g}_j \Delta \bar{s}_j + b_i, \qquad i = t-k, \dots t$$

6: $$s_i^{new} = s_i^{old} + \Delta s_i \qquad i = t-k, \dots t$$

7: **procedure** NONLINEAR MODEL PREDICTIVE CONTROL

8: $\nabla J = \nabla J^{old} + H_s \Delta s_t + H_p \Delta p$ for the NMPC problem

9: Solve the NMPC problem for Δq

$$\min_{q + \Delta q \in C_q} \frac{1}{2} \Delta q^* \mathcal{H} \Delta q + \nabla J^* \Delta q$$

10: $$u^{new} = u^{old} + \partial_p u \Delta p + \sum_{j=1}^{m} z_j \Delta \bar{s}_j + \partial_q u \Delta q$$

11: $$\nabla J^{new} = \nabla J + \mathcal{H} \Delta q$$

5.3 Iteration on the Level B

On the level B we improve the feasibility by taking into account the non-linearity in the model equations. For this purpose we have to solve the PDE over the whole time horizon $[t-k, t+k]$ and compute the residuals b^i, $i = t-k, \dots t+k$, in the multiple shooting nodes. The residuals b^i, $i = t-k, \dots, t$, are inserted into the reduced MHE problem

$$\min_{p \in \mathbb{R}^{n_p}, \Delta s_{\mathcal{R}} \in \mathcal{R}} \left\| \begin{pmatrix} 0 & W_s^{\frac{1}{2}} \\ W_p^{\frac{1}{2}} & 0 \\ J_p & J_g \end{pmatrix} \begin{pmatrix} \Delta p \\ \Delta s_{\mathcal{R}} \end{pmatrix} + \begin{pmatrix} W_s^{\frac{1}{2}}(s_{t-k} - \hat{s}_{t-k}) \\ W_p^{\frac{1}{2}}(p - \hat{p}) \\ F_1 + Jb \end{pmatrix} \right\|^2 .$$

After the iteration on the level A and the corresponding updates of the shooting nodes it holds that

$$\begin{pmatrix} 0 & W_s^{\frac{1}{2}} \\ W_p^{\frac{1}{2}} & 0 \\ JP & JG \end{pmatrix}^* \begin{pmatrix} W_s^{\frac{1}{2}}(s_{t-k} - \hat{s}_{t-k}) \\ W_p^{\frac{1}{2}}(p - \hat{p}) \\ F_1 \end{pmatrix} = 0,$$

hence the reduced MHE problem can be simplified to

$$\min_{p \in \mathbb{R}^{n_p}, \Delta s_{\mathcal{R}} \in \mathcal{R}} \left\| \begin{pmatrix} 0 & W_s^{\frac{1}{2}} \\ W_p^{\frac{1}{2}} & 0 \\ J_p & J_g \end{pmatrix} \begin{pmatrix} \Delta p \\ \Delta s_{\mathcal{R}} \end{pmatrix} + \begin{pmatrix} 0 \\ 0 \\ Jb \end{pmatrix} \right\|^2 .$$

Let us finish this subsection with few comments on the numerical procedure.

The vector spaces $\mathcal{N}$ and $\mathcal{R}$ are orthogonal with respect to the scalar product $< \cdot, \cdot >_{W_s} \; = \; < W_s \cdot, \cdot >_{L^2(\Omega)}$. Hence, the projection $\mathcal{P}_{\mathcal{R},\mathcal{N}}(s_{t-k} - \hat{s}_{t-k})$ can be computed by e.g. the Gram-Schmidt method.

Since the solution of the reduced MHE problem is very fast compared to the solution of the PDE, the estimated states from the MHE problem can be used to update the controls.

The computed states over the prediction horizon are used to update the reduced gradient for the reduced NMPC problem.

For the iteration on the level B we have to solve *online* the original PDE and one linearized PDE for a directional derivative over the estimation horizon $[t-k, t]$. Afterwards, we have to solve a PDE over the prediction horizon $[t, t+k]$ in order to compute the new states. Effort for the solution of the reduced MHE and NMPC problems and update of the states is negligible.

Algorithm 3 Iteration on the level B

1: **procedure** SOLVE PDE OVER THE ESTIMATION HORIZON
2: $b_{t-k} = \mathcal{P}_{\mathcal{R},\mathcal{N}}(s_{t-k} - \hat{s}_{t-k})$
3: $r_{i+1} = S(t_{i+1})(s_i, p, q) - s_{i+1}, \quad i = t-k, \ldots, t-1$
4: $b_{i+1} = \partial_s S(t_{i+1})(s_i, p, q) b_i + r_{i+1}, \quad i = t-k, \ldots, t-1$
5: **procedure** MOVING HORIZON ESTIMATION
6: Solve the MHE problem

$$\min_{p \in \mathbb{R}^{n_p}, \Delta\bar{s} \in \mathbb{R}^m} \left\| J \begin{pmatrix} \Delta p \\ \Delta \bar{s} \end{pmatrix} + \begin{pmatrix} 0 \\ 0 \\ Jb \end{pmatrix} \right\|^2$$

7: $\Delta s_i = P_i \Delta p + \sum_{j=1}^{m} G_i \tilde{g}_j \Delta \bar{s}_j + b_i, \quad i = t-k, \ldots t$
8: $s_i^{new} = s_i^{old} + \Delta s_i \quad i = t-k, \ldots t$
9: **procedure** NONLINEAR MODEL PREDICTIVE CONTROL
10: $\nabla J = \nabla J^{old} + H_s \Delta s_t + H_p \Delta p$
11: Solve the NMPC problem

$$\min_{q \mid \Delta q \in C_q} \frac{1}{2} \Delta q^* \mathcal{H} \Delta q + \nabla J^* \Delta q$$

12: $u^{temp} = u^{old} + \partial_p u \Delta p + \partial_{s_t} u \Delta s_{\mathcal{R}} + \partial_q u \Delta q$
13: $\nabla J^{temp} = \nabla J + \mathcal{H} \Delta q$
14: **procedure** SOLVE PDE OVER THE PREDICTION HORIZON
15: Compute the new states u
16: $\Delta u = u - u^{temp}$
17: **procedure** NONLINEAR MODEL PREDICTIVE CONTROL
18: Update the reduced gradient $\nabla J = \nabla J^{temp} + H_u \Delta u$
19: Solve the NMPC problem

$$\min_{q + \Delta q \in C_q} \frac{1}{2} \Delta q^* \mathcal{H} \Delta q + \nabla J^* \Delta q$$

20: $u^{new} = u + \partial_p u \Delta q$
21: $\nabla J^{new} = \nabla J + \mathcal{H} \Delta q$

5.4 *Iteration on the Level C*

On the level C we improve the optimality of the solution by taking into account the nonlinearity in the cost function. For this purpose we need to solve the adjoint PDE backward in time over the prediction horizon $[t, t+k]$.

Algorithm 4 Iteration on the level C

1: **procedure** SOLVE ADJOINT PDE OVER THE PREDICTION HORIZON
2: Solve

$$-\partial_t \lambda + \partial_u A^*(u, p, q)\,\lambda = (\partial_u f[.], f)_{L^2([t,t+k]\times\Omega)}$$
$$\lambda\big|_{t+k} = E^* E\left(u\big|_{t+k} - \hat{u}_{t+k}\right)$$

3:

$$\nabla f = \begin{pmatrix} \partial_q f(q_1) - < \partial_q A(u, p, q)(q_1), \lambda >_{L^2([t,t+k]\times\Omega)} \\ \vdots \\ \partial_q f(q_{n_q}) - < \partial_q A(u, p, q)(q_{n_q}), \lambda >_{L^2([t,t+k]\times\Omega)} \end{pmatrix}$$

4: $\nabla J = \nabla f$
5: Solve the NMPC problem

$$\min_{q+\Delta q \in C_q} \quad \frac{1}{2}\Delta q^* \mathcal{H} \Delta q + \nabla J^* \Delta q$$

6: $u^{new} = u^{old} + \partial_q u \Delta q$
7: $\nabla J^{new} = \nabla f + \mathcal{H}\Delta q$

6 Numerical Results

We would like to compare the multilevel iterations on the benchmark problem of optimal control for a catalytic plug flow reactor. Catalytic plug flow reactors are used for chemical reactions of gases accelerated by a porous catalyst. Such a reactor consists of several plug flow reactors in one casing. In each plug flow reactor, the gas flows through a fixed bed of porous catalysts. In the fixed bed the gas reacts to the desired product. Outside, in the orifice baffle, there is a cooling fluid. The mathematical model includes nonlinear parabolic PDE models of the mass balance, the energy balance and the momentum balance and ODE models for the chemical reaction.

Each catalytic plug flow reactor is radial symmetric, hence the states and chemical reaction rates depend only on the radial $r \in [0, R]$ and axial coordinates $z \in [0, L]$ of the reactor. This implies, that the catalytic plug flow reactor can be modeled as an pseudo 2D model.

The mass balance leads to a PDE for the concentrations c_i, $i = 1, \ldots, N$, of the involved substances

$$\epsilon \frac{\partial c_i}{\partial t} = -\frac{\partial (u_f c_i)}{\partial z} + D_{eff,z}\frac{\partial^2 c_i}{\partial z^2} + D_{eff,r}\left(\frac{\partial^2 c_i}{\partial r^2} + \frac{1}{r}\frac{\partial c_i}{\partial r}\right) + \frac{\rho_{s,bed}}{\rho_f} R_i.$$

Here, ϵ describes the bed porosity, u_f is the superficial velocity of the gas, D_{eff} is the effective diffusion coefficient of the gas, $\rho_{s,bed}$ is the density of the packed bed, ρ_f is the density of the gas and R_i is the rate of the mass change.

Since the products in the reactor can not permeates the tube wall and the concentration at the inlet equals the concentration of the gas we put into the reactor we obtain the following boundary conditions

$$
\begin{aligned}
c_i(t, r, 0) &= c_i^{inlet}(t, r) && \text{inlet of the tube,} \\
\partial_r\, c_i(t, 0, z) &= 0 && \text{middle of the tube,} \\
\partial_r\, c_i(t, R, z) &= 0 && \text{tube wall.}
\end{aligned}
$$

In the energy balance the fixed bed and the gas are treated as a quasi continuum

$$
\begin{aligned}
(\epsilon\, \rho_f\, c_{p,f} + (1-\epsilon)\rho_{s,bed}\, c_{p,s})\frac{\partial T}{\partial t} = -c_{p,f}&\frac{\partial(u_f\, \rho_f\, T)}{\partial z} + \lambda_{eff,z}\frac{\partial^2 T}{\partial z^2} \\
&+\lambda_{eff,r}\left(\frac{\partial^2 T}{\partial r^2} + \frac{1}{r}\frac{\partial T}{\partial r}\right) \\
&+\rho_{s,bed}\sum_{i=1}^{N}\Delta H_{R,i} R_i .
\end{aligned}
$$

Here T denotes the temperature, $c_{p,f}$, $c_{p,s}$ are the isobar heat capacity of the gas and the packed bed respectively, λ_{eff} is the effective heat conductivity, ΔH_R is the enthalpy of formation.

At the tube wall there will be a temperature loss depending on the heat transmission of the wall and the temperature difference. Effected by the fixed bed the temperature at the inlet will not simply coincide with the temperature of the gas pumped into the reactor. These phenomena are reflected by the boundary conditions

$$
\begin{aligned}
\lambda_{eff,z}\, \partial_z\, T(t, 0, r) &= -u_f\, \rho_f\, c_{p,f}[T_{inlet} - T(t, 0, r)] && \text{inlet of the tube,} \\
\lambda_{eff,r}\, \partial_r\, T(t, z, R) &= k_\alpha[T_{cooling}(t) - T(t, z, R)] && \text{tube wall,} \\
\partial_r\, T(t, z, 0) &= 0 && \text{middle of the tube,}
\end{aligned}
$$

with k_α denoting the heat transmission coefficient.

When the gas flows through the porous catalyst radial and axial accelerations emerge. The turbulences overlap and result in a plug flow. Hence the pressure drop can be modeled using the Ergun equation

$$
-\frac{\partial p}{\partial z} = 150\frac{(1-\epsilon)^2}{\epsilon^3 d_p^2}\eta\, u_f + 1.75\frac{(1-\epsilon)}{\epsilon^3 d_p}\rho_f u_f^2,
$$

where d_p denotes the diameter of the catalyst particles and η is the dynamic viscosity.

In the catalytic plug flow reactor an exothermic reaction of A and B to C plus D takes place ($A + B \longrightarrow C + D$). The influence of the temperature on the reaction rate R_A of the catalyst activity is modeled by the Arrhenius law:

$$R_A = A_0 \exp\left(\frac{E_a}{R_g\, T}\right) c_A\, a_{cat}^2.$$

Here A_0 is the frequency factor of the reaction, E_a is the activation energy, R_g is the gas constant, c_A is the concentration of A, a_{cat} is the activity of the catalyst. The activity of the catalyst satisfies

$$\partial_t a_{cat} = -\, c_A \left(k_{0,des} + A_{0,des} \exp\left(\frac{E_{a,des}}{R_g\, T}\right)\right) a_{cat}^2, \quad a_{cat}\Big|_{t=0} = 1.$$

Here, the subscript des corresponds to the deactivation.

For the optimal steering of the reactor it is crucial that the activity of the catalyst remains as high as possible while the concentration and temperature of the gas stay close to a desired reference trajectory. In order to control the reaction, measurements of the temperature at 16 different points in the middle of the reactor are taken every 10 s. The estimation horizon and the prediction horizon are taken to be 100 s.

The PDE is discretized using the Rothe method: the PDE is first discretized in time with an A-stable scheme, the resulting elliptic problems, for each time point, are solved using finite elements. In the discretized MHE problem 4 parameters and 891 states have to be estimated. The discretized NMPC problem consists of 2 discretized controls and 26730 states.

The aim of the experiments is to compare the performance of multilevel iterations. The time needed for establishing the system as well as the time needed for the iteration on each level is presented in Table 1. The computations were performed on a desktop computer with the following hardware: Intel(R) Core(TM) i7-3770 CPU 3.40 GHz (4 cores), 16 GiB DDR3 RAM 1600 MHz.

Table 1 Computation times

Task	Time (s)
Establishing system	647
Level A	1
Level B	27
Level C	14

7 Conclusions

We have presented for the first time combined NMPC-MHE multilevel iterations for nonstationary PDE models. These are the first steps towards real time feasibility of the NMPC technique for PDE models. According to the results of numerical experiments the iteration on level A computes very fast the control update. This assures the real time feasibility of the level A iterations. The iterations on the next levels still need a lot of time, nevertheless these iterations are faster in magnitudes compared to a Gauss Newton iteration for the state and parameter estimates and an SQP iteration for the control update. For practical applications the multilevel iterations must be accelerated using e.g. reduced order models as proper orthogonal decomposition (POD) or adaptive multigrid methods.

Acknowledgement This work was supported by the BMBF through the programme "Mathematics for Innovation in Industry and Services".

References

1. F. Allgöwer, T.A. Badgwell, J.S. Qin, J.B. Rawlings, and S.J. Wright. Nonlinear predictive control and moving horizon estimation – An introductory overview. In P. M. Frank, editor, *Advances in Control, Highlights of ECC'99*, pages 391–449. Springer, 1999.
2. L. T. Biegler, O. Ghattas, M. Heinkenschloss, D. Keyes, and B. van Bloemen Waanders, editors. *Real-Time PDE-Constrained Optimization*. SIAM, 2007.
3. H. G. Bock. *Randwertproblemmethoden zur Parameteridentifizierung in Systemen nichtlinearer Differentialgleichungen*, volume 183 of *Bonner Mathematische Schriften*. Bonn, 1987.
4. H. G. Bock, M. Diehl, E. Kostina, and J. P. Schlöder. Constrained optimal feedback control for DAE. In L. Biegler, O. Ghattas, M. Heinkenschloss, D. Keyes, and B. van Bloemen Waanders, editors, *Real-Time PDE-Constrained Optimization*, pages 3–24. SIAM, 2007.
5. G. De Nicolao, L. Magni, and R. Scattolini. Stability and robustness of nonlinear receding horizon control. In F. Allgöwer and A. Zheng, editors, *Nonlinear Predictive Control*, volume 26 of *Progress in Systems Theory*, pages 3–23, Basel, 2000. Birkhäuser.
6. M. Diehl, H.G. Bock, and J.P. Schlöder. A real-time iteration scheme for nonlinear optimization in optimal feedback control. *SIAM Journal on Control and Optimization*, 43(5):1714–1736, 2005.
7. M. Diehl, R. Findeisen, S. Schwarzkopf, I. Uslu, F. Allgöwer, H.G. Bock, E.D. Gilles, and J.P. Schlöder. An efficient algorithm for nonlinear model predictive control of large-scale systems. Part I: Description of the method. *Automatisierungstechnik*, 50(12):557–567, 2002.
8. J.V. Frasch, L. Wirsching, S. Sager, and H.G. Bock. Mixed-level iteration schemes for nonlinear model predictive control. In *Proceedings of the IFAC Conference on Nonlinear Model Predictive Control*, 2012.
9. D. Haßkerl, A. Meyer, N. Azadfallah, S. Engell, A. Potschka, L. Wirsching, and H. G. Bock. Study of the performance of the multi-level iteration scheme for dynamic online optimization for a fed-batch reactor example. In *Proceedings of the European Control Conference (ECC)*.
10. H. K. Hesse and G. Kanschat. Mesh adaptive multiple shooting for partial differential equations. Part I: Linear quadratic optimal control problems. *J. Numer. Math.*, 17(3):195–217, 2009.

11. G. Kriwet. An efficient condensing technique suited for parameter estimation problems arising by processes described by partial differential equations. Technical Report, University of Marburg, 2012.
12. M. Hinze, R. Pinnau, M. Ulbrich, and S. Ulbrich. *Optimization with PDE Constraints*. Springer, 2009.
13. D.Q. Mayne. Nonlinear model predictive control: Challenges and opportunities. In F. Allgöwer and A. Zheng, editors, *Nonlinear Predictive Control*, volume 26 of *Progress in Systems Theory*, pages 23–44, Basel, 2000. Birkhäuser.
14. J.B. Rawlings, E.S. Meadows, and K.R. Muske. Nonlinear model predictive control: A tutorial and survey. In *Proc. Int. Symp. Adv. Control of Chemical Processes, ADCHEM*, Kyoto, Japan, 1994.
15. J.P. Schlöder. *Numerische Methoden zur Behandlung hochdimensionaler Aufgaben der Parameteridentifizierung*, volume 187 of *Bonner Mathematische Schriften*. University of Bonn, Bonn, 1988.
16. F. Tröltzsch. *Optimale Steuerung partieller Differentialgleichungen. Theorie, Verfahren und Anwendungen*. Vieweg & Sohn, 2005.
17. S. Ulbrich. Generalized SQP-methods with "parareal" time-domain decomposition for time-dependent PDE-constrained optimization. Technical report, Fachbereich Mathematik TU Darmstadt, 2005.
18. J. Wloka. *Partielle Differentialgleichungen. Sobolevräume und Randwertaufgaben*. Teubner, 1982.

Some Recent Developments in Optimal Control of Multiphase Flows

Michael Hintermüller and Tobias Keil

Abstract The present work serves as a review of some current developments in optimal control of two-phase flows. We discuss the upcoming analytical and numerical challenges and illustrate adequate solution strategies. This includes, among other things, the existence proof of solutions to a coupled Cahn–Hilliard–Navier–Stokes system, the derivation of first order optimality conditions for the associated optimal control problem in the presence of a nonsmooth free energy density. Moreover, we study spatial mesh adaptivity concepts based on a dual weighted residual approach and address future research directions concerned with the derivation of stronger stationarity conditions and/or the design of more efficient numerical solution algorithms.

Keywords Cahn–Hilliard · Navier–Stokes · Mathematical programming with equilibrium constraints · Non-smooth potentials · Optimal control · Stationarity conditions · Semidiscretization in time · Adaptive finite elements method · Dual-weighted residuals

1 Introduction

In this paper we are concerned with the optimal control of multiphase flows including two (or more) immiscible fluids. The behavior of such systems is governed by two main forces: the hydrodynamics of the fluids and the separation process of the different phases. For the mathematical formulation of these processes, we consider a coupled system consisting of an equation of Navier–Stokes type, which captures the hydrodynamic effects, and a phase field model, which takes care of the

---

M. Hintermüller (✉) · T. Keil
Weierstrass-Institut, Berlin, Germany

Humboldt-Universität zu Berlin, Institut für Mathematik, Berlin, Germany
e-mail: hint@math.hu-berlin.de; tkeil@math.hu-berlin.de

V. Schulz, D. Seck (eds.), *Shape Optimization, Homogenization and Optimal Control*, International Series of Numerical Mathematics 169,
https://doi.org/10.1007/978-3-319-90469-6_7

separation process. In contrast to the so-called sharp interface models (where the interface is usually given by the distinct sharp boundaries of the fluid phases with infinitesimal interface width) phase field models contain small (diffuse) interface layers which are described through an order parameter. In this context, the order parameter depicts the concentration of the fluids, attaining extreme values at the pure phases and intermediate values within the interface layer, which is supposed to represent a fluid mixture. It is associated with decreasing/minimizing a suitably chosen energy. Some of the strengths of phase field approaches are due to their ability to overcome both, analytical difficulties of topological changes, such as, e.g., droplet break-ups or the coalescence of interfaces, and numerical challenges in capturing the interface dynamics.

A renowned diffuse interface model is the Cahn–Hilliard system which was first introduced by Cahn and Hilliard in [15]. A first basic model for immiscible, viscous two-phase flows combining the Cahn–Hilliard system with the Navier–Stokes equation was published by Hohenberg and Halperin in [39]. It is however restricted to the case where the two fluids possess nearly identical densities, i.e., matched densities. Recently, Abels, Garcke and Grün [2] obtained the following diffuse interface model for two-phase flows with non-matched densities:

$$\partial_t \varphi + v \nabla \varphi - \operatorname{div}(m(\varphi) \nabla \mu) = 0, \tag{1a}$$

$$-\Delta \varphi + \partial \Psi_0(\varphi) - \mu - \kappa \varphi \ni 0, \tag{1b}$$

$$\begin{aligned}\partial_t (\rho(\varphi) v) + \operatorname{div}(v \otimes \rho(\varphi) v) - \operatorname{div}(2\eta(\varphi) \epsilon(v)) + \nabla \Pi \\ + \operatorname{div}(v \otimes J) - \mu \nabla \varphi = 0,\end{aligned} \tag{1c}$$

$$\operatorname{div} v = 0, \tag{1d}$$

$$v_{|\partial \Omega} = 0, \tag{1e}$$

$$\partial_n \varphi_{|\partial \Omega} = \partial_n \mu_{|\partial \Omega} = 0, \tag{1f}$$

$$(v, \varphi)_{|t=0} = (v_a, \varphi_a), \tag{1g}$$

which is supposed to hold in the space-time cylinder $\Omega \times (0, \infty)$, where $\partial \Omega$ denotes the boundary of Ω. This system is thermodynamically consistent in the sense that it allows for the derivation of local entropy or free energy inequalities.

We introduce the physical quantities of the system in Table 1.

The mobility and viscosity coefficients, and the density of the mixture of the fluids depend on the order parameter, which reflects the mass concentration of the

Table 1 The involved quantities in system (1a)–(1g)

v	Velocity of the fluid	p	Fluid pressure
φ	Order parameter	μ	Chemical potential
$\rho(\varphi)$	Fluid density	$\eta(\varphi)$	Fluid viscosity
$m(\varphi)$	Mobility coefficient	v_a, φ_a	Initial states
κ	Positive constant		

fluid phases. More precisely, the density function is given by

$$\rho(\varphi) = \frac{\rho_1 + \rho_2}{2} + \frac{\rho_2 - \rho_1}{2}\varphi, \tag{2}$$

where φ ranges in the interval $[-1, 1]$, and $0 < \rho_1 \leq \rho_2$ are the given densities of the two fluids under consideration. Furthermore, $\epsilon(v) := \frac{1}{2}(\nabla v + \nabla v^\top)$ denotes the symmetric gradient of v. The Cahn–Hilliard system contains the homogeneous free energy density $\Psi(\varphi) := \Psi_0(\varphi) - \frac{\kappa}{2}\varphi^2$ consisting of a convex part (Ψ_0) and a non-convex contribution ($-\frac{\kappa}{2}\varphi^2$).

The homogeneous free energy density serves the purpose of restricting the order parameter φ to the physically meaningful range $[-1, 1]$ and to capture the spinodal decomposition of the phases. While the latter is typically realized by a negative quadratic energy (here $-\kappa\varphi^2$ for some $\kappa > 0$) there are several choices for confining the order parameter to $[-1, 1]$, i.e. how $\Psi_0(\varphi)$ is chosen depending on the underlying application. In the original work [15], $\Psi_0(\varphi)$ was chosen as a logarithmic barrier function. In [51], Oono and Puri found that in the case of deep quenches of, e.g., binary alloys, the double-obstacle potential $\Psi(\varphi) = I_{[-1,1]}(\varphi) - \frac{\kappa}{2}\varphi^2$ (where $I_{[-1,1]}$ represents the indicator function of the interval $[-1, 1]$) proves to be the best choice for modeling the separation process. A similar observation appears to be true in the case of polymeric membrane formation under rapid wall hardening. However, due to the non-differentiability of the indicator function, the double-obstacle potential gives rise to a variational inequality in (1b) which especially complicates the derivation of stationarity conditions for the optimal control problem (cf. Sect. 4) and its numerical solution (cf. Sect. 5.2).

Regarding physical applications, we point out that the CHNS system is used to model a variety of situations. These range from the aforementioned solidification process of liquid metal alloys, cf. [20], or the simulation of bubble dynamics, as in Taylor flows [4], or pinch-offs of liquid-liquid jets [44], to the formation of polymeric membranes [60] or proteins crystallization, see e.g. [45] and references within. Furthermore, the model can be easily adapted to include the effects of surfactants such as colloid particles at fluid-fluid interfaces in gels and emulsions used in food, pharmaceutical, cosmetic, or petroleum industries [5, 53].

In the literature, the classical case of two-phase flows of liquids with matched densities is well investigated, see e.g. [39]. However, we point out that there exist different approaches to model the case of fluids with non-matched densities. These range from quasi-incompressible models with non-divergence free velocity fields, see e.g. [48], to possibly thermodynamically inconsistent models with solenoidal fluid velocities, cf. [19]. We refer to Boyer [12], Boyer et al. [13], Gal and Grasselli [23] and Aland and Voigt [6] for additional analytical and numerical results for some of these models. In the recent paper [24] the given system (1) with smooth potentials (thus excluding the double-obstacle homogeneous free energy density) is considered in a fully discrete and an alternative semi-discrete in time setting including numerical simulations.

The optimal control problem associated to the Cahn–Hilliard–Navier–Stokes system with matched densities and a non-smooth homogeneous free energy density (double-obstacle potential) has been previously studied in [37, 38]. We also mention the recent articles [22] (which treats the control of a nonlocal Cahn–Hilliard–Navier–Stokes system in two dimensions) and [55]. Apart from these contributions the literature on the optimal control of the coupled CHNS system with non-matched densities is - to the best of our knowledge - essentially void. Nevertheless, we mention that there are numerous publications concerning the optimal control of the phase separation process itself, i.e. the distinct Cahn–Hilliard system, see e.g. [11, 17, 18, 21, 31, 36, 58, 59].

In order to formulate the optimal control problem, we introduce an objective functional $\mathcal{J}$ and a control force u which acts either on the right-hand side of the Navier–Stokes equation (1c) or on the boundary condition (1e). The problem consists of finding the optimal control u which minimizes $\mathcal{J}$ under the condition that the control-version CHNS(u) of the Cahn–Hilliard–Navier–Stokes system is satisfied, i.e.

$$\begin{aligned}&\text{minimize } \mathcal{J}(\varphi, \mu, v, u) \text{ over } (\varphi, \mu, v, u)\\&\text{subject to (s.t.) } u \in U_{ad},\ (\varphi, \mu, v, u) \text{ satisfies CHNS}(u).\end{aligned} \tag{Opt}$$

Here, U_{ad} is a given set of admissible controls. In this paper, we discuss the analytical and numerical challenges connected to the problem (Opt). This covers the difficulties associated with establishing the existence of solutions to the Navier–Stokes equation caused by the non-linear terms contained in (1c), as well as, the different possibilities to formulate stationarity concepts for the first-order primal-dual characterization of solutions due to the presence of a non-smooth homogeneous free energy density associated with the underlying Ginzburg-Landau energy in the Cahn–Hilliard system. More precisely, due to the presence of the variational inequality constraint (1b), classical constraint qualifications (see, e.g., [61]) fail and it is known [33, 37] that the resulting problem falls into the realm of mathematical programs with equilibrium constraints (MPECs) in function space for which stationarity conditions are no longer unique. Moreover, we treat the numerical challenges related to solving an optimal control problem connected to a large-scale nonlinear system upon discretization and discuss options to reduce the computational complexity of the problem as well as the design of appropriate numerical solution algorithms.

We conclude this introduction by introducing some necessary notations and organize the remainder of the paper as follows. In Sect. 2 we discuss the existence of solutions to the Cahn–Hilliard–Navier–Stokes system. Then, Sect. 3 rigorously introduces the optimal control problem and provides a first existence result concerning global solutions. Section 4 is concerned with the various stationarity concepts and the associated relaxation/regularization/penalization techniques. Section 5 investigates the numerical realization of the problem and in Sect. 6 we give a brief outlook on associated future research topics.

1.1 Notation

Let $\Omega \subset \mathbb{R}^n$, $n \in \{2, 3\}$ denote a bounded domain with sufficiently smooth boundary $\partial\Omega$ and outer normal $\mathbf{n}_\Omega$. Let $I = (0, T]$, $T > 0$ denote a time interval.

We use the conventional notation for Sobolev and Hilbert spaces, see e.g. [3]. By $L^p(\Omega)$, $1 \leq p \leq \infty$, we denote the space of measurable functions on Ω, whose modulus to the power p is Lebesgue-integrable. $L^\infty(\Omega)$ denotes the space of measurable functions on Ω, which are essentially bounded. For $p = 2$ we denote by $L^2(\Omega)$ the Hilbert space of square integrable functions on Ω with inner product $(\cdot, \cdot)$ and norm $\|\cdot\|$. For a subset $D \subset \Omega$ and functions $f, g \in L^2(\Omega)$ we denote by $(f, g)_D$ the inner product of f and g restricted to D, and by $\|f\|_D$ the respective norm. By $W^{k,p}(\Omega)$, $k \geq 1, 1 \leq p \leq \infty$, we denote the Sobolev space of functions admitting weak derivatives up to order k in $L^p(\Omega)$. If $p = 2$ we write $H^k(\Omega)$ to acknowledge the Hilbertian structure of the space. The subset $H_0^1(\Omega)$ denotes $H^1(\Omega)$ functions with vanishing boundary trace. For $k \in \mathbb{N}$, we further set

$$H^k_{0,\sigma}(\Omega; \mathbb{R}^n) := \left\{ f \in H^k(\Omega; \mathbb{R}^n) \cap H_0^1(\Omega; \mathbb{R}^n) : \operatorname{div} f = 0, \text{ a.e. on } \Omega \right\};$$

$$\overline{H}^k(\Omega) := H^k_{(0)}(\Omega) := \left\{ f \in H^k(\Omega) : \int_\Omega f dx = 0 \right\};$$

$$\overline{H}^k_{\partial_n}(\Omega) := \left\{ f \in \overline{H}^k(\Omega) : \partial_n f_{|\partial\Omega} = 0 \text{ a.e. on } \partial\Omega \right\}, \quad k \geq 2;$$

where 'a.e.' stands for 'almost everywhere' and the boundary condition is supposed to hold true in the trace sense. The subscript σ here is a common notation representing the solenoidality. Unless otherwise noted, $\langle \cdot, \cdot \rangle := \langle \cdot, \cdot \rangle_{\overline{H}^{-1}, \overline{H}^1}$ represents the duality pairing between $\overline{H}^1(\Omega)$ and its dual $\overline{H}^{-1}(\Omega)$. For $1 \leq p \leq \infty$, $L^p(0, T, X)$ denotes the space of all measurable p-integrable/essentially bounded functions from $(0, T)$ into the Banach space X and $C(0, T, X)$ the set of continuous functions from $(0, T)$ into the Banach space X.

2 Existence of Solutions of the Cahn–Hilliard–Navier–Stokes System

Our main concern in this section is to discuss the well-posedness of the given optimal control problem, i.e. to verify the existence of solutions. This especially includes the very challenging task to establish the existence of solutions to the Cahn–Hilliard–Navier–Stokes system. Hereby, the challenges are primarily imposed by the Navier–Stokes equation (1c) and the contained non-linear convection term $\operatorname{div}(v \otimes \rho(\varphi)v)$.

We point our that for matched densities (i.e. $\rho(\varphi) \equiv \rho_1 = \rho_2$) and a constant viscosity coefficient, equation (1c) reduces to the classical evolutionary non-linear

Navier–Stokes equation

$$\partial_t v + \operatorname{div}(v \otimes v) - 2\eta \operatorname{div}(\epsilon(v)) + \nabla \Pi = f_{ext}, \tag{3}$$

with the external force $f_{ext} := \mu \nabla \varphi + u$.

The proof of existence and uniqueness of weak solutions to the linear Navier–Stokes equation, i.e. without the convection term, is straight-forward based on, e.g., the Faedo-Galerkin method. Hereby, the problem is approximated using a total sequence $\{\psi_n\}_{n\in\mathbb{N}}$ in $H^1_{0,\sigma}(\Omega)$, i.e. a sequence such that $span(\{\psi_1, .., \psi_n, ..\})$ is dense in $H^1_{0,\sigma}(\Omega)$. The solutions to the resulting auxiliary problems in the subspaces $span(\{\psi_1, .., \psi_n\})$ are derived by a fixed point argument and can be shown to be bounded in $L^2(0, T, H^1_{0,\sigma}(\Omega))$ and $L^\infty(0, T, L^2(\Omega))$. Then, the limit point of a weakly convergent subsequence of these solutions can be shown to satisfy (3) in a distributional sense employing (among others) the Rellich-Kondrachov embedding theorem, cf. [3]. The uniqueness proof relies on an interpolation theorem by Lions-Magenes [47] which yields that the solution is in fact equivalent to a continuous function up to a set of measure zero, i.e. it is contained in $C(0, T, L^2(\Omega))$.

However, due to the low regularity of the convection term, these arguments cannot be directly transferred to the non-linear Navier–Stokes equation. In fact, the term $(\operatorname{div}(v_1 \otimes v_2), v_3)$ does only define a trilinear continuous form on $H^1_{0,\sigma}(\Omega) \times H^1_{0,\sigma}(\Omega) \times H^1_{0,\sigma}(\Omega)$, if the space dimension is less or equal to four. Nevertheless, it can be shown – in addition to the conditions for the linear case – that the fractional derivatives of some order $0 < \gamma < \frac{1}{4}$ (which are defined for a function f as the inverse Fourier transform of $(2i\pi\tau)^\gamma \hat{f}(\tau)$ with $\hat{f}$ denoting the Fourier transform of f in this context) of the sequence of approximate solutions remain bounded in $L^2(\mathbb{R}, L^2(\Omega))$. This allows a passage to the limit in the convection term which leads to an existence result for weak solutions of the non-linear Navier–Stokes equation, cf. [56].

While the strong embedding properties for two-dimensional Sobolev spaces allow us to derive the uniqueness of these solutions, this is generally not the case in higher dimensions due to the insufficient regularity of the velocity field v. Even in three dimensions, there is still a gap between the derived regularity of v ($v \in L^{\frac{8}{3}}(0, T, L^4(\Omega))$ with $\partial_t v \in L^{\frac{4}{3}}(0, T, H^{-1}(\Omega))$) and the regularity requirements for the established uniqueness results (e.g., $v \in L^8(0, T, L^4(\Omega))$) for arbitrary initial data $f_{ext} \in L^2(0, T, H^{-1}(\Omega))$ and $v_a \in L^2(\Omega)$. However, if the initial data is more regular and 'sufficiently small', uniqueness can be shown also for the three-dimensional case. If the 'smallness' assumption for the initial data is omitted the uniqueness of a weak solution can still be verified on a small time interval by choosing a specific spatial basis $\{\psi_n\}_{n\in\mathbb{N}}$.

Another approach to derive the existence of weak solutions to the Navier–Stokes equation is based on an implicit discretization in time and a subsequent limiting analysis with respect to the time step size tending to zero, see, e.g., [56]. In the following, we will sketch how this approach can be applied to the Cahn–Hilliard–Navier–Stokes system (1a)–(1g).

2.1 The Semi-Discrete CHNS System

In this subsection, we introduce the semi-discretization in time of the system (1a)–(1g) in its weak formulation. We start by observing that, assuming integrability in time, from (1d), (1a), (1e), and (1f), it follows that

$$\begin{aligned}\int_\Omega \partial_t \varphi dx &= -\int_\Omega v\nabla\varphi dx + \int_\Omega \operatorname{div}(m(\varphi)\nabla\mu)dx \\ &= \int_\Omega \operatorname{div}(v)\varphi dx - \int_{\partial\Omega} v\varphi \mathbf{n}_\Omega dx + \int_{\partial\Omega} m(\varphi)\nabla\mu\mathbf{n}_\Omega dx = 0.\end{aligned}$$

Hence, the integral mean of φ remains constant

$$\frac{1}{|\Omega|}\int_\Omega \varphi dx \equiv: \overline{\varphi_a} \in (-1,1). \tag{4}$$

(Note that (4) excludes the uninteresting case $|\overline{\varphi_a}| = 1$ which corresponds to only one phase being present.) Accordingly, it is sufficient to consider a shifted system (1a)–(1g), where φ is replaced by its projection onto $\overline{L}^2(\Omega)$, with shifted variables such as, e.g. $m(y+\overline{\varphi_a})$, which we again denote by $m(y)$ in a slight misuse of notation.

Our second observation concerns the thermodynamical consistency of the CHNS system. As noted in the introduction, it is possible to derive a (dissipative) energy law by testing (1a), (1b), (1c) and (1d) with μ, $\partial\varphi$, v and Π, respectively, yielding

$$\partial_t E(v,\varphi) + 2\int_\Omega \eta(\varphi)|\epsilon(v)|^2 dx + \int_\Omega m(\varphi)|\nabla\mu|^2 dx \leq 0, \tag{5}$$

where the total energy E is given by the sum of the kinetic and the potential energy, i.e.

$$E(v,\varphi) = \int_\Omega \rho(\varphi)\frac{|v|^2}{2}dx + \int_\Omega \frac{|\nabla\varphi|^2}{2}dx + \Psi(\varphi). \tag{6}$$

Besides mirroring the physical property that the total energy of a closed system is non-increasing, inequality (5) also serves as a very valuable analytical tool, e.g., to secure the boundedness of solutions to (1a)–(1g). Therefore, it is desirable to maintain the energy inequality on the time discrete level which typically requires to preserve the strong coupling of the Cahn–Hilliard system and the Navier–Stokes equation as seen in Definition 1 below. We note, however, that very recently F. Guillén-González and G. Tierra proposed a numerical splitting scheme for the Cahn–Hilliard–Navier–Stokes system which maintains the energy law via introducing a small correction term to the velocity field, cf. [27].

The subsequent assumption rigorously introduces the given data such as the mobility and viscosity coefficients m, η, the density function ρ, the convex part Ψ_0 of the free energy density Ψ and the initial data v_a, φ_a along with some necessary regularity requirements.

Assumption 1

1. The coefficient functions $m, \eta \in C^2(\mathbb{R})$ as well as their derivatives up to second order are bounded, i.e. there exist constants $0 < b_1 \leq b_2$ such that for every $x \in \mathbb{R}$, it holds that $b_1 \leq \min\{m(x), \eta(x)\}$ and

$$\max\{m(x), \eta(x), |m'(x)|, |\eta'(x)|, |m''(x)|, |\eta''(x)|\} \leq b_2.$$

2. The initial state satisfies $(v_a, \varphi_a) \in H^2_{0,\sigma}(\Omega; \mathbb{R}^n) \times \left(\overline{H}^2_{\partial_n}(\Omega) \cap \mathbb{K}\right)$ where

$$\mathbb{K} := \left\{v \in \overline{H}^1(\Omega) : \psi_1 \leq v \leq \psi_2 \text{ a.e. in } \Omega\right\},$$

with $-1 - \overline{\varphi_a} =: \psi_1 < 0 < \psi_2 := 1 - \overline{\varphi_a}$.
3. The density ρ depends on the order parameter φ via

$$\rho(\varphi) = \frac{\rho_1 + \rho_2}{2} + \frac{\rho_2 - \rho_1}{2}(\varphi + \overline{\varphi_a}). \tag{7}$$

4. The functional $\Psi_0 : \overline{H}^1(\Omega) \to \mathbb{R}$ is given by $\Psi_0(\varphi) := \int_\Omega \psi_0(\varphi(x))dx$ where $\psi_0 : \mathbb{R} \to \overline{\mathbb{R}} := \mathbb{R} \cup \{+\infty\}$ represents the double-obstacle potential,

$$\psi_0(z) := i_{[\psi_1;\psi_2]} := \begin{cases} +\infty & \text{if } z < \psi_1, \\ 0 & \text{if } \psi_1 \leq z \leq \psi_2, \\ +\infty & \text{if } z > \psi_2. \end{cases}$$

We point out that since the double-obstacle potential restricts the order parameter to the physically relevant interval $[\psi_1, \psi_2]$, the density remains always positive which is important for deriving appropriate energy estimates.

With these assumptions we now state the control-version of the semi-discrete Cahn–Hilliard Navier–Stokes system with a distributed control u on the right-hand side of the Navier–Stokes equation. The problem with boundary control is formulated analogously. Here, $\tau > 0$ denotes the time step-size and $K \in \mathbb{N}$ the total number of time instants.

Definition 1 (Semi-Discrete CHNS System) Fixing $(\varphi_{-1}, v_0) = (\varphi_a, v_a)$ we say that a triple

$$(\varphi, \mu, v) = ((\varphi_i)_{i=0}^{K-1}, (\mu_i)_{i=0}^{K-1}, (v_i)_{i=1}^{K-1})$$

in $\overline{H}^2_{\partial_n}(\Omega)^K \times \overline{H}^2_{\partial_n}(\Omega)^K \times H^1_{0,\sigma}(\Omega;\mathbb{R}^n)^{K-1}$ solves the semi-discrete CHNS system with respect to a given control $u = (u_i)_{i=1}^{K-1} \in L^2(\Omega;\mathbb{R}^n)^{K-1}$, denoted by $(\varphi,\mu,v) \in S_\Psi(u)$, if it holds for all $\phi \in \overline{H}^1(\Omega)$ and $\psi \in H^1_{0,\sigma}(\Omega;\mathbb{R}^n)$ that

$$\left\langle \frac{\varphi_{i+1}-\varphi_i}{\tau}, \phi \right\rangle + \langle v_{i+1}\nabla\varphi_i, \phi\rangle + \big(m(\varphi_i)\nabla\mu_{i+1}, \nabla\phi\big) = 0, \tag{8}$$

$$(\nabla\varphi_{i+1}, \nabla\phi) + \langle\partial\Psi_0(\varphi_{i+1}), \phi\rangle - \langle\mu_{i+1}, \phi\rangle - \langle\kappa\varphi_i, \phi\rangle \ni 0, \tag{9}$$

$$\begin{aligned}\left\langle \frac{\rho(\varphi_i)v_{i+1}-\rho(\varphi_{i-1})v_i}{\tau}, \psi \right\rangle_{H^{-1}_{0,\sigma},H^1_{0,\sigma}} &- \big(v_{i+1}\otimes\rho(\varphi_{i-1})v_i, \nabla\psi\big) \\ + \left(v_{i+1}\otimes\frac{\rho_2-\rho_1}{2}m(\varphi_{i-1})\nabla\mu_i, \nabla\psi\right) &+ (2\eta(\varphi_i)\epsilon(v_{i+1}), \epsilon(\psi)) \\ &- \langle\mu_{i+1}\nabla\varphi_i, \psi\rangle_{H^{-1}_{0,\sigma},H^1_{0,\sigma}} = \langle u_{i+1}, \psi\rangle_{H^{-1}_{0,\sigma},H^1_{0,\sigma}}.\end{aligned} \tag{10}$$

The first two equations are supposed to hold for every $0 \le i+1 \le K-1$ and the last equation holds for every $1 \le i+1 \le K-1$.

In the above system the boundary conditions specified in (1e) and (1f) are incorporated in the respective function spaces and the definition already includes the inherent regularity properties of φ and μ which anticipates the results of Lemma 2 below. Moreover, the semi-discrete CHNS system involves three time instants $(i-1, i, i+1)$ and (φ_0, μ_0) is characterized in an initialization step by the (decoupled) Cahn–Hilliard system only.

At the subsequent time instants, the strong coupling of the Cahn–Hilliard and Navier–Stokes system is maintained. As discussed above, this gives rise to the following (dissipative) energy law with respect to the semi-discrete equivalent of the total energy introduced in (6) (cf. [32])

$$E(v,\varphi,\varphi_{-1}) = \int_\Omega \rho(\varphi_{-1})\frac{|v|^2}{2}dx + \int_\Omega \frac{|\varphi|^2}{2}dx + \Psi(\varphi). \tag{11}$$

Lemma 1 (Energy Estimate for a Single Time Step) *Let $\varphi_i, \varphi_{i-1} \in \overline{H}^1(\Omega)\cap\mathbb{K}$, $\mu_i \in \overline{H}^1(\Omega)$, $v_i \in H^1_{0,\sigma}(\Omega;\mathbb{R}^n)$ be the state of the system at time step i and $u_{i+1} \in (H^1_{0,\sigma}(\Omega;\mathbb{R}^n))^*$ a given external force.*

Then, if $(\varphi_{i+1}, \mu_{i+1}, v_{i+1}) \in \overline{H}^1(\Omega) \times \overline{H}^1(\Omega) \times H^1_{0,\sigma}(\Omega; \mathbb{R}^n)$ *solves the system* (8)–(10) *for one time step, the corresponding total energy is bounded by*

$$
\begin{aligned}
E(v_{i+1}, \varphi_{i+1}, \varphi_i) &+ \int_\Omega \rho(\varphi_{i-1}) \frac{|v_{i+1} - v_i|^2}{2} dx + \int_\Omega \frac{|\nabla\varphi_{i+1} - \nabla\varphi_i|^2}{2} dx \\
&+ \tau \int_\Omega 2\eta(\varphi_i)\, |\varepsilon(v_{i+1})|^2 \, dx + \tau \int_\Omega m(\varphi_i) \left|\nabla\mu_{i+1}\right|^2 dx + \int_\Omega \kappa \frac{(\varphi_{i+1} - \varphi_i)^2}{2} \\
&\leq E(v_i, \varphi_i, \varphi_{i-1}) + \langle u_{i+1}, v_{i+1} \rangle_{H^{-1}_{0,\sigma}, H^1_{0,\sigma}} \, .
\end{aligned} \tag{12}
$$

Note that, due to the positivity of the density and the coefficients m, η, all the terms of the left-hand side of the inequality are always non-negative such that Lemma 1 indeed ensures that the energy of the next time step is non-increasing if the external force u_{i+1} is absent.

2.2 Existence of Solutions

In this subsection, we establish the existence of feasible points for the semi-discrete Cahn–Hilliard Navier–Stokes system. The proof is based on Schaefer's fixed point theorem, also called the Leray-Schauder principle, and uses some arguments from monotone operator theory. In order to apply the theorem, it is necessary to guarantee a boundedness condition for the solutions of a slightly modified system (8)–(10). This can be done with the help of the energy estimate (12) and yields the following result concerning the solvability of the semi-discrete system (8)–(10) for single time steps. For rigorous proofs of the results of this section we refer to Hintermüller et al. [32].

Theorem 1 (Existence of Solutions to the CHNS System for a Single Time Step) *Let the assumptions of Lemma 1 be satisfied. Then the system* (8)–(10) *admits a solution* $(\varphi_{i+1}, \mu_{i+1}, v_{i+1}) \in \overline{H}^1(\Omega) \times \overline{H}^1(\Omega) \times H^1_{0,\sigma}(\Omega; \mathbb{R}^n)$ *for one time step.*

As noted above, it is possible to derive higher regularity properties for the solution of the Navier–Stokes system if the initial data is slightly more regular. A similar observation holds true for the solution of variational inequalities, see, e.g., [46]. Combining these arguments using a bootstrap argument, it is possible to prove the following lemma.

Note that for the actual applications it is natural to consider the control force u_{i+1} to be an element of $L^2(\Omega; \mathbb{R}^n)$, later on, in order to permit a point-wise interpretation almost everywhere on Ω.

Lemma 2 (Regularity of Solutions) *Let the assumptions of Lemma 1 be satisfied, and suppose additionally that* $\varphi_i \in H^2(\Omega)$ *and* $u_{i+1} \in L^2(\Omega; \mathbb{R}^n)$*. Moreover,* *let* $(\varphi_{i+1}, \mu_{i+1}, v_{i+1}) \in \overline{H}^1(\Omega) \times \overline{H}^1(\Omega) \times H^1_{0,\sigma}(\Omega; \mathbb{R}^n)$ *be a solution to the system* (8)–(10)

Then it holds that $\varphi_{i+1}, \mu_{i+1} \in \overline{H}^2_{\partial_n}(\Omega)$, $\varphi_{i+1} \in \mathbb{K}$ *and* $v_{i+1} \in H^2(\Omega; \mathbb{R}^n)$, *and there exists a constant* $C = C(N, \Omega, b_1, b_2, \tau, \kappa) > 0$ *such that*

$$\begin{aligned} &\left\|\varphi_{i+1}\right\|_{H^2} + \left\|\mu_{i+1}\right\|_{H^2} + \|v_{i+1}\|_{H^2} \\ &\quad \leq C(\left\|\varphi_{i+1}\right\| + \left\|\mu_{i+1}\right\| + \left\|\varphi_i\right\| + \|v_{i+1}\|_{H^1} \left\|\varphi_i\right\|_{H^2}). \end{aligned} \tag{13}$$

This also ensures that the system (8)–(10) is well-posed for each subsequent time step. Thus, applying Theorem 1 and Lemma 2 repeatedly for each time step $i = 0, .., K-2$ directly verifies the existence of solutions to the semi-discrete Cahn–Hilliard–Navier–Stokes system given in Definition 1.

Proposition 1 (Existence of Feasible Points) *Let* $u \in L^2(\Omega; \mathbb{R}^n)^{K-1}$ *be given. Then the semi-discrete CHNS system admits a solution* $(\varphi, \mu, v) \in \overline{H}^2_{\partial_n}(\Omega)^K \times \overline{H}^2_{\partial_n}(\Omega)^K \times H^2_{0,\sigma}(\Omega; \mathbb{R}^n)^{K-1}$.

From here onwards, the existence of solutions to the time-continuous CHNS system can be established via a limiting process with respect the total number of time instances $K \to \infty$ and the time step size $\tau := \frac{T}{K} \to 0$. For this purpose, one considers certain step functions f^K_{step} with respect to the time t which are equal to the time discrete solutions $f^K_{step}(t) = f_i$ on each interval $t \in [(i-1)\tau, i\tau)$; $i = 1, .., K-1$ for $f \in \{v, \varphi, \mu\}$. Employing the energy estimate (12) these functions can be bounded in the spaces $v^K_{step} \in L^2(0, T, H^1(\Omega; \mathbb{R}^n))$, $\varphi^K_{step} \in L^\infty(0, T, H^1(\Omega))$, $\mu^K_{step} \in L^2(0, T, H^1(\Omega))$. Then, the limit point of an appropriate subsequence can be shown to satisfy the system (1a)–(1g). For more details, we refer to, e.g., [1], where this method has been successfully applied to a similar system where the free energy density is defined through the logarithmic potential $\Psi(\varphi) = (1+\varphi)\ln(1+\varphi) + (1-\varphi)\ln(1-\varphi) - \frac{\kappa}{2}\varphi^2$.

However, since the same arguments cannot be applied to the adjoint system associated with the optimal control problem later on, we restrict our subsequent investigations on the semi-discrete system.

3 The Optimal Control Problem

This section finally presents the optimal control problem for the semi-discrete CHNS system. For this purpose, let $U_{ad} \subset L^2(\Omega; \mathbb{R}^n)^{K-1}$ and $\mathcal{J} : \mathcal{X} \to \mathbb{R}$ be a Fréchet differentiable function, with

$$\mathcal{X} := \overline{H}^1(\Omega)^K \times \overline{H}^1(\Omega)^K \times H^1_{0,\sigma}(\Omega; \mathbb{R}^n)^{K-1} \times L^2(\Omega; \mathbb{R}^n)^{K-1}.$$

Definition 2 The optimal control problem is given by

$$\begin{aligned} &\min \mathscr{J}(\varphi, \mu, v, u) \text{ over } (\varphi, \mu, v, u) \in \mathscr{X} \\ &\text{s.t. } u \in U_{ad}, \ (\varphi, \mu, v) \in S_\Psi(u). \end{aligned} \tag{P_Ψ}$$

Further requirements on U_{ad} and $\mathscr{J}$ are made explicit in connection with Theorem 2, below. However, we point out that a tracking-type functional, like, e.g.,

$$\mathscr{J}(\varphi, \mu, v, u) := \frac{1}{2} \left\|\varphi_{K-1} - \varphi_d\right\|^2 + \frac{\xi}{2} \|u\|^2_{(L^2)^{(K-1)}}, \ \xi > 0, \tag{14}$$

with desired states $\varphi_d \in L^2(\Omega)$, which is used in many applications, satisfies these assumptions.

Due to the results from the previous section, the feasible set of problem (P_Ψ) is non-empty. The existence of globally optimal points can be verified via standard arguments from optimization theory, if some classical assumptions on the objective functional and the constraint set U_{ad} are imposed, cf. [32].

Theorem 2 (Existence of Global Solutions) *Suppose that the objective functional* $\mathscr{J} : \overline{H}^2_{\partial_n}(\Omega)^K \times \overline{H}^2_{\partial_n}(\Omega)^K \times H^1_{0,\sigma}(\Omega; \mathbb{R}^n)^{K-1} \times L^2(\Omega; \mathbb{R}^n)^{K-1} \to \mathbb{R}$ *is convex and weakly lower-semi-continuous and* U_{ad} *is non-empty, closed and convex. Assume that either* U_{ad} *is bounded or* $\mathscr{J}$ *is partially coercive, i.e. for every sequence* $\left\{(\varphi^{(k)}, \mu^{(k)}, v^{(k)}, u^{(k)})\right\}_{k\in\mathbb{N}}$ *which satisfies* $\lim_{k\to\infty} \left\|u^{(k)}\right\| = +\infty$ *it holds true that* $\lim_{k\to\infty} \mathscr{J}(\varphi^{(k)}, \mu^{(k)}, v^{(k)}, u^{(k)}) = +\infty$.

Then the optimization problem (P_Ψ) *admits a global solution.*

Although the previous theorem ensures the existence of solutions to problem (P_Ψ), it is our goal to directly compute such solutions or candidates thereof efficiently. Therefore, the next section explores different ways to characterize these solutions.

4 Stationarity Conditions

In order to compute solutions of the optimal control problem it is desirable to establish stationarity conditions for the problem, which consist of a set of equations involving the derivatives of the objective functional and the constraint mappings connected via certain (Lagrange) multipliers and can be solved numerically.

Due to the presence of the double-obstacle potential in the Cahn–Hilliard system the constraint system of the optimal control problem includes a variational inequality constraint. This imposes a severe analytical challenge as the solution operator to a variational inequality is in general not differentiable (even in finite dimensions). To illustrate this, we note that the variational inequality (1b) is equivalent to the problem of finding $\varphi \in \mathbb{K}$ and $a \in \partial\Psi_0(\varphi)$ such that

$$-\Delta\varphi + a - \mu - \kappa\varphi = 0, \tag{15}$$

which can be reformulated as

$$\langle -\Delta\varphi - \mu - \kappa\varphi, \phi - \varphi\rangle = \langle -a, \phi - \varphi\rangle \geq 0, \ \forall \phi \in \mathbb{K}. \tag{16}$$

This can be related to the first-order condition of a constrained optimization problem over $\phi \in \mathbb{K}$, where a, respectively, a^+ and a^- with $a = a^+ - a^-$, act as Lagrange multipliers associated to the inequality constraints $\varphi \geq \psi_1$ and $\varphi \leq \psi_2$. Condition (16) can be further transformed into the following complementarity problem

$$-\Delta\varphi + a^+ - a^- - \mu - \kappa\varphi = 0, \tag{17}$$

$$\varphi \geq \psi_1, \quad a^- \geq 0, \quad \langle a^-, \varphi - \psi_1\rangle = 0, \tag{18}$$

$$\varphi \leq \psi_2, \quad a^+ \geq 0, \quad \langle a^+, \varphi - \psi_2\rangle = 0. \tag{19}$$

Note that, in general, a is an element of the dual space $H^{-1}(\Omega)$. However, employing the properties of the Laplace operator it is possible to show that $\varphi \in H^2(\Omega)$ and a is contained in $L^2(\Omega)$ as we have seen it in the previous section which allows us to impose the sign conditions in (18), (19) a.e. on Ω and to understand the respective duality pairings as L^2-inner products. Then, it can be seen that a solution φ either meets one of the bounds (which defines the so-called active sets)

$$\varphi = \psi_1, \quad (a^+ = 0, \ a^- \geq 0), \tag{20}$$

$$\varphi = \psi_2, \quad (a^+ \geq 0, \ a^- = 0), \tag{21}$$

or it satisfies the partial differential equation (which happens on the inactive set)

$$-\Delta\varphi - \mu - \kappa\varphi = 0, \quad (a^+ = a^- = 0). \tag{22}$$

This structure of the solution set invokes the aforementioned non-differentiability of the solution operator which specifically occurs at the transition from active to the inactive parts of the domain, which includes the so-called biactive sets where the partial differential equation is satisfied as well as one of the constraints (20), (21).

As a consequence, classical constraint qualifications (see, e.g., [61]) fail which prevents the application of Karush-Kuhn-Tucker (KKT) theory in Banach spaces for the first-order characterization of an optimal solution by (Lagrange) multipliers. As a result, stationarity conditions for this problem class are no longer unique (even in finite dimensions, cf. [49, 52]) but rather depend on the underlying problem structure and/or on the chosen analytical approach; compare [33, 34] in function space and, e.g., [54] in finite dimensions.

In [33], different stationarity concepts for mathematical programs with equilibrium conditions are discussed on the basis of the optimal control of a standard elliptic variational inequality: C-stationarity and strong stationarity. They involve multipliers associated to the partial differential equations which in our case are represented by (1a), (22) and (1c). The corresponding multipliers are called adjoint states and will be subsequently denoted by (p, r, q). Furthermore, the concepts contain additional multipliers (subsequently denoted by λ^- and λ^+) which are connected to the state constraints involving ψ_1 and ψ_2. The main difference is caused by the complicated interaction of these constraints explained above. While C-stationarity just imposes conditions of the type '$r/\lambda^-/\lambda^+ = 0$ if $a \neq 0/\varphi \neq \psi_1/\varphi \neq \psi_2$', strong stationarity conditions include additional sign conditions for r, λ^-, λ^+ on the biactive set. Another challenge comes from the fact that state-constrained optimal control problems typically feature low multiplier regularity (see, e.g., [16]), i.e. λ^-, λ^+ are just contained in $H^{-1}(\Omega)$. As a consequence, the conditions of the type '$\lambda^-/\lambda^+ = 0$ if $\varphi \neq \psi_1/\varphi \neq \psi_2$' cannot be interpreted pointwise almost everywhere. Instead, there are different possibilities to transfer the condition to an infinite-dimensional setting. This leads to three different versions of C- and strong stationarity, respectively, summarized in the hierarchical structure below, cf. [33]:

strong stationarity $\Rightarrow$	almost strong stationarity $\Rightarrow$	ε-almost strong stationarity
$\Downarrow$	$\Downarrow$	$\Downarrow$
C-stationarity $\Rightarrow$	almost C-stationarity $\Rightarrow$	ε-almost C-stationarity

Here, the notion of ε-almost refers to the use of Egorov's theorem in the corresponding proof.

Typical approaches to establish these stationarity concepts rely on the relaxation, regularization and/or penalization of the degeneracy of the lower-level problem. In the sequel, we sketch how a Yosida regularization technique with a subsequent passage to the limit with the Yosida parameter can be used to derive conditions of ε-almost C-stationarity type for the optimal control problem (P_Ψ). For more details on the proofs of the results of this section we refer to Hintermüller et al. [32].

For this purpose, we introduce the following sequence of double-well type potentials which will be used to approximate the double-obstacle potential in the lower-level problem. Here, γ denotes the subdifferential of the indicator function of $[\psi_1, \psi_2]$, i.e. $\gamma := \partial i_{[\psi_1,\psi_2]}$.

Definition 3 Let a mollifier $\zeta \in C^1(\mathbb{R})$ with supp $\zeta \subset [-1, 1]$, $\int_{\mathbb{R}} \zeta = 1$ and $0 \leq \zeta \leq 1$ a.e. on $\mathbb{R}$, and a function $\theta : \mathbb{R}^+ \to \mathbb{R}^+$, with $\theta(\alpha) > 0$ and $\frac{\theta(\alpha)}{\alpha} \to 0$ as $\alpha \to 0$, be given. For the Yosida approximation γ_α with parameter $\alpha > 0$ of γ define

$$\zeta_\alpha(s) := \tfrac{1}{\alpha}\zeta\left(\tfrac{s}{\alpha}\right), \quad \widetilde{\gamma}_\alpha := \gamma_\alpha * \zeta_{\theta(\alpha)}, \quad \psi_{0,\alpha}(s) := \textstyle\int_0^s \widetilde{\gamma}_\alpha(x)\, dx,$$

$$\Psi_{0,\alpha}(\varphi) := \textstyle\int_\Omega (\psi_{0,\alpha} \circ \varphi)(x)\, dx.$$

Moreover, we set $\alpha_k := k^{-1}$, $\Psi_0^{(k)} := \Psi_{0,\alpha_n}$.

In the sequel, we approximate the problem (P_Ψ) by auxiliary problems $(P_\Psi^{(k)})$ where the double-obstacle Ψ_0 is replaced by $\Psi_0^{(k)}$. We then derive necessary first-order optimality conditions for the auxiliary problems and establish a stationarity system for (P_Ψ) by considering the limit process for $\alpha \to 0$.

We point out that this method requires the verification of the existence of globally optimal solutions to the auxiliary problems. Hereby, the arguments of Sect. 2 cannot be directly transferred to the smooth case since ρ is no longer guaranteed to be non-negative if we maintain the affine connection of the order parameter and the density given in (7). However, the line of argumentation can be saved by manually enforcing the non-negativity of the density function, i.e. $\rho(\varphi) := \max\left\{\frac{\rho_1+\rho_2}{2} + \frac{\rho_2-\rho_1}{2}(\varphi + \overline{\varphi_a}), 0\right\}$, with the help of the following theorem.

Theorem 3 *Let $u \in L^2(\Omega;\mathbb{R}^n)^{K-1}$ be given and let $\left\{(\varphi^{(k)}, \mu^{(k)}, v^{(k)})\right\}_{k\in\mathbb{N}}$ be a sequence of solutions to the systems* (8)–(10) *with $\Psi_0 = \Psi_0^{(k)}$. Then*

$$\left\|\max(-\varphi^{(k)} + \psi_1, 0)\right\|_{L^\infty} \to 0, \text{ as } k \to \infty.$$

Theorem 3 ensures that the order parameter of a solution to the system (8)–(10) for the double-well type potentials under consideration is always greater than $\psi_1 - \varepsilon$ for a small $\varepsilon > 0$ if k is sufficiently large. Consequently, the corresponding density is positive.

The next theorem verifies the consistency of Moreau–Yosida type approximations, i.e. the convergence of a sequence of solutions to the auxiliary problems to a solution of (P_Ψ) for $k \to \infty$.

Theorem 4 (Consistency of the Regularization) *Let the assumptions of Theorem 2 be fulfilled and let $\mathcal{J} : \overline{H}^1(\Omega)^K \times \overline{H}^1(\Omega)^K \times H^1_{0,\sigma}(\Omega;\mathbb{R}^n)^{K-1} \times L^2(\Omega;\mathbb{R}^n)^{K-1} \to \mathbb{R}$ be upper-semicontinuous.*

Then a sequence $\left\{(\varphi^{(k)}, \mu^{(k)}, v^{(k)}, u^{(k)})\right\}_{k\in\mathbb{N}}$ of global solutions to $(P_{\Psi^{(k)}})$ in $\overline{H}^2(\Omega)^K \times \overline{H}^2(\Omega)^K \times H^1_{0,\sigma}(\Omega;\mathbb{R}^n)^{K-1} \times U_{ad}$ converges to a global solution of $(P_{\overline{\Psi}})$, provided that $\left\{\mathcal{J}(\varphi^{(k)}, \mu^{(k)}, v^{(k)}, u^{(k)})\right\}_{k\in\mathbb{N}}$ is assumed bounded, whenever U_{ad} is unbounded.

At this point, we turn our attention to the derivation of stationarity conditions for the optimal control problem. For the smooth potential functions, first-order optimality conditions can be directly derived using a classical result from Zowe and Kurcyusz, [61, Theorem 4.1] which leads to the following theorem.

Theorem 5 (First-Order Optimality Conditions for Smooth Potentials) *Let $\mathcal{J} : \overline{H}^1(\Omega)^K \times \overline{H}^1(\Omega)^K \times H^1_{0,\sigma}(\Omega;\mathbb{R}^n)^{K-1} \times L^2(\Omega;\mathbb{R}^n)^{K-1} \to \mathbb{R}$ be Fréchet differentiable and let $\overline{z} := (\overline{\varphi}, \overline{\mu}, \overline{v}, \overline{u})$ be a minimizer of the auxiliary problem $(P_\Psi^{(k)})$.*

Then there exist $(p, r, q) \in \overline{H}^1(\Omega)^K \times \overline{H}^1(\Omega)^{K-1} \times H^1_{0,\sigma}(\Omega; \mathbb{R}^n)^{K-1}$*, with* $p = (p_{-1}, \dots p_{K-2})$, $r = (r_{-1}, \dots r_{K-2})$, $q = (q_0, \dots q_{K-2})$*, such that*

$$\begin{aligned}
-\frac{1}{\tau}(p_i - p_{i-1}) + m'(\varphi_i)\nabla\mu_{i+1} \cdot \nabla p_i - \operatorname{div}(p_i v_{i+1}) - \Delta r_{i-1}& \\
+\Psi_0''(\varphi_i)^* r_{i-1} - \kappa r_{i+1} - \frac{1}{\tau}\rho'(\varphi_i) v_{i+1} \cdot (q_{i+1} - q_i)& \\
-(\rho'(\varphi_i) v_{i+1} - \frac{\rho_2 - \rho_1}{2} m'(\varphi_i)\nabla\mu_{i+1})(Dq_{i+1})^\top v_{i+2}& \\
+2\eta'(\varphi_i)\epsilon(v_{i+1}) : Dq_i + \operatorname{div}(\mu_{i+1} q_i) &= \frac{\partial \mathcal{J}}{\partial \varphi_i}(\overline{z}),
\end{aligned} \tag{23}$$

$$\begin{aligned}
-r_{i-1} - \operatorname{div}(m(\varphi_{i-1})\nabla p_{i-1}) - \operatorname{div}(\frac{\rho_2 - \rho_1}{2} m(\varphi_{i-1})(Dq_i)^\top v_{i+1})& \\
-q_{i-1} \cdot \nabla\varphi_{i-1} &= \frac{\partial \mathcal{J}}{\partial \mu_i}(\overline{z}),
\end{aligned} \tag{24}$$

$$\begin{aligned}
-\frac{1}{\tau}\rho(\varphi_{j-1})(q_j - q_{j-1}) - \rho(\varphi_{j-1})(Dq_j)^\top v_{j+1}& \\
-(Dq_{j-1})(\rho(\varphi_{j-2}) v_{j-1} - \frac{\rho_2 - \rho_1}{2} m(\varphi_{j-2})\nabla\mu_{j-1})& \\
-\operatorname{div}(2\eta(\varphi_{j-1})\varepsilon(q_{j-1})) + p_{j-1}\nabla\varphi_{j-1} &= \frac{\partial \mathcal{J}}{\partial v_j}(\overline{z}),
\end{aligned} \tag{25}$$

$$\left(\frac{\partial \mathcal{J}}{\partial u_k}(\overline{z}) - q_{k-1}\right)_{k=1}^{K-1} \in \left[\mathbb{R}_+(U_{ad} - \bar{u})\right]^+, \tag{26}$$

for all $i = 0, \dots, K-1$ *and* $j = 1, \dots, K-1$*. Here,* $\left[\mathbb{R}_+(U_{ad} - \bar{u})\right]^+$ *denotes the polar cone of the set* $\{r(w-u) | w \in U_{ad},\ r \in \mathbb{R}^+\}$*. Furthermore, we use the convention that* p_i, r_i, q_i *are equal to* 0 *for* $i \geq K-1$ *along with* q_{-1} *and* φ_i, μ_i, v_i *for* $i \geq K$*.*

In order to pass to the limit with respect to $k \to \infty$ it is necessary to ensure that the adjoint state (p, r, q) is bounded independently of the regularization parameter. This leads to the following theorem which states the adjoint system for the optimal control problem (P_Ψ), cf. [32].

Theorem 6 (Stationarity Conditions) *Suppose that the following assumptions are satisfied.*

1. *$\mathcal{J}'$ is a bounded mapping from $\overline{H}^1(\Omega)^K \times \overline{H}^1(\Omega)^K \times H^1_{0,\sigma}(\Omega;\mathbb{R}^n)^{K-1} \times U_{ad}$ into the space $(\overline{H}^1(\Omega)^K \times \overline{H}^1(\Omega)^K \times H^1_{0,\sigma}(\Omega;\mathbb{R}^n)^{K-1} \times L^2(\Omega;\mathbb{R}^n)^{K-1})^*$ and $\frac{\partial \mathcal{J}}{\partial u}$ satisfies the following weak lower-semicontinuity property*

$$\Big\langle \frac{\partial \mathcal{J}}{\partial u}(\hat{z}), \hat{u}\Big\rangle \leq \liminf_{n\to\infty} \Big\langle \frac{\partial \mathcal{J}}{\partial u}(\hat{z}^{(k)}), \hat{u}^{(k)}\Big\rangle,$$

for $\hat{z}^{(k)} = (\hat{\varphi}^{(k)}, \hat{\mu}^{(k)}, \hat{v}^{(k)}, \hat{u}^{(k)})$ converging weakly in $\overline{H}^2_{\partial_n}(\Omega)^K \times \overline{H}^2_{\partial_n}(\Omega)^K \times H^1_{0,\sigma}(\Omega;\mathbb{R}^n)^{K-1} \times U_{ad}$ to $\hat{z} = (\hat{\varphi}, \hat{\mu}, \hat{v}, \hat{u})$.
2. *Let $(\varphi^{(k)}, \mu^{(k)}, v^{(k)}, u^{(k)}) \in \overline{H}^2_{\partial_n}(\Omega)^K \times \overline{H}^2_{\partial_n}(\Omega)^K \times H^1_{0,\sigma}(\Omega;\mathbb{R}^n)^{K-1} \times U_{ad}$ be a minimizer for $(P_{\Psi^{(k)}})$ and let further $(p^{(k)}, r^{(k)}, q^{(k)}) \in \overline{H}^1(\Omega)^K \times \overline{H}^1(\Omega)^K \times H^1_{0,\sigma}(\Omega;\mathbb{R}^n)^{K-1}$ be given as in Theorem 5.*

Then there exists an element $(\varphi, \mu, v, u, p, r, q)$ and a subsequence denoted by $\{(\varphi^{(m)}, \mu^{(m)}, v^{(m)}, u^{(m)}, p^{(m)}, r^{(m)}, q^{(m)})\}_{m\in\mathbb{N}}$ with

$$\varphi^{(m)} \to \varphi \text{ weakly in } \overline{H}^2_{\partial_n}(\Omega)^K,\ \mu^{(m)} \to \mu \text{ weakly in } \overline{H}^2_{\partial_n}(\Omega)^{K-1},$$

$$v^{(m)} \to v \text{ weakly in } H^2(\Omega;\mathbb{R}^n)^{K-1},\ u^{(m)} \to u \text{ weakly in } L^2(\Omega;\mathbb{R}^n)^{K-1},$$

$$p^{(m)} \to p \text{ weakly in } \overline{H}^1(\Omega)^K,\ r^{(m)} \to r \text{ weakly in } \overline{H}^1(\Omega)^{K-1},$$

$$q^{(m)} \to q \text{ weakly in } H^1_{0,\sigma}(\Omega;\mathbb{R}^n)^{K-1},\ \Psi_0^{(m)''}(\varphi^{(m)}_{i+1})^* r^{(m)}_i \to \lambda_i \text{ weakly in } \overline{H}^1(\Omega)^*,$$

for all $i = -1, \ldots, K-2$ which satisfies (23)–(26) where $\Psi_0''(\varphi_i)^* r_{i-1}$ is replaced by λ_{i-1}.

If the set U_{ad} is bounded, Theorem 6 holds also true for a sequence of stationary points for $(P_{\Psi^{(k)}})$. If it is unbounded, then the result can still be transferred to sequences of stationary points by assuming that the sequence $\{u^{(k)}\}_{k\in\mathbb{N}}$ is bounded in $L^2(\Omega;\mathbb{R}^n)^{K-1}$.

The aforementioned multiplier conditions on r and λ are derived through a careful limiting analysis of the corresponding terms.

Theorem 7 (Limiting ε-almost C-stationarity) *Let $\Psi_0^{(m)}$, $m \in \mathbb{N}$ be the functionals of Definition 3, and let the tuples $(\varphi^{(m)}, \mu^{(m)}, v^{(m)}, u^{(m)}, p^{(m)}, r^{(m)}, q^{(m)})$, $(\varphi, \mu, v, u, p, r, q)$ and $\mathcal{J}$ be as in Theorem 6. Moreover, let $\Lambda : \mathbb{R} \to \mathbb{R}$ be a Lipschitz function with $\Lambda(\psi_1) = \Lambda(\psi_2) = 0$. For*

$$a_i^{(m)} := \Psi_0^{(m)'}(\varphi_i^{(m)}), \quad \lambda_i^{(m)} := \Psi_0^{(m)''}(\varphi_i^{(m)})^* r_{i-1}^{(m)} \tag{27}$$

for $i = 0, \dots, K$, *and for* a_i *denoting the limit of* $a_i^{(m)}$, *it holds that*

$$(a_i, \Lambda(\varphi_i))_{L^2} = 0, \qquad \langle \lambda_i, \Lambda(\varphi_i) \rangle = 0, \tag{28}$$

$$(a_i, r_{i-1})_{L^2} = 0, \qquad \liminf (\lambda_i^{(m)}, r_{i-1}^{(m)})_{L^2} \geq 0. \tag{29}$$

Moreover, for every $\varepsilon > 0$ *there exist a measurable subset* M_i^ε *of* $M_i := \{x \in \Omega : \psi_1 < \varphi_i(x) < \psi_2\}$ *with* $|M_i \setminus M_i^\varepsilon| < \varepsilon$ *and*

$$\langle \lambda_i, v \rangle = 0 \qquad \forall v \in \overline{H}^1(\Omega), \;\; v|_{\Omega \setminus M_i^\varepsilon} = 0.$$

In combination with the adjoint system from Theorem 6, the last theorem states stationarity conditions corresponding to a function space version of limiting ε-almost C-stationarity type. For the underlying problem class, this is currently the most (and, to the best of our knowledge, only) selective stationarity system available.

4.1 Boundary Control

Since a distributed control as introduced in Definition 2 might be difficult to realize in some applications, we briefly discuss the case of boundary control in this subsection. Hereby, the homogeneous boundary condition (1e) is replaced by

$$v_{|\partial\Omega} = u. \tag{30}$$

In this case, the control u is an element of the space $H_{tr} := Tr(H_\sigma^1(\Omega; \mathbb{R}^n))$, where Tr denotes the zero-order trace operator, cf., e.g., [3]. Due to the embedding properties of Sobolev spaces, H_{tr} is contained in $H^{\frac{1}{2}}(\partial\Omega; \mathbb{R}^n)$. Moreover, it is a Hilbert space and the trace operator regarded from $H_\sigma^1(\Omega; \mathbb{R}^n)$ into H_{tr} is a linear, bounded and surjective mapping between Hilbert spaces. Hence, there exists a right inverse operator $B_{tr} : H_{tr} \to H_\sigma^1(\Omega; \mathbb{R}^n)$ such that $Tr \circ B_{tr}$ equals the identity operator on H_{tr}, cf. [7, 28].

The operator can be employed to reduce the inhomogeneous Navier–Stokes system to the problem with homogeneous Dirichlet boundary conditions, which is used in [37], to derive the existence of solutions to the CHNS system via similar arguments as in Sect. 2 (namely Brouwer's fix point theorem and monotone operator theory). More precisely, a boundary-control equivalent of the problem (P_Ψ) is studied with a tracking-type functional for matched densities. Furthermore, the constraint set U_{ad} is assumed to be a closed, linear subspace of H_{tr} and an additional compatibility condition on the given data is imposed.

Though the involved trace spaces require a careful (embedding) analysis as sketched above, similar arguments as for the distributed control can be cited to derive an analogous stationarity system due to the linearity of the trace operator, cf. [37].

5 Numerical Solution Methods

Based on the analytical results of the previous section, we now discuss how to solve the optimal control problem (P_Ψ) numerically. Hereby, we benefit from the fact that the specific time-discretization of Sect. 2 represents a first step towards a numerical investigation/realization of the problem.

Due to the non-differentiability of the control-to-state operator, there exist various approaches to design numerical solution algorithms for the problem. The different approaches can be typically linked to a specific derivation of the stationarity conditions and (like those) either rely on the relaxation, regularization and/or penalization of the degeneracy of the lower-level problem and a suitable adjustment of the corresponding (relaxation) parameter or on a direct characterization/calculation of some generalized derivative of the control-to-state operator. In this section, we illustrate a regularizing algorithm which corresponds to the presented analytical approach. However, we briefly discuss a different numerical solution method in Sect. 6.

One of the main computational challenges is imposed by the Navier–Stokes system which typically causes an immense numerical expense. Furthermore, the special structure of the phase field that models the spatial distribution suggests the utilization of a non-uniform spatial mesh. For this reason, the subsequent subsection deals with suitable adaptation processes for the underlying space mesh. Here, we incorporate the fact that, for optimal control problems, one is usually interested in an accurate estimation of the target quantity, i.e., the objective functional.

We note that there is also a wealth of literature on how to deal with the numerical challenges imposed by the Cahn–Hilliard system itself including, e.g., multigrid solvers (see, e.g., in [43, 44]), and residual based error estimation (see, e.g., [8, 9, 30, 31]. We also refer to [6] for a collection of benchmark computations for a couple of the diffuse interface models cited in Sect. 1.

5.1 Adaptive Mesh Refinement Techniques

Common adaptive refinement concepts for phase field models are based on the order parameter φ or its derivative to take different values on the interfacial layers ($-1 < \varphi < 1$ and $|\nabla\varphi| > 0$) than on the bulk phases ($|\varphi| = 1$ and $\nabla\varphi = 0$), see, e.g., [6, 10, 42] and [26, 31], respectively. Such a refinement strategy however ignores the contributions of the velocity field v (which is included for residual based methods) and, in the presence of the optimal control problem, the contribution of the adjoint variables to the total discretization error.

In this subsection, we derive an adaptive error estimator which consists of dual-weighted primal residuals, primal-weighted dual residuals and complementarity errors following an approach presented in [14]. It is based on the notion of a modified Lagrangian associated with the MPEC and uses the associated saddle-point condition for optimal points to characterize the error in the objective function between the continuous solution and a fully discretized problem.

For this purpose, we suppose that the stationarity conditions (23)–(26) are discretized in space based on Taylor-Hood finite elements which are known to be LBB-stable in case of the Navier–Stokes equation, cf., e.g., [25, 57]. More precisely, the phase field and the chemical potential are discretized via piecewise linear and globally continuous finite elements, whereas the discretization of the velocity field utilizes piecewise quadratic and globally continuous finite elements. For more details on the chosen discretization approach we refer the reader to [29]. Furthermore, we consider the objective functional of tacking type given in (14).

The subsequent definition characterizes the MPCC-Lagrangian of the optimal control problem (P_Ψ), which is defined on the product function space

$$\begin{aligned}\mathscr{Y} :=&\overline{H}^1(\Omega)^K \times \overline{H}^1(\Omega)^K \times H^1_{0,\sigma}(\Omega;\mathbb{R}^n)^{K-1} \times \overline{L}^2(\Omega)^K \times L^2(\Omega;\mathbb{R}^n)^{K-1}\\ &\times \overline{H}^1(\Omega)^K \times \overline{H}^1(\Omega)^K \times H^1_{0,\sigma}(\Omega;\mathbb{R}^n)^{K-1}\\ &\times \overline{H}^1(\Omega)^K \times \left(\overline{H}^1(\Omega)^*\right)^K \times \left(\overline{H}^1(\Omega)^*\right)^K.\end{aligned}$$

In contrast to the classical Lagrange function, the MPCC-Lagrangian does not include a multiplier for the complementarity condition. It rather corresponds to the Lagrange function of certain tightened nonlinear problems associated to the MPEC, cf., e.g., [49, 54].

Definition 4 The MPCC-Lagrangian $L : \mathscr{Y} \to \mathbb{R}$ corresponding to (P_Ψ) is given by

$$\begin{aligned}&L(\varphi,\mu,v,a,u,p,r,q,\pi,\lambda^+,\lambda^-) := \mathscr{J}(\varphi,\mu,v,u)\\ &+\sum_{i=-1}^{K-2}\left[\left\langle\frac{\varphi^{i+1}-\varphi^i}{\tau},p^{i+1}\right\rangle+\left\langle v^{i+1}\nabla\varphi^i,p^{i+1}\right\rangle\right.\\ &\left.-\left\langle \operatorname{div}(m(\varphi^i)\nabla\mu^{i+1}),p^{i+1}\right\rangle\right]\\ &+\sum_{i=-1}^{K-2}\left[\left\langle-\Delta\varphi^{i+1},r^{i+1}\right\rangle+\left\langle a^{i+1},r^{i+1}\right\rangle-\left\langle\mu^{i+1},r^{i+1}\right\rangle\right.\\ &\left.-\left\langle\kappa\varphi^i,r^{i+1}\right\rangle\right]\\ &+\sum_{i=0}^{K-2}\left[\left\langle\frac{\rho(\varphi^i)v^{i+1}-\rho(\varphi^{i+1})v^i}{\tau},q^{i+1}\right\rangle_{H^{-1},H^1_0}\right.\end{aligned}$$

$$
\begin{aligned}
&+\left\langle \operatorname{div}(v^{i+1} \otimes \rho(\varphi^{i+1})v^i), q^{i+1}\right\rangle_{H^{-1},H_0^1} \\
&-\left\langle \operatorname{div}(v^{i+1} \otimes \frac{\rho_2-\rho_1}{2} m(\varphi^{i+1})\nabla\mu^i), q^{i+1}\right\rangle_{H^{-1},H_0^1} \\
&+(2\eta(\varphi^i)\epsilon(v^{i+1}), \varepsilon(q^{i+1})) \\
&-\left\langle \mu^{i+1}\nabla\varphi^i, q^{i+1}\right\rangle_{H^{-1},H_0^1} - \left\langle u^{i+1}, q^{i+1}\right\rangle_{H^{-1},H_0^1}\Big] \\
&-\sum_{i=0}^{K-1}\left\langle a^i, \pi^i\right\rangle + \sum_{i=0}^{K-1}\left\langle (\lambda^i)^+, \varphi^i - \psi_2\right\rangle - \sum_{i=0}^{K-1}\left\langle (\lambda^i)^-, \varphi^i - \psi_1\right\rangle.
\end{aligned}
$$

In the sequel, we denote the state of the optimal control problem by $y := (\varphi, \mu, a, v)$ and the adjoint variables by $\Phi := (p, r, q)$. Moreover, $\mathscr{Y}_h$ denotes the discrete equivalent to $\mathscr{Y}$.

Remark 1 Note that if (y, u) is an ε-almost C-stationary point of (P_Ψ) with adjoints $(\Phi, \pi, \lambda^+, \lambda^-)$ then

$$
L(y, u, \Phi, \pi, \lambda^+, \lambda^-) = \mathscr{J}(\varphi, \mu, v, u). \tag{31}
$$

Based on the MPCC-Lagrangian we provide an estimation of the difference of the objective values at stationary points of the semi-discrete and the fully discretized problem, cf. [29]. Hereby, the index δ denotes the difference of the discrete and the continuous variables, e.g. $(y_\delta, u_\delta, \Phi_\delta) := (y_h, u_h, \Phi_h) - (y, u, \Phi)$.

Theorem 8 *Let $(y, u, \Phi, \pi, \lambda^+, \lambda^-)$ be a stationary point of the optimal control problem (P_Ψ) and assume that $(y_h, u_h, \Phi_h, \pi_h, \lambda_h^+, \lambda_h^-) \in \mathscr{Y}_h$ satisfy the discretized stationarity system. Then it holds that*

$$
\begin{aligned}
\mathscr{J}(\varphi_h, \mu_h, v_h, u_h) - \mathscr{J}(\varphi, \mu, v, u) &= \frac{1}{2}\left(\sum_{i=0}^{K-1}\left\langle a_h^i, \pi^i\right\rangle - \sum_{i=0}^{K-1}\left\langle a^i, \pi_h^i\right\rangle\right) \\
&+\frac{1}{2}\left(\sum_{i=0}^{K-1}\left\langle (\lambda^i)^+, \varphi_h^i - \psi_2\right\rangle - \sum_{i=0}^{K-1}\left\langle (\lambda_h^i)^+, \varphi^i - \psi_2\right\rangle\right) \\
&+\frac{1}{2}\left(\sum_{i=0}^{K-1}\left\langle (\lambda^i)^-, \varphi_h^i - \psi_1\right\rangle - \sum_{i=0}^{K-1}\left\langle (\lambda_h^i)^-, \varphi^i - \psi_1\right\rangle\right) \\
&+\frac{1}{2}\nabla_x L(y_h, u_h, \Phi_h, \pi_h, \lambda_h^+, \lambda_h^-)((y_h, u_h, \Phi_h) - (y, u, \Phi)) \\
&+O(\|(y_h, u_h, \Phi_h) - (y, u, \Phi)\|^3),
\end{aligned} \tag{32}
$$

where the error function O is continuous such that $|O(x)| \leq C|x|$; $C \in \mathbb{R}^+$.

The penultimate term on the right-hand side of Eq. (32) assembles the weighted dual and primal residuals. Whereas the previous terms display the mismatch in the complementarity between the discretized solution and the original one.

In other words, the discretization error with respect to the objective function can be estimated by

$$\begin{aligned}\mathcal{J}(\varphi_h, \mu_h, v_h, u_h) - \mathcal{J}(\varphi, \mu, v, u) \\ \approx \sum_{i=0}^{K-1} (\eta_{CM1,i} + \eta_{CM2,i} + \eta_{CM3,i} + \eta_{CM4,i} + \eta_{CH1,i} \\ + \eta_{CH2,i} + \eta_{NS,i} + \eta_{AD\varphi,i} + \eta_{AD\mu,i} + \eta_{ADv,i}),\end{aligned} \tag{33}$$

where the complementarity error terms $\eta_{CM1,i}, .., \eta_{CM4,i}$, the weighted primal residuals $\eta_{CH1,i}, \eta_{CH2,i}, \eta_{NS,i}$ and the weighted dual residuals $\eta_{AD\varphi,i}, \eta_{AD\mu,i}, \eta_{ADv,i}$ are defined as in [29, Section 4]. We point out that the integral structure of these error terms allows a patchwise evaluation on the underlying mesh. Apart from the weights φ_δ^i, μ_δ^i and v_δ^i and p_δ^i, q_δ^i, r_δ^i, respectively, the primal-dual-weighted error estimators only contain discrete quantities. In order to obtain a fully a-posteriori error estimator the weights are approximated involving a local higher-order approximation based on the respective discrete variables.

5.2 *Numerical Results*

In this subsection, we briefly sketch how the above error estimator can be converted into a numerical solution algorithm for the optimal control problem and showcase the numerical results for a specific example.

In accordance with the previous subsection, the algorithm utilizes linear finite elements for φ, μ, and p and quadratic finite elements for v and operates on a sequence of meshes $(\mathcal{T}^i)_{i=1}^K$. Mirroring the analytical approach from Theorem 6, we solve the resulting fully discrete optimization problem for a sequence $\alpha \to 0$, where we approximate Ψ_0 by the following expression

$$\Psi_{0,\alpha}(\varphi) := \frac{1}{2\alpha}\left(\max(0, \varphi - 1)^2 + \min(\varphi + 1)^2\right), \ \alpha > 0.$$

If a solution to these regularized problems is successfully calculated by a steepest descent method employing the characterization of the derivative of $\mathcal{J}$ in Theorem 5, then α is decreased until an approximate optimal control on the current sequence of grids is found that solves the original optimal control problem sufficiently well in the sense that it satisfies the complementarity conditions (28), (29) up to a given tolerance tol_c. Here, we define the multipliers a, λ based on the relation (27). The algorithm is enhanced by an outer adaptation loop which is

Data: Initial data: $\varphi_{-1}, \varphi_0, v_0$ and $u_0 = 0$;
1 **repeat**
2 **repeat**
3 solve the regularized problem (P_{Ψ_α}) using a steepest descent method;
4 decrease α;
5 **until** complementarity conditions (28), (29) are satisfied up to a tolerance tol_c;
6 calculate the error indicators and identify the sets $\mathcal{M}_r, \mathcal{M}_c$ of cells to refine/coarsen;
7 adapt $(\mathcal{T}^i)_{i=1}^K$ based on $\mathcal{M}_r$ and $\mathcal{M}_c$;
8 **until** $\sum_{i=1}^K |\mathcal{T}^i| > \mathcal{A}_{\max}$;

Algorithm 1: The overall solution procedure

based on the error estimator given in (33). The complete procedure is sketched in Algorithm 1.

The mesh refinement of the grids $(\mathcal{T}^i)_{i=1}^K$ relies on the Dörfler marking procedure. More precisely, the error indicators from (33) are evaluated for all time steps i and for all cells $T \in \mathcal{T}^i$ individually and a set $\mathcal{M}_r$ of cells to be refined is chosen as the set with the smallest cardinality which satisfies

$$\sum_{T \in \mathcal{M}_r} \eta_T \geq \theta^r \sum_{i=1}^{K} \sum_{T \in \mathcal{T}^i} \eta_T,$$

for a given parameter $0 < \theta^r < 1$. Due to the movement of the interface, we also select cells for coarsening if the calculated error indicator is smaller than a certain fraction of the mean error, i.e.

$$M_c := \left\{ T \in (\mathcal{T}^i)_{i=1}^K \mid \eta_T \leq \frac{\theta^c}{\mathcal{A}} \sum_{i=1}^{K} \sum_{T \in \mathcal{T}^i} \eta_T \right\},$$

where $0 < \theta^c < 1$ is fixed and $\mathcal{A} := \sum_{i=1}^K |\mathcal{T}^i|$. The mesh refinement process is terminated if a desired total number of cells $\mathcal{A}_{\max}$ is exceeded.

Finally, we briefly illustrate the performance of the proposed Algorithm 1. In the subsequent example, we aim to form a curved channel out of a ring-shaped initial region. For this purpose, 16 locally supported Ansatz functions of the control are distributed over the two-dimensional domain as depicted in Fig. 1. The corresponding objective functional is defined as in (14) with $\xi = 1e - 11$.

The associated fluid parameters are given by $\rho_1 = 1000$, $\rho_2 = 100$, $\eta_1 = 10$, $\eta_2 = 1$, and $\sigma = 24.5 \cdot \frac{2}{\pi}$ and are taken from a benchmark problem for rising bubble dynamics in [40]. Furthermore, we incorporate a gravitational acceleration $g = 0.981$ in the vertical direction and set $\varepsilon = 0.04$, $m(\varphi) \equiv \frac{1}{12500}$. The time horizon is set to $T = 1.0$ and the time step size is $\tau = 0.00125$.

For the marking procedure we use the parameters $\theta^r = 0.7$ and $\theta^c = 0.01$. Furthermore, the stopping criteria use the tolerance $tol_c = 1e - 3$ for the

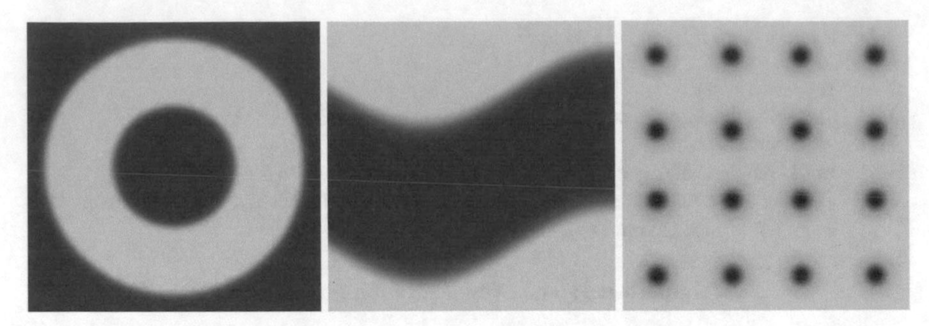

Fig. 1 The initial state (left), the desired state (middle) and the control (right) of the optimal control problem

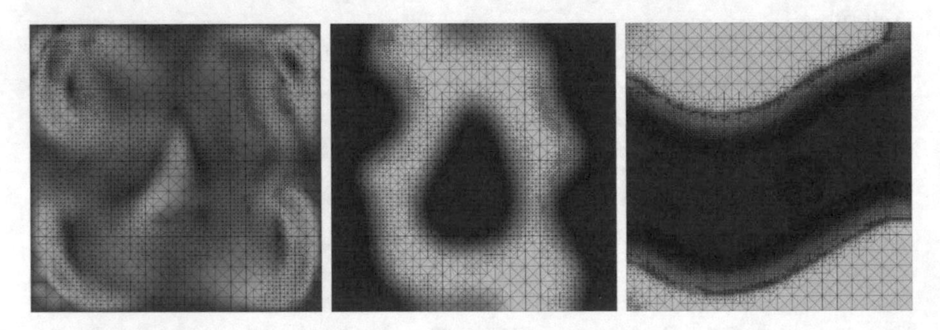

Fig. 2 The velocity field at time $t = 0.375$ (left), the concentration at $t = 0.375$ (middle) and the obtained final state at $t = 1$ (right)

Table 2 Number of gradient steps

Adaptation step	1			2	3	4	5	6	7	8	$\sum$
Regularization parameter	1e−07	5e−15	1e−15	1e−15	1e−15	1e−15	1e−15	1e−15	1e−15	1e−15	
Gradient steps	33	111	33	10	14	11	32	4	22	11	281

complementarity conditions and the maximum amount of cells $\mathscr{A}_{\max} = 8e6$ for the adaptation process, which relates to $1e4$ cells in average per time instance.

In Fig. 2, we present the obtained final state of the algorithm which closely matches the desired state. During the evolution, the ring-shaped initial region is deformed, where the upper part of the ring is pushed towards the top of the domain and the lower part is pushed towards the bottom as it can be seen on the intermediate state of the process given in Fig. 2. As a result the phase splits into two separate regions towards the end of the evolution.

The algorithm terminated after 281 gradient steps and 8 adaptation steps, cf. Table 2. The decrease of the objective functional is shown in Fig. 3 with respect to a logarithmic scaling. The small increases in the objective are related to the mesh

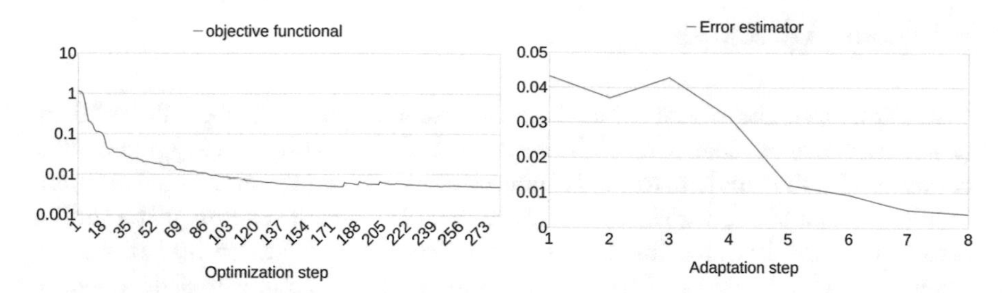

Fig. 3 The value of the objective functional at each gradient step (left) and the estimated error over all time steps η_{total} at each adaptation step (right)

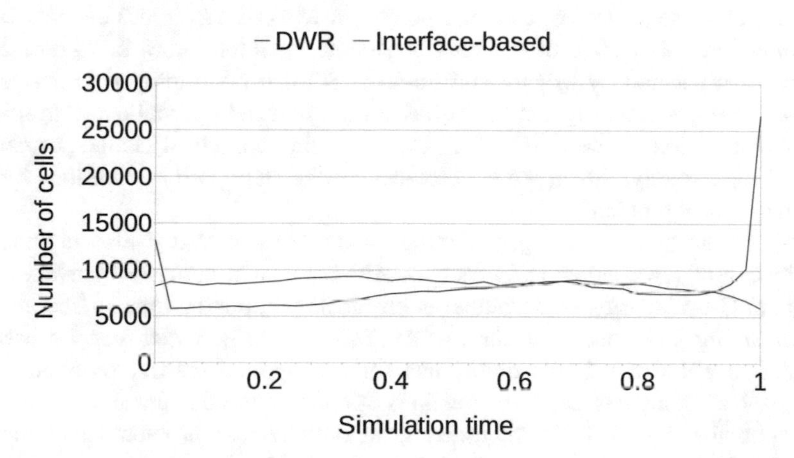

Fig. 4 The number of cells at each time step for the goal-oriented method and for an adaptation based on the order parameter φ

adaptation, which essentially imposes a different optimization problem in the finite dimensional space.

The right diagram of Fig. 3 further shows the decrease of the total error estimator as given in (33) with respect to the adaptation steps. As it can be seen exemplary in Fig. 2, the error indicator is dominated by the numerical error connected to the order parameter, which leads to a small resolution of the interfacial boundaries similar to conventional adaptation techniques based on the gradient of φ. However, we additionally observe refinements related to the velocity field outside of the interface, if we compare, e.g., the left and the middle picture of Fig. 2 which depict the velocity and the order parameter at an intermediate time step along with the corresponding mesh. Furthermore, the goal-oriented error indicator incorporates the structure of the optimization problem which leads to comparatively more refinements at the end of the evolution process and (to a smaller extend) at its beginning. This is illustrated in Fig. 4, where we compare the distribution of cells over the simulation time for the dual-weighted residuals method and a conventional adaptive method based on φ.

6 Future Directions

In this article, we showcased a C-stationarity system for the optimal control problem (P_Ψ). As stated in Sect. 2 it might be beneficial to establish strong stationarity conditions for the problem. Recently, Jarusek et al. verified the directional differentiability of a parabolic variational inequality, cf. [41], which suggests that a similar result might be derivable for the coupled Cahn–Hilliard–Navier–Stokes system. Their method utilizes the Lipschitz continuity of the control-to-state operator S_Ψ, which can be established for the Cahn–Hilliard–Navier–Stokes system, in order to bound certain difference quotients. This enables the extraction of weakly convergent subsequences whose limit points prove to be auspicious candidates for the directional derivative of the control-to-state operator. From this point on, a technique pioneered by Mignot and Puel in [50] can be employed to derive the desired strong stationarity conditions based on a reduced optimal control problem (after elimination of the state). If all the upcoming analytical challenges can be treated successfully, this approach provides strong stationarity conditions for the optimal control problem.

One of the main advantages of strong stationarity is that it also permits the application of novel numerical concepts. Although most numerical solvers target a type of C-stationarity, Hintermüller et al. recently applied a bundle-free implicit programming technique to a class of MPECs, see [35], which can be used to the design efficient solution algorithms for the optimal control problem (P_Ψ). The method computes descent directions for the reduced control problem at a non-optimal point $(\overline{u}, \overline{y})$ by minimizing an optimization problem involving the directional derivative of the control-to-state operator S_Ψ, i.e.

$$\min_h J(\overline{u}; h) = D_u J(\overline{u}, \overline{y})h + D_y J(\overline{u}, \overline{y}) S'_\Psi(\overline{u}; h) + \frac{1}{2} b(h, h). \tag{34}$$

In the presence of a non-empty biactive set, the problem becomes a very non-linear optimization problem and a smoothing algorithm connected to the regularization approach from Sect. 4 is applied. The method results in a globally convergent numerical solver, which computes at least a C-stationary point, and in many cases even a strongly stationary point.

Acknowledgements The authors gratefully acknowledge the support of the DFG through the priority program 1506 “Transport processes at fluidic interfaces” under the grant HI 1466/2-1 and the DFG-AIMS Workshop in Mbour, Sénégal. This research was further supported by the Research Center MATHEON through project C-SE5 and D-OT1 funded by the Einstein Center for Mathematics Berlin. In addition, this research was partly supported by the Berlin Mathematical School.

References

1. Abels, H., Depner, D., Garcke, H.: Existence of weak solutions for a diffuse interface model for two-phase flows of incompressible fluids with different densities. J. Math. Fluid Mech. **15**(3), 453–480 (2013). https://doi.org/10.1007/s00021-012-0118-x.
2. Abels, H., Garcke, H., Grün, G.: Thermodynamically consistent, frame indifferent diffuse interface models for incompressible two-phase flows with different densities. Math. Models Methods Appl. Sci. **22**(3), 1150013, 40pp (2012). https://doi.org/10.1142/S0218202511500138.
3. Adams, R.A., Fournier, J.J.F.: Sobolev spaces, *Pure and Applied Mathematics (Amsterdam)*, vol. 140, second edn. Elsevier/Academic Press, Amsterdam (2003).
4. Aland, S., Boden, S., Hahn, A., Klingbeil, F., Weismann, M., Weller, S.: Quantitative comparison of Taylor flow simulations based on sharp-interface and diffuse-interface models. International Journal for Numerical Methods in Fluids **73**(4), 344–361 (2013)
5. Aland, S., Lowengrub, J., Voigt, A.: Particles at fluid-fluid interfaces: A new Navier-Stokes-Cahn-Hilliard surface-phase-field-crystal model. Physical Review E **86**(4 Pt 2), 046321 (2012)
6. Aland, S., Voigt, A.: Benchmark computations of diffuse interface models for two-dimensional bubble dynamics. International Journal for Numerical Methods in Fluids **69**, 747–761 (2012)
7. Aubin, J.P.: Applied functional analysis. John Wiley & Sons, New York-Chichester-Brisbane (1979). Translated from the French by Carole Labrousse, With exercises by Bernard Cornet and Jean-Michel Lasry
8. Bañas, L., Nürnberg, R.: Adaptive finite element methods for Cahn–Hilliard equations. Journal of Computational and Applied Mathematics **218**, 2–11 (2008)
9. Bañas, L., Nürnberg, R.: A posteriori estimates for the Cahn–Hilliard equation. Mathematical Modelling and Numerical Analysis **43**(5), 1003–1026 (2009)
10. Blank, L., Butz, M., Garcke, H.: Solving the Cahn–Hilliard variational inequality with a semi-smooth Newton method. ESAIM: Control, Optimisation and Calculus of Variations **17**(4), 931–954 (2011)
11. Blowey, J.F., Elliott, C.M.: The Cahn–Hilliard gradient theory for phase separation with non-smooth free energy. Part I: Mathematical analysis. European Journal of Applied Mathematics **2**, 233–280 (1991)
12. Boyer, F.: A theoretical and numerical model for the study of incompressible mixture flows. Computers & Fluids **31**(1), 41–68 (2002)
13. Boyer, F., Chupin, L., Fabrie, P.: Numerical study of viscoelastic mixtures through a Cahn-Hilliard flow model. Eur. J. Mech. B Fluids **23**(5), 759–780 (2004). https://doi.org/10.1016/j.euromechflu.2004.03.001
14. Brett, C., Elliott, C.M., Hintermüller, M., Löbhard, C.: Mesh adaptivity in optimal control of elliptic variational inequalities with point-tracking of the state. Interfaces Free Bound. **17**(1), 21–53 (2015). https://doi.org/10.4171/IFB/332
15. Cahn, J.W., Hilliard, J.E.: Free energy of a nonuniform system. I. Interfacial free energy. The Journal of Chemical Physics **28**(2), 258–267 (1958)
16. Casas, E.: Control of an elliptic problem with pointwise state constraints. SIAM J. Control Optim. **24**(6), 1309–1318 (1986). https://doi.org/10.1137/0324078
17. Colli, P., Farshbaf-Shaker, M.H., Gilardi, G., Sprekels, J.: Optimal boundary control of a viscous Cahn-Hilliard system with dynamic boundary condition and double obstacle potentials. SIAM J. Control Optim. **53**(4), 2696–2721 (2015). https://doi.org/10.1137/140984749
18. Colli, P., Gilardi, G., Sprekels, J.: A boundary control problem for the viscous cahn–hilliard equation with dynamic boundary conditions. Applied Mathematics & Optimization **73**(2), 195–225 (2016). https://doi.org/10.1007/s00245-015-9299-z
19. Ding, H., Spelt, P.D., Shu, C.: Diffuse interface model for incompressible two-phase flows with large density ratios. Journal of Computational Physics **226**(2), 2078–2095 (2007)
20. Eckert, S., Nikrityuk, P.A., Willers, B., Räbiger, D., Shevchenko, N., Neumann-Heyme, H., Travnikov, V., Odenbach, S., Voigt, A., Eckert, K.: Electromagnetic melt flow control during solidification of metallic alloys. The European Physical Journal Special Topics **220**(1), 123–137 (2013)

21. Elliott, C.M., Songmu, Z.: On the Cahn-Hilliard equation. Arch. Rational Mech. Anal. **96**(4), 339–357 (1986). https://doi.org/10.1007/BF00251803
22. Frigeri, S., Rocca, E., Sprekels, J.: Optimal distributed control of a nonlocal Cahn-Hilliard/Navier-Stokes system in two dimensions. SIAM J. Control Optim. **54**(1), 221–250 (2016). https://doi.org/10.1137/140994800
23. Gal, C.G., Grasselli, M.: Asymptotic behavior of a Cahn-Hilliard-Navier-Stokes system in 2D. Ann. Inst. H. Poincaré Anal. Non Linéaire **27**(1), 401–436 (2010). https://doi.org/10.1016/j.anihpc.2009.11.013
24. Garcke, H., Hinze, M., Kahle, C.: A stable and linear time discretization for a thermodynamically consistent model for two-phase incompressible flow. Applied Numerical Mathematics **99**, 151–171 (2016). https://doi.org/10.1016/j.apnum.2015.09.002; http://www.sciencedirect.com/science/article/pii/S0168927415001324
25. Girault, V., Raviart, P.A.: Finite element methods for Navier-Stokes equations: Theory and algorithms, *Springer Series in Computational Mathematics*, vol. 5. Springer, Berlin (1986). https://doi.org/10.1007/978-3-642-61623-5.
26. Grün, G., Klingbeil, F.: Two-phase flow with mass density contrast: Stable schemes for a thermodynamic consistent and frame indifferent diffuse interface model. Journal of Computational Physics **257**(A), 708–725 (2014)
27. Guillén-Gonzáles, F., Tierra, G.: Splitting schemes for a Navier–Stokes–Cahn–Hilliard model for two fluids with different densities. Journal of Computational Mathematics **32**(6), 643–664 (2014)
28. Héron, B.: Quelques propriétés des applications de trace dans des espaces de champs de vecteurs à divergence nulle. Comm. Partial Differential Equations **6**(12), 1301–1334 (1981). https://doi.org/10.1080/03605308108820212
29. Hintermüller, M., Hinze, M., Kahle, C., Keil, T.: A goal-oriented dual-weighted adaptive finite element approach for the optimal control of a nonsmooth Cahn–Hilliard–Navier–Stokes system. Optimization and Engineering (2018). https://doi.org/10.1007/s11081-018-9393-6
30. Hintermüller, M., Hinze, M., Kahle, C.: An adaptive finite element Moreau–Yosida-based solver for a coupled Cahn–Hilliard/Navier–Stokes system. Journal of Computational Physics **235**, 810–827 (2013)
31. Hintermüller, M., Hinze, M., Tber, M.H.: An adaptive finite-element Moreau-Yosida-based solver for a non-smooth Cahn-Hilliard problem. Optim. Methods Softw. **26**(4–5), 777–811 (2011). https://doi.org/10.1080/10556788.2010.549230
32. Hintermüller, M., Keil, T., Wegner, D.: Optimal control of a semidiscrete Cahn-Hilliard-Navier-Stokes system with nonmatched fluid densities. SIAM J. Control Optim. **55**(3), 1954–1989 (2017). https://doi.org/10.1137/15M1025128
33. Hintermüller, M., Kopacka, I.: Mathematical programs with complementarity constraints in function space: C- and strong stationarity and a path-following algorithm. SIAM J. Optim. **20**(2), 868–902 (2009). https://doi.org/10.1137/080720681
34. Hintermüller, M., Mordukhovich, B.S., Surowiec, T.M.: Several approaches for the derivation of stationarity conditions for elliptic MPECs with upper-level control constraints. Math. Program. **146**(1–2, Ser. A), 555–582 (2014). https://doi.org/10.1007/s10107-013-0704-6
35. Hintermüller, M., Surowiec, T.: A bundle-free implicit programming approach for a class of elliptic MPECs in function space. Mathematical Programming **160**(1), 271–305 (2016). https://doi.org/10.1007/s10107-016-0983-9
36. Hintermüller, M., Wegner, D.: Distributed optimal control of the Cahn-Hilliard system including the case of a double-obstacle homogeneous free energy density. SIAM J. Control Optim. **50**(1), 388–418 (2012). https://doi.org/10.1137/110824152
37. Hintermüller, M., Wegner, D.: Optimal control of a semidiscrete Cahn-Hilliard-Navier-Stokes system. SIAM J. Control Optim. **52**(1), 747–772 (2014). https://doi.org/10.1137/120865628
38. Hintermüller, M., Wegner, D.: Distributed and boundary control problems for the semidiscrete Cahn-Hilliard/Navier-Stokes system with nonsmooth Ginzburg-Landau energies. In: U. Langer, H. Albrecher, H. Engl, R. Hoppe, K. Kunisch, H. Niederreiter, C. Schmeisser (eds.) Topological Optimization and Optimal Transport, *Radon Series on Computational and Applied Mathematics*, vol. 17. De Gruyter (2017)

39. Hohenberg, P.C., Halperin, B.I.: Theory of dynamic critical phenomena. Reviews of Modern Physics **49**(3), 435 (1977)
40. Hysing, S., Turek, S., Kuzmin, D., Parolini, N., Burman, E., Ganesan, S., Tobiska, L.: Quantitative benchmark computations of two-dimensional bubble dynamics. International Journal for Numerical Methods in Fluids **60**(11), 1259–1288 (2009). https://doi.org/10.1002/fld.1934; http://onlinelibrary.wiley.com/doi/10.1002/fld.1934/abstract
41. Jarušek, J., Krbec, M., Rao, M., Sokołowski, J.: Conical differentiability for evolution variational inequalities. J. Differential Equations **193**(1), 131–146 (2003). https://doi.org/10.1016/S0022-0396(03)00136-0
42. Kay, D., Styles, V., Welford, R.: Finite element approximation of a Cahn–Hilliard–Navier–Stokes system. Interfaces and Free Boundaries **10**(1), 15–43 (2008). http://www.ems-ph.org/journals/show_issue.php?issn=1463-9963&vol=10&iss=1
43. Kay, D., Welford, R.: A multigrid finite element solver for the Cahn–Hilliard equation. Journal of Computational Physics **212**, 288–304 (2006)
44. Kim, J., Kang, K., Lowengrub, J.: Conservative multigrid methods for Cahn-Hilliard fluids. J. Comput. Phys. **193**(2), 511–543 (2004). https://doi.org/10.1016/j.jcp.2003.07.035
45. Kim, J., Lowengrub, J.: Interfaces and multicomponent fluids. Encyclopedia of Mathematical Physics pp. 135–144 (2004)
46. Kinderlehrer, D., Stampacchia, G.: An introduction to variational inequalities and their applications, *Classics in Applied Mathematics*, vol. 31. Society for Industrial and Applied Mathematics (SIAM), Philadelphia, PA (2000). https://doi.org/10.1137/1.9780898719451. Reprint of the 1980 original
47. Lions, J., Magenes, E.: Non-Homogeneous Boundary Value Problems and Applications. Springer, Berlin (1972). https://doi.org/10.1007/978-3-642-65161-8
48. Lowengrub, J., Truskinovsky, L.: Quasi-incompressible Cahn-Hilliard fluids and topological transitions. R. Soc. Lond. Proc. Ser. A Math. Phys. Eng. Sci. **454**(1978), 2617–2654 (1998). https://doi.org/10.1098/rspa.1998.0273
49. Luo, Z.Q., Pang, J.S., Ralph, D.: Mathematical programs with equilibrium constraints. Cambridge University Press, Cambridge (1996). https://doi.org/10.1017/CBO9780511983658
50. Mignot, F.: Contrôle dans les inéquations variationelles elliptiques. J. Functional Analysis **22**(2), 130–185 (1976)
51. Oono, Y., Puri, S.: Study of phase-separation dynamics by use of cell dynamical systems. I. Modeling. Physical Review A **38**(1), 434 (1988)
52. Outrata, J., Kočvara, M., Zowe, J.: Nonsmooth approach to optimization problems with equilibrium constraints: Theory, applications and numerical results, *Nonconvex Optimization and its Applications*, vol. 28. Kluwer Academic Publishers, Dordrecht (1998). https://doi.org/10.1007/978-1-4757-2825-5.
53. Praetorius, S., Voigt, A.: A phase field crystal model for colloidal suspensions with hydrodynamic interactions. arXiv preprint arXiv:1310.5495 (2013)
54. Scheel, H., Scholtes, S.: Mathematical programs with complementarity constraints: stationarity, optimality, and sensitivity. Math. Oper. Res. **25**(1), 1–22 (2000). https://doi.org/10.1287/moor.25.1.1.15213
55. Tachim Medjo, T.: Optimal control of a Cahn-Hilliard-Navier-Stokes model with state constraints. J. Convex Anal. **22**(4), 1135–1172 (2015)
56. Temam, R.: Navier-Stokes equations. Theory and numerical analysis. North-Holland Publishing Co., Amsterdam-New York-Oxford (1977). Studies in Mathematics and its Applications, Vol. 2
57. Verfürth, R.: A posteriori error analysis of space-time finite element discretizations of the time-dependent Stokes equations. Calcolo **47**(3), 149–167 (2010). https://doi.org/10.1007/s10092-010-0018-5
58. Wang, Q.F., Nakagiri, S.I.: Weak solutions of Cahn-Hilliard equations having forcing terms and optimal control problems. Sūrikaisekikenkyūsho Kōkyūroku (1128), 172–180 (2000). Mathematical models in functional equations (Japanese) (Kyoto, 1999)

59. Yong, J.M., Zheng, S.M.: Feedback stabilization and optimal control for the Cahn-Hilliard equation. Nonlinear Anal. **17**(5), 431–444 (1991). https://doi.org/10.1016/0362-546X(91)90138-Q
60. Zhou, B., et al.: Simulations of polymeric membrane formation in 2D and 3D. Ph.D. thesis, Massachusetts Institute of Technology (2006)
61. Zowe, J., Kurcyusz, S.: Regularity and stability for the mathematical programming problem in Banach spaces. Appl. Math. Optim. **5**(1), 49–62 (1979). https://doi.org/10.1007/BF01442543

Localized Model Reduction in PDE Constrained Optimization

Mario Ohlberger, Michael Schaefer, and Felix Schindler

Abstract We present efficient localized model reduction approaches for PDE constraint optimization or optimal control. The first approach focuses on problems where the underlying PDE is given as a locally periodic elliptic multiscale problem. The second approach is more universal and focuses on general underlying multiscale or large scale problems. Both methods make use of reduced basis techniques and rely on efficient a posteriori error estimation for the approximation of the underlying parameterized PDE. The methods are presented and numerical experiments are discussed.

Keywords Localized model reduction · Reduced basis methods · Optimal control · PDE constrained optimization · LRBMS · Heterogeneous multiscale method

Mathematics Subject Classification (2010). Primary 65K10; Secondary 65N30

1 Introduction

In this contribution we are concerned with efficient approximation schemes for the following class of multiscale or large scale PDE constrained optimization problems.

M. Ohlberger (✉) · M. Schaefer · F. Schindler
Applied Mathematics Muenster, Münster, Germany
e-mail: mario.ohlberger@uni-muenster.de; michael.schaefer@uni-muenster.de; felix.schindler@uni-muenster.de

V. Schulz, D. Seck (eds.), *Shape Optimization, Homogenization and Optimal Control*, International Series of Numerical Mathematics 169,
https://doi.org/10.1007/978-3-319-90469-6_8

$$\left.\begin{aligned} \text{Find} \quad & \mu^* = \arg\min J\big(u(\mu), \mu\big) \\ \text{subject to} \quad & C_j\big(u(\mu), \mu\big) \leq 0 \quad \forall j = 1, \ldots, m, \\ & \mu \in \mathcal{P} \end{aligned}\right\} \tag{1.1}$$

with a compact *parameter set* $\mathcal{P} \subset \mathbb{R}^P$ for $P \in \mathbb{N}$. In (1.1), the *state variable* $u(\mu)$ is given as the solution of the following (parameterized) multiscale problem:

$$\left.\begin{aligned} -\nabla \cdot \big(A(\mu)\nabla u(\mu)\big) &= f(\mu) \quad && (\text{in } \Omega), \\ u(\mu) &= 0 && (\text{on } \partial\Omega). \end{aligned}\right\} \tag{1.2}$$

In (1.2), $\Omega \subset \mathbb{R}^d$ for $d = 1, 2, 3$ is a bounded domain and A denotes a diffusion tensor. We make use of the short notation $u(\mu) := u(\cdot; \mu)$ and will use analogue expressions for all functions that depend on both spatial variables and parameters.

We are particularly interested in multiscale or large scale applications in the sense that the diffusion tensor A has a rich structure that would lead to very high dimensional approximation spaces for the state space when approximated with classical finite element type methods. For parameter independent multiscale problems of this type there has been a tremendous development of suitable numerical multiscale methods in the last two decades including the multiscale finite element method (MsFEM) [14, 15, 22, 24], the heterogeneous multiscale method (HMM) [1, 3, 12, 13, 20, 21, 35], the variational multiscale method (VMM) [25, 26, 28–31] or the more recent local orthogonal decomposition (LOD) [19, 33].

For parameterized partial differential equations, among others, reduced basis methods (RBM) have seen a great development in the last decade [8, 23, 41]. Meanwhile, also several applications of RBM in the context of multiscale methods have been proposed [2, 9, 10, 36–39].

In this contribution we first derive a general framework for localized model reduction of multiscale or large scale PDE constrained optimization problems. We then focus on PDE constrained optimization, where the underlying multiscale problem has locally periodic structure, i.e. $A(x) = A^{\varepsilon}(x) = \hat{A}(x, \frac{x}{\varepsilon})$ and present an efficient combined RB-HMM approximation scheme. PDE constrained optimization for locally periodic structures has particular applications in shape optimization as, e.g., presented in [5, 17]. Finally, we also provide a localized approach for general multiscale or large scale problems based on the localized reduced basis multiscale method with online enrichment [38, 39]. This approach leads to a novel paradigm towards optimal complexity in PDE constrained optimization as recently introduced in [40].

2 Weak Formulation for the Parameterized Multiscale Problem and Non-conforming Reference Approximation

Definition 2.1 (Weak Solution of the Multiscale Problem) We call $u(\mu) \in H_0^1(\Omega)$ weak solution of (1.2), if

$$b\big(u(\mu), v; \mu\big) = L(v; \mu) \qquad \text{for all } v \in H_0^1(\Omega). \tag{2.1}$$

Here, the bilinear form b and the right hand side L are given as

$$b(v, w; \mu) := \int_\Omega A(\mu)\nabla v \cdot \nabla w, \quad L(v; \mu) := \int_\omega f v.$$

In order to derive a suitable formulation for (non-)conforming weak reference approximations for our model reduction approach, we first assume that a non-overlapping decomposition of the underlying domain Ω is given by a coarse grid $\mathcal{T}_H$ with cells $T_j \in \mathcal{T}_H, j = 1, \dots N_H$. Furthermore, each macro cell T_j is further decomposed by a local fine resolution triangulation $\tau_h(T_j)$, that resolves all fine scale features of the multiscale problem. We then define the global fine scale partition τ_h as the union of all its local contributions, i.e., $\tau_h = \bigcup_{j=1}^{N_H} \tau_h(T_j)$. Let $H^p\big(\tau_h(\omega)\big) := \big\{v \in L^2(\omega) \bigm| v|_t \in H^p(t) \;\; \forall t \in \tau_h(\omega)\big\}$ for a triangulation $\tau_h(\omega)$ of some $\omega \subset \Omega$ denote the broken Sobolev space of order $p \in \mathbb{N}$ on τ_h, then $H^p(\tau_h)$ naturally inherits the decomposition $H^p(\tau_h) = \bigoplus_{j=1}^{N_H} H^p(\tau_h(T_j))$.

Definition 2.2 ((Non-)conforming Approximations of the Multiscale Problem in Broken Spaces) Let $V(\tau_h) \subset H^2(\tau_h)$ denote any approximate subset of the broken Sobolev space.

We call $u_h(\mu) \in V(\tau_h)$ an approximate weak reference solution of (1.2), if

$$\mathcal{A}_{\mathrm{DG}}\big(u_h(\mu), v; \mu\big) = L_{\mathrm{DG}}(v; \mu) \qquad \text{for all } v \in V(\tau_h). \tag{2.2}$$

Here, the DG bilinear form $\mathcal{A}_{\mathrm{DG}}$ and the right hand side L_{DG} are given as

$$\mathcal{A}_{\mathrm{DG}}(v, w; \mu) := \sum_{t\in\tau_h} \int_t A(\mu)\nabla v \cdot \nabla w + \sum_{e\in\mathcal{F}(\tau_h)} \mathcal{A}_{\mathrm{DG}}^e(v, w; \mu)$$

$$L_{\mathrm{DG}}(v; \mu) := \sum_{t\in\tau_h} \int_t f v,$$

where $\mathcal{F}(\cdot)$ denotes the set of all faces of a triangulation and the DG coupling bilinear form $\mathcal{A}_{\mathrm{DG}}^{e}$ for a face e is given by

$$\mathcal{A}_{\mathrm{DG}}^{e}(v, w; \mu) := \int_{e} \langle A(\mu)\nabla v \cdot \mathbf{n_e}\rangle[w] + \langle A(\mu)\nabla w \cdot \mathbf{n_e}\rangle[v] + \frac{\sigma_e(\mu)}{|e|^{\beta}}[v][w].$$

For any triangulation $\tau_h(\omega)$ of some $\omega \subseteq \Omega$, we assign to each face $e \in \mathcal{F}(\tau_h(\omega))$ a unique normal $\mathbf{n_e}$ pointing away from the adjacent cell t^-, where an inner face is given by $e = t^- \cap t^+$ and a boundary face is given by $e = t^- \cap \partial\omega$, for appropriate cells $t^{\pm} \in \tau_h(\omega)$. In the above, the average and jump of a two-valued function $v \in H^2(\tau_h(\omega))$ are given by $\langle v\rangle := \frac{1}{2}(v|_{t^-} + v|_{t^+})$ and $[v] := v|_{t^-} - v|_{t^+}$ for an inner face and by $\langle v\rangle := [v] := v$ for a boundary face, respectively. The parametric penalty function $\sigma_e(\mu)$ and the parameter β need to be chosen appropriately to ensure coercivity of $\mathcal{A}_{\mathrm{DG}}$ and may involve A. For simplicity, we restrict ourselves to the above symmetric interior penalty DG scheme; other DG variants can be easily accommodated and we refer to [16, 39] and the references therein for further details.

Note that Definition 2.2 contains both, continuous Galerkin finite element approximations, if $V(\tau_h) \subset H^2(\tau_h) \cap H_0^1(\Omega)$ and discontinuous Galerkin finite elements if $V(\tau_h) \subset H^2(\tau_h)$, $V(\tau_h) \not\subset H_0^1(\Omega)$. In the continuous Galerkin case, we naturally have $\mathcal{A}_{\mathrm{DG}} \equiv b$ and $L_{\mathrm{DG}} \equiv L$.

3 A General Non-conforming Weak Formulation for Numerical Multiscale Methods

We assume that for each quadrature point x_T there exists open environments M_T, O_T with $x_T \in M_T \subset O_T$. We call M_T the local reconstruction region and O_T the local oversampling region. The subsets M_T and O_T are further decomposed by local fine resolution grids $\tau_h(M_T)$, $\tau_h(O_T)$, which resolve all fine scale features of the multiscale problem. For sake of simplicity we assume that $\tau_h(M_T) \subset \tau_h(O_T) \subset \tau_h$.

Let $V^c := V(\mathcal{T}_H) \subset H^2(\mathcal{T}_H)$ denote a global finite dimensional coarse scale space, i.e., a finite dimensional subset of the broken Sobolev space $H^2(\mathcal{T}_H)$. Furthermore, let $V^f(O_T) := V(\tau_h(O_T)) \subset H^2(\tau_h(O_T))$ denote a suitable finite dimensional local fine scale space, that resolves all features of the underlying multiscale problem. We define $V^f(M_T) := V(\tau_h(M_T)) := V(\tau_h(O_T))|_{M_T} \subset H^1(\tau_h(M_T))$.

Definition 3.1 (General Non-conforming Multiscale Approximation) The macroscopic part of the multiscale approximation $u^c \subset V^c$ is defined as the solution of

$$\text{find } u^c(\mu) \in V^c: \qquad \mathcal{A}_{\mathrm{DG}}^{c}(u^c(\mu), \Phi; \mu) = L_{\mathrm{DG}}(\Phi; \mu) \quad \text{for all } \Phi \in V^c.$$

Here, the coarse DG bilinear form $\mathcal{A}^c_{\text{DG}} : V^c \times V^c \times \mathcal{P} \to \mathbb{R}$ is given as

$$\mathcal{A}^c_{\text{DG}}(v, w; \mu) := \sum_{T \in \mathcal{T}_H} \frac{|T|}{|M_T|} \Bigg\{ \sum_{t \in \tau_h(M_T)} \int_t A(\mu) \nabla \mathcal{R}^\mu(v) \cdot \nabla w + \sum_{e \in \mathcal{F}(\tau_h(M_T))} \mathcal{A}^e_{\text{DG}}\big(\mathcal{R}^\mu(v), w; \mu\big) \Bigg\}.$$

Furthermore, for any $\Phi \in V^c, T \in \mathcal{T}_H$, the local reconstruction operators $\mathcal{R}^\mu|_T : V^c \to V^c|_{O_T} + V^f(O_T)$ are defined through the solution of the following local problem on the oversampling domain O_T:

find $\mathcal{R}^\mu|_T(\Phi) \in \Phi|_{O_T} + V^f(O_T)$ such that

$$\mathcal{A}_{O_T}(\mathcal{R}^\mu|_T(\Phi), \phi; \mu) = F_{O_T}(\phi) \quad \forall \phi \in V^f(O_T),$$

where the fine scale bilinear form $\mathcal{A}_{O_T}$ is defined as

$$\mathcal{A}_{O_T}(v, w; \mu) := \sum_{t \in \tau_h(O_T)} \int_t A(\mu) \nabla v \cdot \nabla w + \sum_{e \in \mathcal{F}(\tau_h(O_T))} \mathcal{A}^e_{\text{DG}}(v, w; \mu)$$

and $F_{O_T} \subset V^f(O_T)'$ is a suitable fine scale right hand side that depends on the particular realization of the method.

Note that for $M_T := T$ for all $T \in \mathcal{T}_H$, we get from the definition above $\mathcal{A}^c_{\text{DG}}(v, w : \mu) = \mathcal{A}_{\text{DG}}(\mathcal{R}^\mu(v), w; \mu)$.

Particular instances of multiscale methods that fit in the general framework of Definition 3.1 are the Heterogeneous Multiscale Method (HMM) and the Discontinuous Galerkin Multiscale Finite Element Method (DG-MsFEM) with oversampling. These methods are obtained with the following specifications:

Definition 3.2 The HMM with oversampling is given through Definition 3.1 with the following choices of domains and spaces:

$V^c := V^1(\mathcal{T}_H) \cap H^1_0(\Omega) \subset H^1_0(\Omega)$,
$M_T := Y_{T,\varepsilon} := \big\{ y \in \mathbb{R}^n \mid |||y - x_T||_\infty \le \varepsilon \big\}$,
$O_T := Y_{T,\delta} := \big\{ y \in \mathbb{R}^n \mid ||y - x_T||_\infty \le \delta \}, \varepsilon \le \delta$,
$V^f(M_T) := V^1_\#(\tau_h(M_T)) \subset \tilde{H}^1_\#(M_T)$,
$V^f(O_T) := V^1_\#(\tau_h(O_T)) \cap \tilde{H}^1_\#(O_T) \subset \tilde{H}^1_\#(O_T)$,
$F_{O_T}(\phi) \equiv 0$.

Here $\tilde{H}^1_\#(O_T)$ denotes the space of H^1-functions on O_T with periodic boundary values and zero mean and $V^1(\tau_h(\omega))$ ($V^1_\#(\tau_h(\omega))$) denotes the standard conforming piecewise linear Lagrange finite element space on ω (with mean zero and periodic boundary condition).

Definition 3.3 The DG-MsFEM with oversampling is given through Definition 3.1 with the following choices of domains and spaces:

$$
\begin{aligned}
&V^c := Q^1(\mathcal{T}_H) \subset H^2(\mathcal{T}_H),\\
&M_T := T,\\
&O_T := U_l(T),\\
&V^f(M_T) := Q^1(\tau_h(T)) \subset H^2(\tau_h(T)),\\
&V^f(O_T) := Q^1(\tau_h(U_l(T))) \subset H^2(\tau_h(U_l(T))),\\
&F_{O_T}(\phi) \equiv \textstyle\int_\Omega f\phi.
\end{aligned}
$$

Here $Q^1(\mathcal{T}_H)$, $Q^1(\tau_h(\omega))$ denote the standard non-conforming piecewise linear discontinuous Galerkin finite element space on $\mathcal{T}_H$, $\tau_h(\omega)$ respectively. Furthermore, $U_l(T)$ denotes an environment of T, consisting of l additional layers of fine scale elements $t \in \tau_h$.

4 A General Framework for Localized Model Reduction Based on Numerical Multiscale Methods with Oversampling

Based on the general framework for numerical multiscale methods from Definition 3.1, we define general localized reduced basis multiscale methods (LRBMS) as follows.

Definition 4.1 (Localized Model Reduction Multiscale Methods) Let $V_N^c \subset V^c$, $V_N^f(O_T) \subset V^f(O_T), T \in \mathcal{T}_H$ denote suitably defined local reduced coarse scale and fine scale subspaces that are obtained e.g. by a greedy algorithm from snapshots $u^c(\mu)$, $(\mathcal{R}^\mu|_T(u^c) - u^c)(\mu)$ for a number of suitably chosen parameters μ. Furthermore, define $V_N^f(M_T) := \{v|_{M(T)} | v \in V^f(O_T)\}$, such that $V_N^f(M_T) \subset V^f(M_T)$. Then the corresponding localized reduced basis multiscale method is defined as the DG-Galerkin projection onto these subspaces as follows.

The macroscopic part of the localized multiscale model reduction approximation $u_N^c \subset V_N^c$ is defined as the solution of

$$\text{find } u_N^c(\mu) \in V_N^c : \qquad \mathcal{A}_{\mathrm{DG}}^c(u_N^c(\mu), \Phi; \mu) = L_{\mathrm{DG}}(\Phi; \mu) \quad \text{for all } \Phi \in V_N^c.$$

Furthermore, for any $\Phi \in V_N^c, T \in \mathcal{T}_H$ the local reconstruction operators $\mathcal{R}^\mu|_T : V_N^c \to V_N^c|_T + V_N^f(O_T)$ are defined through the solution of the following local problem on the oversampling domain O_T:

$$\text{find } \mathcal{R}^\mu|_T(\Phi) \in \Phi|_T + V_N^f(O_T) \text{ such that}$$

$$\mathcal{A}_{O_T}(\mathcal{R}^\mu|_T(\Phi), \phi; \mu) = F_{O_T}(\phi) \quad \forall \phi \in V_N^f(O_T).$$

In the next two sections we will study in some more detail two localized model reduction methods that can be seen as reduced approximations of the RB-HMM on the one hand and of the DG-MsFEM with oversampling on the other hand.

5 The Reduced Basis Heterogeneous Multiscale Method (RB-HMM)

In this section we first review the RB-HMM method from [37] in the given optimization context and finally present some new numerical experiments.

The RB-HMM is defined through Definition 4.1 with the HMM choice given in Definition 3.2 and a greedy construction of the corresponding reduced subspaces. In the case of locally periodic homogenization problems, i.e. $A^\varepsilon(\mu) = A(x, \frac{x}{\varepsilon}; \mu)$, where $A(x, y; \mu)$ is 1-periodic with respect to y, it has been shown in [37], that the resulting method for $\delta = \varepsilon$ can be equivalently formulated as a reduced Galerkin approximation of the two-scale homogenized limit equations with quadrature. To this end, let us first introduce the continuous two scale solution space

$$\mathcal{H} := H_0^1(\Omega) \times L^2(\Omega; \tilde{H}_\#^1(Y))$$

which is a Hilbert space equipped with the scalar product

$$(u, v)_\mathcal{H} := ((u^0, u^1), (v^0, v^1)) := \int_\Omega \nabla u^0 \cdot \nabla v^0 \, \mathrm{d}x + \int_\Omega \int_Y \nabla_y u^1 \cdot \nabla_y v^1 \, \mathrm{d}y \, \mathrm{d}x.$$

Definition 5.1 (Discrete Two Scale Homogenized Bilinear Form and HMM Approximation) We first define a piecewise constant approximation of the coefficient function $A(x, y; \mu)$ through

$$\mathcal{B}_h(x, y; \mu)\big|_{T\times t} := A(x_T, y_t; \mu) \quad \forall\ T \times t \in \mathcal{T}_H \times \tau_h(Y).$$

where y_t is the center of gravity of t. We then define the discrete two scale homogenized bilinear form $\mathcal{B}_h : \mathcal{H} \times \mathcal{H} \times \mathcal{P} \to \mathbb{R}$ as

$$\mathcal{B}_h(u, v; \mu) := \int_\Omega \int_Y \mathcal{B}_h(\mu) \left(\nabla u^0 + \nabla_y u^1\right) \cdot \left(\nabla v^0 + \nabla_y v^1\right) \mathrm{d}y \, \mathrm{d}x.$$

With $\mathcal{H}_H := V^1(\mathcal{T}_H) \times V^0(\Omega; V^1(\tau_h(Y))) \subset \mathcal{H}$ it has been shown in [35] that the HMM approximation $u^c \in V^1(\mathcal{T}_H)$ from Definitions 3.1, 3.2 can be equivalently defined as the solution of the following discrete two scale limit equation.

Find $u_H(\mu) = (u^c(\mu), u^f(\mu)) \in \mathcal{H}_H$ such that for all $v_H := (v^c, v^f) \in \mathcal{H}_H$ we have

$$\mathcal{B}_h(u_H(\mu), v_H; \mu) = L(v^c; \mu). \tag{5.1}$$

Finally we obtain the following equivalent formulation of the RB-HMM method defined above.

Definition 5.2 (RB-HMM in the Two Scale Limit Formulation) Consider a *sample set* $\mathcal{S} = \{\mu_1, \ldots, \mu_N\} \subset \mathcal{P}$ chosen, e.g., from a greedy algorithm. We define the *RB-HMM space* $\mathcal{H}_N \subset \mathcal{H}_H$ as

$$\mathcal{H}_N = \text{span}\{u_H(\mu_i), i = 1, \ldots, N\} \tag{5.2}$$

with $u_H(\mu_i) \in \mathcal{H}_H$ solving (5.1) for $\mu_i \in \mathcal{S}$. The RB-HMM approximation $u_N(\mu) \in \mathcal{H}_N$ of $u_H(\mu)$ for any $\mu \in \mathcal{P}$ is then defined as the Galerkin projection of (5.1) onto $\mathcal{H}_N$, i.e.

$$\mathcal{B}_h(u_N(\mu), v_N; \mu) = L(v_N) \qquad (v_N \in \mathcal{H}_N). \tag{5.3}$$

5.1 A Posteriori Error Estimation

Based on the Galerkin framework it is hence straight forward to derive a robust and efficient a posteriori error estimation that allows for an efficient offline/online decomposition as has been detailed in [37]. To this end, we introduce the *residual* $\text{Res} : \mathcal{P} \to \mathcal{H}^{-1}$ via

$$-\langle \text{Res}(\mu), v \rangle_{\mathcal{H}^{-1}\times\mathcal{H}} := L(v; \mu) - \mathcal{B}_h(u_N(\mu), v; \mu). \tag{5.4}$$

With the Hilbert space structure of $\mathcal{H}_H$ we further define its Riesz representative $v_H^{\text{Res}}(\mu) \in \mathcal{H}_H$ through

$$\left(v_H^{\text{Res}}(\mu), v_H\right)_{\mathcal{H}} = \langle \text{Res}(\mu), v_H \rangle_{\mathcal{H}^{-1}\times\mathcal{H}}$$

for all $v_H \in \mathcal{H}_H$. If we assume $A(x, y; \mu)\xi \cdot \xi \geq \alpha(\mu)\, |\xi|^2$ for all $(x, y, \mu) \in \Omega \times Y \times \mathcal{P}$ and $\xi \in \mathbb{R}^d$ with $\inf_{\mu\in\mathcal{P}} \alpha(\mu) > 0$, the bilinear form $\mathcal{B}_h$ is coercive and we get the following theorem.

Theorem 5.3 (A Posteriori Error Estimate)

$$||u_H(\mu) - u_N(\mu)||_{\mathcal{H}} \leq \frac{||v_H^{Res}(\mu)||_{\mathcal{H}}}{\alpha(\mu)} =: \Delta_N(\mu). \tag{5.5}$$

It is this a posteriori error estimator that we use in the greedy construction of the RB-HMM space.

5.2 Optimization and Computation of Derivative Information

Following [37], the optimization problem (1.1) will be approximated in our RB-HMM setting as follows. We define the RB-HMM approximations of the functionals J and C_j through

$$J_N(\mu) := J(u_N^c(\mu), \mu),$$
$$C_{N,j}(\mu) := C_j(u_N^c(\mu), \mu).$$

The *reduced optimization problem* then reads

$$\begin{aligned} \text{Find} \quad & \mu_N^* = \arg\min J_N(\mu) \\ \text{subject to} \quad & C_{N,j}(\mu) \leq 0 \quad \forall j = 1, \dots, m, \\ & \mu \in \mathcal{P}. \end{aligned} \tag{5.6}$$

Note that the reduced functionals involve only the macroscopic part of $u_N(\mu)$. For an efficient optimization process, the involved functionals must allow an offline/online splitting. We thus assume that J_N and $C_{N,j}$ allow a representation as follows

$$J_N(\mu) = \sum_{q=1}^{Q_J} \sigma_J^q(\mu) J_N^q \left(\varphi_1^N, \dots, \varphi_N^N\right),$$

$$C_{N,j}(\mu) = \sum_{q=1}^{Q_{C_j}} \sigma_{C_j}^q(\mu) C_{N,j}^q \left(\varphi_1^N, \dots, \varphi_N^N\right),$$

where the mappings J_N^q and $C_{N,j}^q$ can (in principle) be arbitrary. These expansions ensure that we can separate the parts of the functionals that depend on the basis functions from those depending on parameters which is necessary to achieve the offline/online splitting.

The only thing that is still missing are the parameter derivatives of the state $u_N(\mu)$. Equations for these quantities are established by differentiating the defining Eq. (5.3) of $u_N(\mu)$ with respect to μ_i. This results in a weak formulation for $\partial_{\mu_i} u_N(\mu) \in \mathcal{H}_N$:

$$\mathcal{B}_h(\partial_{\mu_i} u_N(\mu), v_N; \mu) = -\partial_{\mu_i} \mathcal{B}_h(u_N(\mu), v_N; \mu) \tag{5.7}$$

for all $v_N \in \mathcal{H}_N$, if the functional L does not depend on μ. Since $\mathcal{B}_h$ is affinely decomposable, we have

$$\partial_{\mu_i} \mathcal{B}_h(u, v; \mu) = \sum_{q=1}^{Q_A} \partial_{\mu_i} \sigma_q^A(\mu) \mathcal{B}_h^q(u, v).$$

Hence, we can reuse both the same reduced spaces and precomputed reduced system matrices as for the approximation of $u_N(\mu)$. Thus the computational costs for the parameter derivatives are negligible. Higher order derivatives can be specified by further differentiation of (5.7).

5.3 Numerical Experiments

As a benchmark for the RB-HMM in a homogenization setting we look at the following model problem. For some scale parameter $\varepsilon > 0$ let $A^\varepsilon : \Omega \times \mathcal{P} \to \mathbb{R}^{d\times d}$ be a parameterized rapidly oscillating diffusion tensor, $\eta, g_N : \Gamma_N \to \mathbb{R}$ and f some right hand side. For $\mu \in \mathcal{P}$ we seek $u^\varepsilon(\mu)$ as solution of

$$\begin{aligned} -\nabla \cdot \left(A^\varepsilon(\mu)\nabla u(\mu)\right) &= f && \text{in } \Omega, \\ u^\varepsilon(\mu) &= 0 && \text{on } \Gamma_D, \\ -A^\varepsilon(\mu)\nabla u(\mu)\cdot n &= \eta\,(u^\varepsilon(\mu) - g_N) && \text{on } \Gamma_N, \end{aligned} \tag{5.8}$$

where $n(x)$ is the outer normal on Γ_D.

In particular we choose $\Omega \equiv (0, 0.6)\times(0, 0.2) \subset \mathbb{R}^2$ and define $\Gamma_{N_1} \equiv [0, 0.2]\times\{0.2\}$, $\Gamma_{N_2} \equiv [0.4, 0.6] \times \{0.2\}$ and $\Gamma_N \equiv \partial\Omega \setminus \left(\Gamma_{N_1} \cup \Gamma_{N_2}\right)$ (see Fig. 1). With $r_1 = r_2 = 0.025$ and $x_1 = (0.15, 0.1), x_2 = (0.45, 0.1) \in \mathbb{R}^2$ we introduce the source term $f : \Omega \to \mathbb{R}$, the Neumann value function $g_N : \partial\Omega \to \mathbb{R}$ and the heat

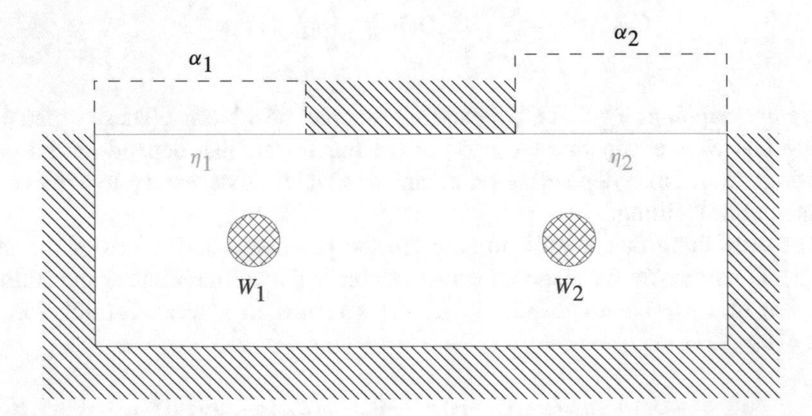

Fig. 1 Geometry of the model problem: Two heat sources (red), no flow (blue) and Robin (green) boundary conditions. The head transfer coefficients and surrounding temperatures are $\eta_1 = 0.3$, $\eta_2 = 0.4$ resp. $\alpha_1 = \alpha_2 = 300$

transfer coefficient function $\eta : \partial\Omega \to \mathbb{R}$ as

$$f(x) \equiv 500 \cdot \mathbf{1}_{B_{r_1}(x_1)}(x) + 800 \cdot \mathbf{1}_{B_{r_2}(x_2)}(x), \tag{5.9}$$

$$g_N(x) \equiv 300 \cdot \mathbf{1}_{\Gamma_{N_1} \cup \Gamma_{N_2}}(x), \tag{5.10}$$

$$\eta(x) \equiv 0.3 \cdot \mathbf{1}_{\Gamma_{N_1}}(x) + 0.4 \cdot \mathbf{1}_{\Gamma_{N_2}}(x). \tag{5.11}$$

Additionally, we define the parameter set $\mathcal{P} \equiv [0.001, 1]^2$ and the parameterized diffusion tensor as

$$A^{\varepsilon}(\mu) = A(x, \frac{x}{\varepsilon}; \mu) \text{ with } A(x, y; \mu) = 16y_1^2(1 - y_1)^2(\mu_2 - \mu_1) + \mu_1.$$

As an objective functional for the optimization we choose

$$J(u, \mu) = \int_{\Omega} (u - u_{\text{ref}})^2 \, \mathrm{d}x + \frac{\alpha}{2} ||\mu||^2$$

with a reference solution $u_{\text{ref}} = u^0(\mu)$, $\mu = (0.9112, 0.0062)$.

Note that since the diffusion tensor does not depend on y_2, the cell solutions $\chi^2(\mu)$ vanish. Samples of the macroscopic and the first cell solution are displayed in Fig. 2 (the macroscopic solution is also used as target function for the optimization problems). Additionally, we impose the restriction $\mu \in \mathcal{P}$.

The convergence behavior of the a posteriori error estimator is shown in Fig. 3. Figure 4 shows the parameters chosen by the greedy algorithm in the offline phase for the reduced basis construction. Table 1 shows a runtime comparisons. For that we have run the detailed and reduced optimization procedure for different underlying grid widths. The detailed optimization scales more or less quadratically with the grid width while in the reduced setting it stays constant. In the detailed optimization, gradient and Hessian information are approximated with finite differences, while in the reduced setting they can be easily generated directly form the corresponding reduced model problem. This makes the detailed approach more sensitive to the discretization, resulting in the necessity of more iterations and function evaluations which slows down the whole process significantly. Finally, in Fig. 4 we display the convergence of the reduced objective functional and the computed optimal parameter value with respect to the high dimensional reference computation in dependence of the basis size of the RB-HMM.

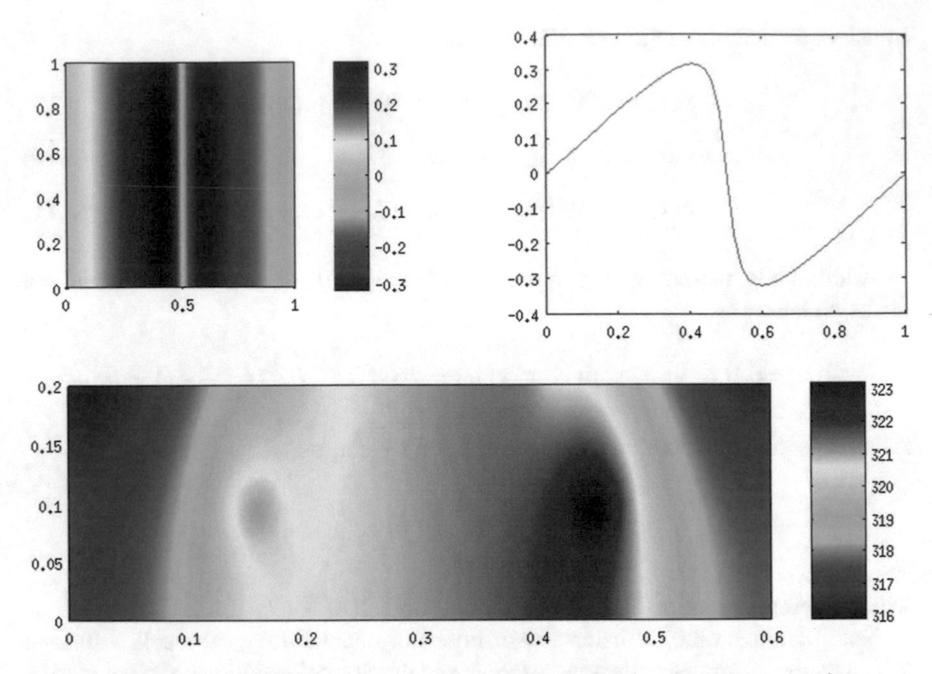

Fig. 2 Example of a solution of the test problem for $\mu = (0.9112, 0.0062)$. Top left: $\chi_H^1(\mu)$, top right: cross-section plot of $\chi_H^1(\mu)$ in y_1-direction; bottom: $u_H^0(\mu)$. The effective diffusion tensor is $\mathcal{B}_h^0(\mu) = \mathrm{diag}(0.069, 0.429)$

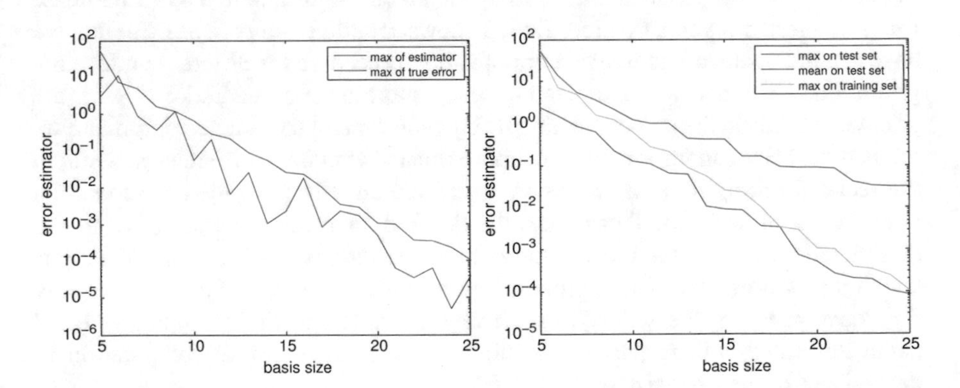

Fig. 3 Convergence of the error estimator. Left: Estimator and true error on training set. Right: Estimator on test set

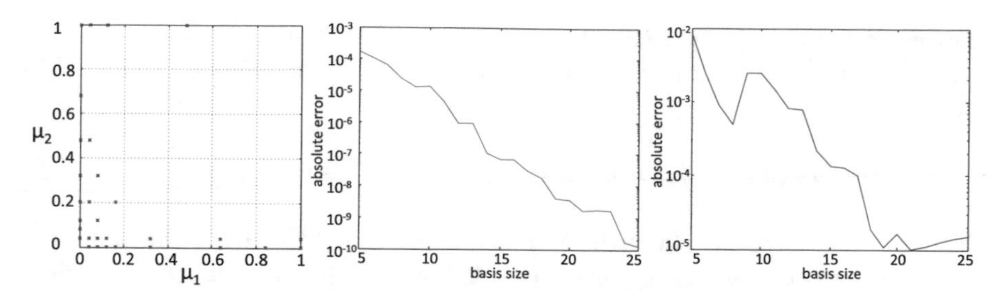

Fig. 4 Parameters from the training set chosen by the greedy algorithm (left). Convergence of the reduced objective functional (middle) and optimal parameter value (right) with respect to the high dimensional reference optimization problem

Table 1 Runtime comparison between detailed and reduced simulations for a single forward solve and the complete optimization problem

	Detailed (s)	Reduced (s)	Speed-up	
Single solve	0.5	0.1	~ 5	$(h = 4 \times 10^{-3})$
	1.5	0.1	~ 15	$(h = 2 \times 10^{-3})$
	6.4	0.1	~ 64	$(h = 1 \times 10^{-3})$
Optimization	341	24	~ 14	$(h = 4 \times 10^{-3})$
	999	24	~ 42	$(h = 2 \times 10^{-3})$

6 The Localized Reduced Basis Multiscale Method (LRBMS) with Online Enrichment

In this section we first review the LRBMS method with online enrichment from [38, 39] in the given optimization context. The LRBMS is defined through Definition 4.1 with the DG-MsFEM choice given in Definition 3.3. The construction of the corresponding local reduced subspaces is done using a localized a posteriori error estimator and a greedy online enrichment strategy during the optimization loop. This procedure follows a new paradigm that we recently introduced in [40].

The online enrichment procedure that we describe in Sect. 6.2 below constructs appropriate low dimensional local approximation spaces $U_N^j \subset V^c|_{T_j} \oplus V^f(T_j) \subset H^1(\tau_h^j)$ of local dimensions N_j that form the global reduced solution space via

$$U_H^N = \bigoplus_{j=1}^{N_H} U_N^j, \qquad N := \dim(U_H^N) = \sum_{j=1}^{N_H} N_j. \tag{6.1}$$

Once such a reduced approximation space is constructed, the LRBMS approximation is defined as follows.

Definition 6.1 (The Localized Reduced Basis Multiscale Method) We call $u_N(\mu) \in U_H^N$ a localized reduced basis multiscale approximation of (2.2) if it satisfies

$$\mathcal{A}_{\mathrm{DG}}(u_N(\mu), v_N; \mu) = L_{\mathrm{DG}}(\mu; v_N) \qquad \text{for all } v_N \in U_H^N. \tag{6.2}$$

Note that $u_N(\mu)$ defined through (6.2) corresponds to $\mathcal{R}^\mu(u_N^c(\mu))$, where $u_N^c(\mu)$ is defined in 4.1. It solves a globally coupled reduced problem, where all arising quantities can nevertheless be locally computed w.r.t. the local reduced spaces U_N^j.

6.1 A Posteriori Error Estimation

In recent contributions [4, 10, 27, 39] we discussed several possibilities to construct local reduced spaces U_N^j from global or localized snapshot computations. Thereby, in the concept presented above, it is possible to use finite volume, DG or conforming finite element approximations on the underlying fine partition τ_h or restrictions thereof to a local neighborhood of the macro elements $T_j \in \mathcal{T}_H$. In what follows, we consider the iterative construction of reduced approximation spaces and related surrogate models based on localized a posteriori error control and local enrichment as recently introduced in [39]. In these circumstances we obtain the following estimate on the error w.r.t. the unknown weak solution of (1.2).

Theorem 6.2 (Localizable a Posteriori Error Estimate) *With the assumptions and the notation of [39, Cor. 4.5], the following estimate on the full approximation error in the energy semi-norm* $|||v|||_{\overline{\mu}} := \sum_{t\in\tau_h} \int_t (A^\varepsilon(\overline{\mu})\nabla v)\cdot\nabla v$ *holds for arbitrary* $\mu, \overline{\mu}, \hat{\mu} \in \mathcal{P}$,

$$|||u(\mu) - u_N(\mu)|||_{\overline{\mu}} \leq \eta\big(u_N(\mu)\big) := C(\mu, \overline{\mu}, \hat{\mu}) \Bigg[\sum_{j=1}^{N_H} \Big(\eta_j^{nc}(u_N(\mu))^2\Big)^{1/2} + \sum_{j=1}^{N_H} \Big(\eta_j^{r}(u_N(\mu))^2\Big)^{1/2} + \sum_{j=1}^{N_H} \Big(\eta_j^{df}(u_N(\mu))^2\Big)^{1/2} \Bigg]$$

with a computable constant $C(\mu, \overline{\mu}, \hat{\mu}) > 0$ *and fully computable local indicators* η_j^{nc}, η_j^r *and* η_j^{df} *corresponding to the local non-conformity errors, residual errors, and diffusive flux reconstruction errors, respectively.*

We refer to [39] for a more detailed presentation and derivation of this result and the definition of the corresponding local indicators. Note that a similar result holds for a full norm comprising the above semi-norm and a DG-jump semi-norm.

6.2 Optimization and Online Enrichment

Within an optimization loop, typically only parameters along a path towards the optimal parameter are depicted. As introduced in [40], we thus suggest a new iterative procedure to successively build up or enhance the surrogate model (6.2) by using the concept of local enrichment from [39]. Hence, only localized snapshot computations for the parameters that are selected during the optimization loop are computed. The resulting approach is thus tailored towards the specific optimization problem in an a posteriori manner.

Algorithm 6.1 Parameter optimization with adaptive enrichment

---

Require: $\boldsymbol{\mu}^{(0)} \in \mathcal{P}$, initial local bases $\Phi_j{}^{(0)}$, $\Delta_{\text{model}}, \Delta_{\text{opt}} > 0$, a marking strategy `MARK` and an orthonormalization procedure `ONB` (see [39, Sec. 5]), an optimization routine `OPT` (returning a new parameter and status of convergence).

$n \leftarrow 0$, $U_H^{N(0)} \leftarrow \bigoplus_{j=1}^{N_H} \text{span}\big(\Phi_j{}^{(0)}\big)$
repeat
  $m \leftarrow n$
  Solve (6.2) *for* $u_N^{(m)}(\boldsymbol{\mu}^{(n)}) \in U_H^{N(n)}$.
  while $\eta(\boldsymbol{\mu}^{(n)}) > \Delta_{\text{model}}$ **do**
    for all $j = 1, \ldots, N_H$ **do**
      Compute local error indicators $\eta_j(\boldsymbol{\mu}^{(n)})$ *according to* [39, Cor. 4.5].
    end for
    $\tilde{\mathcal{T}}_H \leftarrow \text{MARK}\big(\mathcal{T}_H, \{\eta_j(\boldsymbol{\mu}^{(n)})\}_{j=1}^{N_H}\big)$
    for all $T_j \in \tilde{\mathcal{T}}_H$ **do**
      Solve locally on O_{T_j} *for enhanced local snapshot* $u_h^j(\boldsymbol{\mu}^{(n)}) \in H^1(\tau_h^j)$.
      $\Phi_j{}^{(m+1)} \leftarrow \text{ONB}\big(\{\Phi_j{}^{(m)}, u_h^j(\boldsymbol{\mu}^{(n)})\}\big)$
    end for
    $U_H^{N(m+1)} \leftarrow \bigoplus_{T_j \in \tilde{\mathcal{T}}_H} \text{span}\big(\Phi_j{}^{(m+1)}\big) \oplus \bigoplus_{T_j \in \mathcal{T}_H \setminus \tilde{\mathcal{T}}_H} \text{span}\big(\Phi_j{}^{(m)}\big)$
    Solve (6.2) *for* $u_N^{(m+1)}(\boldsymbol{\mu}^{(n)}) \in U_H^{N(m+1)}$.
    $m \leftarrow m + 1$
  end while
  $(\boldsymbol{\mu}^{(n+1)}, \texttt{success}) \leftarrow \text{OPT}\big(\boldsymbol{\mu}^{(n)}, \Delta_{\text{opt}}, u_N^{(m)}(\boldsymbol{\mu}_n)\big)$
  $n \leftarrow n + 1$
until `success`
return optimal parameter $\boldsymbol{\mu}^{(n)}$ and state $u_N^{(m)}(\boldsymbol{\mu}^{(n)})$

In more detail, in a first step we initialize the local reduced spaces U_N^j with a classical polynomial coarse scale DG basis of prescribed order, thus ensuring that any reduced solution of the state equation is at least as good as a DG solution on the coarse grid $\mathcal{T}_H$. During the following optimization loop, given any $\boldsymbol{\mu} \in \mathcal{P}$ from the optimization algorithm, we compute a reduced solution $u_N(\boldsymbol{\mu}) \in U_H^N$ and efficiently assess its quality using the localized a posteriori error estimator $\eta(\boldsymbol{\mu}) := \big(\sum_{j=1}^{N_H} \eta_j(\boldsymbol{\mu})^2\big)^{1/2}$ derived in [38, 39]. If the estimated error is above a prescribed tolerance, $\Delta_{\text{model}} > 0$, we start an intermediate local enrichment phase to enhance the surrogate model in the SEMR (solve → estimate → mark

→ refine) spirit of adaptive mesh refinement. We refer to Algorithm 6.1 and [39] for a detailed description and evaluation of this enrichment procedure that only involves local snapshot computations for the given parameter $\boldsymbol{\mu}$ on the local neighborhoods O_{T_j}, $T_j \in \mathcal{T}_H$ with Dirichlet boundary values obtained from the insufficient previous reduced surrogate. The algorithm calls a routine `OPT` that performs one optimization step with a descent method based on the old parameter value and the corresponding state with respect to the parameters. It returns the new parameter value and `success=true`, if the optimization criteria have been met.

6.3 Numerical Experiment

To investigate the performance of the online enrichment, we consider (1.2) on $\Omega := [-1, 1]^2$ with right hand side $f(x, y) := \frac{1}{2}\pi^2 \cos(\frac{1}{2}\pi x)\cos(\frac{1}{2}\pi y)$, a parameter space $\mathcal{P} := [0, \pi]^2$ and a parametric scalar diffusion $A(\mu) := \sum_{\xi=1}^{2} \theta_\xi(\mu) A_\xi$ with coefficients $A_1 := \chi_{\Omega\setminus\omega}$, $A_2 := \chi_\omega$ and parameter functionals $\theta_1(\mu) := 1.1 + \sin(\mu_0)\mu_1$, $\theta_2(\mu) := 1.1 + \sin(\mu_1)$, where χ denotes an indicator function for the given domain and $\omega := [-\frac{2}{3}, -\frac{1}{3}]^2 \cup ([-\frac{2}{3}, -\frac{1}{3}] \times [\frac{1}{3}, \frac{2}{3}])$, compare Fig. 5, top left. We are interested in minimizing the compliant quantity of interest (QoI) $J(u(\mu); \mu) := \int_\Omega f u(\mu)$ over $\mathcal{P}$. While this problem does not contain any multiscale features, it may serve as a model problem to compare model reduction using standard global RB methods with localized RB methods in the context of PDE-constrained optimization.

We discretize Ω by a triangulation τ_h with 2018 simplices and approximate the solution of (1.2) using a P^1-SWIPD scheme [16] (similar to Definition 2.2) with 6144 unknowns and use an L-BFGS-B algorithm [11] with a finite difference approximation of the objectives derivatives as optimization routine. We compare four different scenarios (compare Table 2), which we discuss further below: (i) using only the reference discretization ("standard FEM"); (ii) using a reduced order model (ROM) based on a single reduced basis with global support ("standard RB"); (iii) using a ROM based on a local reduced basis on each subdomain, containing only the constant 1 ("localized RB: Q^0-basis"); (iv) same as (iii), but with adaptive online enrichment of the local bases according to Algorithm 6.1 ("localized RB: adaptive").

Using the standard FEM approach (i) and starting from an initial guess of $\mu^{(0)} = (0.25, 0.5)$ we obtain the reference minimizer $\mu_h^* \approx (\frac{\pi}{2}, \pi)$ after 7 iterations of the optimization routine, with an additional 32 evaluations of J for the approximation of its derivatives (resulting in 39 evaluations of the reference discretization).

In the standard RB approach (ii), we employ a weak greedy algorithm (using the standard residual based a posteriori error estimate on the model reduction error w.r.t. the energy product induced by the SWIPDG bilinear form for a fixed parameter $\overline{\mu} = (0, 0)$) to build a single reduced basis with global support, requiring 18 evaluations of the reference discretization to reach a model reduction error of $1.77 \cdot 10^{-11}$ over the training set of 625 equally distributed parameters. We did not

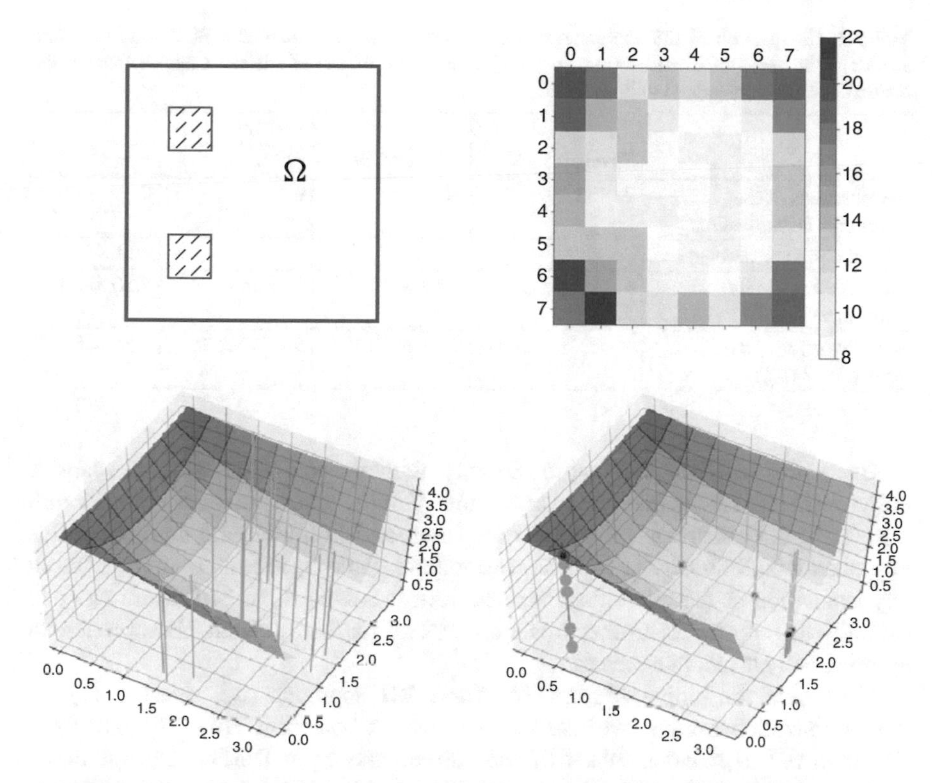

Fig. 5 Physical domain and coarse grid (top row) and QoI J (purple surface, computed with a reference discretization) over parameter space (bottom row). Top left: Ω and ω (shaded regions). Top right: $\mathcal{T}_H$ with 8×8 subdomains and sizes of the local reduced bases (between 8 and 22) at the end of the computation. Bottom left: selected parameters (orange bars) during greedy basis generation of a standard RB approximation. Bottom right: selected parameters (orange bars) during optimization, intermediate evaluations of the QoI $J\big(u_N^{(m)}(\mu^{(n)}), \mu^{(n)}\big)$ during the adaptive enrichment (orange circles) and those given to OPT (purple dots)

employ an a posteriori error estimate on the quantity of interest, since the estimate on the state is an equivalent one in the present compliant setting (compare [18]). As we observe in Table 2, third column, using this ROM during the optimization yields very satisfactory results (thus justifying this choice). Compared to the standard FEM approach, we require only half of the evaluations of the reference discretization (since the finite difference approximation of the objectives derivative is now performed using the ROM). However, the purpose of the greedy algorithm is to build a reduced basis that is equally valid for the whole parameter space and the reference discretization is thus evaluated over a large part of the parameter space (compare Fig. 5, bottom left) that is not required for the optimization (although a certain symmetry in the parameterization is detected). One could thus argue that too many evaluations of the reference discretization are required, compared to the (unknown) trajectory of the optimizer through the parameter space.

Table 2 Comparison of the computational effort (in terms of global/local PDE solutions) and accuracy (in terms of relative errors in the QoI and minimizer) of different approaches in the context of PDE-constrained optimization

	Standard FEM	Standard RB	Localized RB	
			Q^0-basis	Adaptive
#evaluations of the reference discretization	$7+32$	18	0	0
#local corrector problems	–	–	–	709
Relative error in the minimizer w.r.t. μ_h^*	–	4.53×10^{-5}	9.18×10^{-5}	4.50×10^{-5}
Relative error in the QoI w.r.t. $J(\mu_h^*)$	–	9.01×10^{-9}	9.47×10^{-1}	1.36×10^{-5}

Using the localized RB approach only with the Q^0-basis without adaptive enrichment (iii) can be interpreted as a Finite Volume scheme on an 8×8 grid with a higher order quadrature. If one is only interested in finding the minimizer, this approach would already be sufficient (compare Table 2, fourth column), without any evaluation of the reference discretization. It is thus subject of future work to establish an a posteriori error estimate on the QoI for the localized RB approach, to automatically detect this scenario.

Finally, in the online adaptive localized RB approach (iv), in each step of the optimization routine, we use the localized a posteriori error estimate from Theorem 6.2 to select a subset of the subdomains by a Dörfler and age-based marking strategy and enrich the corresponding local reduced bases by solutions to local corrector problems posed on the neighborhood of each of these subdomains (containing all adjacent subdomains), see [39] and the references therein. Using this approach, we obtain a satisfactory approximation of the minimum as well as the minimizer (compare Table 2, fifth column), without requiring any evaluation of the reference discretization. However, we did require the solution of 709 local corrector problems.

It is clear that for the simple example at hand, the adaptive localized RB approach does not pay off in terms of computational time when compared to the standard FEM or standard RB approach. However, if applied to a large real-world multiscale problem, the solution of which requires the use of large computing clusters, the localized nature of this approach should show significant benefits over the other approaches.

We used the generic discretization toolbox `dune-gdt`,[1] based on the `dune-xtensions` [32] and the `DUNE` software framework [6, 7], together with the model reduction package `pyMOR` [34]. To reproduce the experiments follow the instructions on:

https://github.com/ftschindler-work/proceedings-mbour-2017-lrbms-control

References

1. A. Abdulle. The finite element heterogeneous multiscale method: a computational strategy for multiscale PDEs. In *Multiple scales problems in biomathematics, mechanics, physics and numerics*, volume 31 of *GAKUTO Internat. Ser. Math. Sci. Appl.*, pages 133–181. Gakkōtosho, Tokyo, 2009.
2. A. Abdulle and P. Henning. A reduced basis localized orthogonal decomposition. *J. Comput. Phys.*, 295:379–401, 2015.
3. A. Abdulle and A. Nonnenmacher. Adaptive finite element heterogeneous multiscale method for homogenization problems. *Comput. Methods Appl. Mech. Engrg.*, 200(37–40):2710–2726, 2011.
4. F. Albrecht, B. Haasdonk, S. Kaulmann, and M. Ohlberger. The localized reduced basis multiscale method. In *Proceedings of Algoritmy 2012, Conference on Scientific Computing, Vysoke Tatry, Podbanske, September 9–14, 2012*, pages 393–403. Slovak University of Technology in Bratislava, Publishing House of STU, 2012.
5. G. Allaire, E. Bonnetier, G. Francfort, and F. Jouve. Shape optimization by the homogenization method. *Numer. Math.*, 76(1):27–68, 1997.
6. P. Bastian, M. Blatt, A. Dedner, C. Engwer, R. Klöfkorn, R. Kornhuber, M. Ohlberger, and O. Sander. A generic grid interface for parallel and adaptive scientific computing. II. Implementation and tests in DUNE. *Computing*, 82(2–3):121–138, 2008.
7. P. Bastian, M. Blatt, A. Dedner, C. Engwer, R. Klöfkorn, M. Ohlberger, and O. Sander. A generic grid interface for parallel and adaptive scientific computing. I. Abstract framework. *Computing*, 82(2–3):103–119, 2008.
8. P. Benner, A. Cohen, M. Ohlberger, and K. Willcox. *Model Reduction and Approximation: Theory and Algorithms*, volume 15 of *Computational Science and Engineering*. SIAM Publications, Philadelphia, PA, 2017.
9. S. Boyaval. Reduced-basis approach for homogenization beyond the periodic setting. *Multiscale Model. Simul.*, 7(1):466–494, 2008.
10. A. Buhr, C. Engwer, M. Ohlberger, and S. Rave. ArbiLoMod, a Simulation Technique Designed for Arbitrary Local Modifications. *SIAM J. Sci. Comput.*, 39(4):A1435–A1465, 2017.
11. R. H. Byrd, P. Lu, J. Nocedal, and C. Y. Zhu. A limited memory algorithm for bound constrained optimization. *SIAM J. Sci. Comput.*, 16(5):1190–1208, 1995.
12. W. E and B. Engquist. The heterogeneous multiscale methods. *Commun. Math. Sci.*, 1(1):87–132, 2003.
13. W. E and B. Engquist. The heterogeneous multi-scale method for homogenization problems. In *Multiscale methods in science and engineering*, volume 44 of *Lect. Notes Comput. Sci. Eng.*, pages 89–110. Springer, Berlin, 2005.
14. Y. Efendiev, T. Hou, and V. Ginting. Multiscale finite element methods for nonlinear problems and their applications. *Commun. Math. Sci.*, 2(4):553–589, 2004.

[1] https://github.com/dune-community/dune-gdt.

15. Y. Efendiev and T. Y. Hou. *Multiscale finite element methods*, volume 4 of *Surveys and Tutorials in the Applied Mathematical Sciences*. Springer, New York, 2009. Theory and applications.
16. A. Ern, A. F. Stephansen, and P. Zunino. A discontinuous galerkin method with weighted averages for advection–diffusion equations with locally small and anisotropic diffusivity. *IMA J. Numer. Anal.*, 29(2):235–256, 2009.
17. B. Geihe and M. Rumpf. A posteriori error estimates for sequential laminates in shape optimization. *Discrete Contin. Dyn. Syst. Ser. S*, 9(5):1377–1392, 2016.
18. P. Haasdonk. *Reduced Basis Methods for Parametrized PDEsA Tutorial Introduction for Stationary and Instationary Problems*, In P. Benner, A. Cohen, M. Ohlberger and K. Willcox (editors) *Model reduction and approximation. Theory and algorithms.*, 65136, Comput. Sci. Eng., 15, SIAM, Philadelphia, PA, 2017. chapter 2, pages 65–136.
19. P. Henning, A. Malqvist, and D. Peterseim. A localized orthogonal decomposition method for semi-linear elliptic problems. *ESAIM Math. Model. Numer. Anal.*, 48(5):1331–1349, 2014.
20. P. Henning and M. Ohlberger. The heterogeneous multiscale finite element method for elliptic homogenization problems in perforated domains. *Numer. Math.*, 113(4):601–629, 2009.
21. P. Henning and M. Ohlberger. The heterogeneous multiscale finite element method for advection-diffusion problems with rapidly oscillating coefficients and large expected drift. *Netw. Heterog. Media*, 5(4):711–744, 2010.
22. P. Henning, M. Ohlberger, and B. Schweizer. An adaptive multiscale finite element method. *Multiscale Model. Simul.*, 12(3):1078–1107, 2014.
23. J. S. Hesthaven, G. Rozza, and B. Stamm. *Certified reduced basis methods for parametrized partial differential equations*. SpringerBriefs in Mathematics. Springer, Cham; BCAM Basque Center for Applied Mathematics, Bilbao, 2016. BCAM SpringerBriefs.
24. T. Y. Hou and X.-H. Wu. A multiscale finite element method for elliptic problems in composite materials and porous media. *J. Comput. Phys.*, 134(1):169–189, 1997.
25. T. J. R. Hughes. Multiscale phenomena: Green's functions, the Dirichlet-to-Neumann formulation, subgrid scale models, bubbles and the origins of stabilized methods. *Comput. Methods Appl. Mech. Engrg.*, 127(1–4):387–401, 1995.
26. T. J. R. Hughes, G. R. Feijóo, L. Mazzei, and J.-B. Quincy. The variational multiscale method - a paradigm for computational mechanics. *Comput. Methods Appl. Mech. Engrg.*, 166(1–2):3–24, 1998.
27. S. Kaulmann, M. Ohlberger, and B. Haasdonk. A new local reduced basis discontinuous Galerkin approach for heterogeneous multiscale problems. *C. R. Math. Acad. Sci. Paris*, 349(23–24):1233–1238, 2011.
28. M. G. Larson and A. Malqvist. Adaptive variational multiscale methods based on a posteriori error estimation: duality techniques for elliptic problems. In *Multiscale methods in science and engineering*, volume 44 of *Lect. Notes Comput. Sci. Eng.*, pages 181–193. Springer, Berlin, 2005.
29. M. G. Larson and A. Malqvist. Adaptive variational multiscale methods based on a posteriori error estimation: energy norm estimates for elliptic problems. *Comput. Methods Appl. Mech. Engrg.*, 196(21–24):2313–2324, 2007.
30. M. G. Larson and A. Malqvist. An adaptive variational multiscale method for convection-diffusion problems. *Comm. Numer. Methods Engrg.*, 25(1):65–79, 2009.
31. M. G. Larson and A. Malqvist. A mixed adaptive variational multiscale method with applications in oil reservoir simulation. *Math. Models Methods Appl. Sci.*, 19(7):1017–1042, 2009.
32. T. Leibner, R. Milk, and F. Schindler. Extending dune: The dune-xt modules. *Archive of Numerical Software*, 5:193–216, 2017.
33. A. Malqvist and D. Peterseim. Localization of elliptic multiscale problems. *Math. Comp.*, 83(290):2583–2603, 2014.
34. R. Milk, S. Rave, and F. Schindler. pyMOR – generic algorithms and interfaces for model order reduction. *SIAM Journal on Scientific Computing*, 38(5):S194–S216, jan 2016.

35. M. Ohlberger. A posteriori error estimates for the heterogeneous multiscale finite element method for elliptic homogenization problems. *Multiscale Model. Simul.*, 4(1):88–114, 2005.
36. M. Ohlberger and M. Schaefer. A reduced basis method for parameter optimization of multiscale problems. In *Proceedings of Algoritmy 2012, Conference on Scientific Computing, Vysoke Tatry, Podbanske, September 9–14, 2012*, pages 272–281, september 2012.
37. M. Ohlberger and M. Schaefer. Error control based model reduction for parameter optimization of elliptic homogenization problems. In Yann Le Gorrec, editor, *1st IFAC Workshop on Control of Systems Governed by Partial Differential Equations, CPDE 2013; Paris; France; 25 September 2013 through 27 September 2013; Code 103235*, volume 1, pages 251–256. International Federation of Automatic Control (IFAC), 2013.
38. M. Ohlberger and F. Schindler. A-posteriori error estimates for the localized reduced basis multi-scale method. In J. Fuhrmann, M. Ohlberger, and C. Rohde, editors, *Finite Volumes for Complex Applications VII-Methods and Theoretical Aspects*, volume 77 of *Springer Proceedings in Mathematics & Statistics*, pages 421–429. Springer International Publishing, 2014.
39. M. Ohlberger and F. Schindler. Error control for the localized reduced basis multi-scale method with adaptive on-line enrichment. *SIAM J. Sci. Comput.*, 37(6):A2865–A2895, 2015.
40. M. Ohlberger and F. Schindler. Non-conforming localized model reduction with online enrichment: Towards optimal complexity in PDE constrained optimization. In *Finite volumes for complex applications VIIIhyperbolic, elliptic and parabolic problems*, 357–365, Springer Proc. Math. Stat., 200, Springer, Cham, 2017.
41. A. Quarteroni, A. Manzoni, and F. Negri. *Reduced basis methods for partial differential equations*, volume 92 of *Unitext*. Springer, Cham, 2016. An introduction, La Matematica per il 3+2.

Numerical Simulation for a Dimensionless Coupled System of Shallow Water Equation with Long Term Dynamic of Sand Dunes Equation

Mouhamadou A. M. T. Baldé and Diaraf Seck

Abstract In this paper we aim to do numerical simulation of the dimensionless coupled system (SW-LTDD) of Shallow Water equations (SWE) and Long term dynamic of sand dunes (LTDD) equation with a small parameter ϵ, presented in our previous paper (Baldé and Seck, Discrete Contin Dyn Syst Ser S 9(5):1521–1551, 2016) to model the erosion phenomenon. The small parameter ϵ was for the first time introduced in the paper Faye et al. (J Nonlinear Anal Appl 2016(2):82–105, 2016). In Baldé and Seck (Discrete Contin Dyn Syst Ser S 9(5):1521–1551, 2016), we have proved existence and uniqueness of a solution of the coupled system. Now we give a numerical scheme of the system based on finite volume method. The stability of the scheme is proved and some numerical tests are performed.

Keywords Shallow water equations (SWE) · Long term dynamic of dunes of sand (LTDD) · Numerical simulation · Finite volume · PDE · Models

Mathematics Subject Classification (2010). Primary 65N08; Secondary 35Q35

1 Introduction and Model

The system we shall study has been presented in the paper [1] to model the erosion phenomenon. In the paper [1] we have presented the model and given existence and uniqueness results of solution of the system. In [1] the system has been obtained after scaling and Nondimensionalization of a coupled system of Shallow Water Equation (SWE) and an equation of dunes of sand. The dimensionless system was obtained by rewriting the system according to a small parameter ϵ which was, for

M. A. M. T. Baldé (✉) · D. Seck
Laboratoire de Mathématiques de la Décision et d'Analyse Numérique (L.M.D.A.N) F.A.S.E.G, Université Cheikh Anta Diop de Dakar, Dakar, Senegal
e-mail: mouhamadouamt.balde@ucad.edu.sn; diaraf.seck@ucad.edu.sn

V. Schulz, D. Seck (eds.), *Shape Optimization, Homogenization and Optimal Control*, International Series of Numerical Mathematics 169,
https://doi.org/10.1007/978-3-319-90469-6_9

the first time, defined in the paper [3]. ϵ corresponds to the ratio of a 1-month tide period over a long 16-years observation period of the tide, so $\epsilon = 1/192$. Because of the long observation period of the tide the equation of dunes of sand presented in [3] was named Long term dynamic of dunes of sand (LTDD). For simplicity, in the following we will always note the coupled system by SW-LTDD.

The system is expressed as follows:

$$\begin{cases} \dfrac{\partial v^\epsilon}{\partial t} + \dfrac{1}{\epsilon^4}\displaystyle\sum_{j=1}^{2}\dfrac{\partial}{\partial x_j}F_j(v^\epsilon) = \dfrac{1}{\epsilon}h(v^\epsilon) - \dfrac{1}{\epsilon^4}P(v^\epsilon, z^\epsilon) \\ \dfrac{\partial z^\epsilon}{\partial t} - \dfrac{1}{\epsilon^2}\nabla\cdot[\mathcal{A}^\epsilon\nabla z^\epsilon] \quad = \dfrac{1}{\epsilon^2}\nabla\cdot C^\epsilon \end{cases} \tag{1.1}$$

With $v^\epsilon = \begin{pmatrix} m^\epsilon \\ q_1^\epsilon \\ q_2^\epsilon \end{pmatrix}$, $h(v^\epsilon) = -\frac{f}{\epsilon}v^{\epsilon\,\perp} + \frac{1}{\epsilon}H(v^\epsilon)$ and

$$F(v^\epsilon) = (F_1, F_2) = \begin{pmatrix} a_1q_1 & a_1q_2 \\ \dfrac{a^1q_1^2}{m^\epsilon + b^1} + \dfrac{c^1}{2}(m^\epsilon + b^1)^2 & \dfrac{a^1q_1^\epsilon q_2^\epsilon}{m^\epsilon + b^1} \\ \dfrac{a^1q_1^\epsilon q_2^\epsilon}{m^\epsilon + b^1} & \dfrac{a^1q_2^2}{m^\epsilon + b^1} + \dfrac{c^1}{2}(m^\epsilon + b^1)^2 \end{pmatrix}$$

with

$$F_1(v^\epsilon) = \begin{pmatrix} F_1^1 \\ F_1^2 \\ F_1^3 \end{pmatrix} = \begin{pmatrix} a_1q_1 \\ \dfrac{a^1q_1^2}{m^\epsilon + b^1} + \dfrac{c^1}{2}(m^\epsilon + b^1)^2 \\ \dfrac{a^1q_1^\epsilon q_2^\epsilon}{m^\epsilon + b^1} \end{pmatrix},$$

$$F_2(v^\epsilon) = \begin{pmatrix} F_2^1 \\ F_2^2 \\ F_2^3 \end{pmatrix} = \begin{pmatrix} a_1q_2 \\ \dfrac{a^1q_1^\epsilon q_2^\epsilon}{m^\epsilon + b^1} \\ \dfrac{a^1q_2^2}{m^\epsilon + b^1} + \dfrac{c^1}{2}(m^\epsilon + b^1)^2 \end{pmatrix},$$

$$H(v^\epsilon) = \begin{pmatrix} 0 \\ -\dfrac{kq_1}{m^\epsilon + b^1} \\ -\dfrac{kq_2}{m^\epsilon + b^1} \end{pmatrix}, \quad P(v^\epsilon, z^\epsilon) = \begin{pmatrix} 0 \\ d^1(m + b^1)\dfrac{\partial z^\epsilon}{\partial x} \\ d^1(m + b^1)\dfrac{\partial z^\epsilon}{\partial y} \end{pmatrix}$$

$$\mathcal{A}^\epsilon = \frac{a(1 - b\epsilon m^\epsilon)|q^\epsilon|^3}{|m^\epsilon + b^1|^3} \quad \text{and} \quad C^\epsilon = \frac{c(1 - b\epsilon m^\epsilon)|q^\epsilon|^2 q^\epsilon}{(m^\epsilon + b^1)^3}$$

The equation

$$\frac{\partial v^\epsilon}{\partial t} + \frac{1}{\epsilon^4}\sum_{j=1}^{2}\frac{\partial}{\partial x_j}F_j(v^\epsilon) = \frac{1}{\epsilon}h(v^\epsilon) - \frac{1}{\epsilon^4}P(v^\epsilon, z^\epsilon) \tag{1.2}$$

is a first order hyperbolic system of balance law and while

$$\frac{\partial z^\epsilon}{\partial t} - \frac{1}{\epsilon^2}\nabla\cdot[\mathcal{A}^\epsilon\nabla z^\epsilon] = \frac{1}{\epsilon^2}\nabla\cdot C^\epsilon \tag{1.3}$$

is a parabolic equation, which may become singular or degenerate when ϵ take particular values: for example when $\epsilon = 0$ we are in front of a singular case, and when $\mathcal{A}^\epsilon = 0$ for some values of ϵ the Eq. (1.3) is degenerated.
If we suppose that v^ϵ is smooth enough at least in the sense of distributions, then:

$$\frac{\partial}{\partial x_j}F_j(v^\epsilon) = D_v F_j(v^\epsilon)\cdot\frac{\partial v^\epsilon}{\partial x_j},\ \forall\, j = 1, 2.$$

And the system becomes:

$$\frac{\partial v^\epsilon}{\partial t} + \frac{1}{c^4}\sum_{j=1}^{2} D_v F_j(v^\epsilon)\cdot\frac{\partial v^\epsilon}{\partial x_j} + \frac{1}{\epsilon}\tilde{A}_0 v^{\epsilon\,\perp} = \frac{1}{\epsilon}H(v^\epsilon) - \frac{1}{\epsilon^4}P(v^\epsilon, z^\epsilon) \tag{1.4}$$

with $D_v F_j(v^\epsilon) = \begin{pmatrix} \frac{\partial}{\partial m^\epsilon}F_j^1 & \frac{\partial}{\partial q_1^\epsilon}F_j^1 & \frac{\partial}{\partial q_2^\epsilon}F_j^1 \\ \frac{\partial}{\partial m^\epsilon}F_j^2 & \frac{\partial}{\partial q_1^\epsilon}F_j^2 & \frac{\partial}{\partial q_2^\epsilon}F_j^2 \\ \frac{\partial}{\partial m^\epsilon}F_j^3 & \frac{\partial}{\partial q_1^\epsilon}F_j^3 & \frac{\partial}{\partial q_2^\epsilon}F_j^3 \end{pmatrix}$

After calculations, we get:

$$D_v F_1(v^\epsilon) = \begin{pmatrix} 0 & a^1 & 0 \\ -\frac{a^1 q_1^{\epsilon\,2}}{(m^\epsilon + b^1)^2} + c^1(m^\epsilon + b^1) & \frac{2a^1 q_1}{(m^\epsilon + b^1)} & 0 \\ -\frac{a^1 q_1 q_2}{(m^\epsilon + b^1)^2} & \frac{a^1 q_2}{(m^\epsilon + b^1)} & \frac{a^1 q_1}{(m^\epsilon + b^1)} \end{pmatrix}$$

$$D_v F_2(v^\epsilon) = \begin{pmatrix} 0 & 0 & a^1 \\ -\frac{a^1 q_1 q_2}{(m^\epsilon + b^1)^2} & \frac{a^1 q_2}{(m^\epsilon + b^1)} & \frac{a^1 q_1}{(m^\epsilon + b^1)} \\ -\frac{a^1 q_2^{\epsilon\,2}}{(m^\epsilon + b^1)^2} + c^1(m^\epsilon + b^1) & 0 & \frac{2a^1 q_2}{(m^\epsilon + b^1)} \end{pmatrix}$$

Setting $A^j(v^\epsilon) = D_v F_j(v^\epsilon), \forall\, j = 1, 2$, we have:

$$\frac{\partial v^\epsilon}{\partial t} + \frac{1}{\epsilon^4}\sum_{j=1}^{2} A^j \cdot \frac{\partial v^\epsilon}{\partial x_j} = \frac{1}{\epsilon} h(v^\epsilon) - \frac{1}{\epsilon^4} P(v^\epsilon, z^\epsilon) \tag{1.5}$$

It is well known that standard methods that solve correctly systems of conservation laws can fail in solving systems of balance laws, specially when approaching equilibria or near to equilibria solutions. Moreover, they can produce unstable methods when they are applied to coupled systems of conservation or balance laws. Many authors have studied well-balanced numerical schemes for balance laws or coupled systems (see [4] and [5]).

One of these techniques consists on discretizing the conservative terms of the system by using a well-known solver for homogeneous conservative systems and studying a discretization of the source term and/or the non-conservative terms. To do the numerical simulations we use a splitting scheme which consists on solving separately the shallow water equation and the LTDD equation during a same time step Δt, but using the same approximation of v^ϵ and z^ϵ in both equations.

2 Finite Volume Scheme for the Dimensionless Coupled System (SW-LTDD)

We consider a two-dimensional admissible mesh V of the torus $\mathbb{T}^2$ or a square domain that means there is a gravity center x_i^c in each control volume V_i such that (x_i^c, x_j^c) is orthogonal with the ridge $V_i|V_j = \overline{V}_i \cap \overline{V}_j = \Gamma_{ij} = \sigma$ and let's set $d_{ij} = \left\| x_i^c - x_j^c \right\|$, $j \in J_i$. This condition is necessary to obtain a consistent approximation of the diffusion flux in the parabolic diffusion equation LTDD.
$\bigcup_j \Gamma_{ij}$ is the boundary of V_i and n_{ij} is the normal vector to Γ_{ij} pointing outside to V_i.
$\partial V_i = \bigcup_{j\in J_i} \Gamma_{ij} = \bigcup_{\sigma\in\xi_i} \Gamma_{ij}$, where J_i is the set of ridge of V_i and $\xi_i = \xi_i^{int} \cup \xi_i^{ext}$, with ξ_i^{int} being the set of interior boundary of V_i and ξ_i^{ext} the set of exterior boundary of V_i.
We consider $h = (\sup_{V_i\in V} |V_i|)^{1/2}$ or $h = (\sup_{V_i\in V} diam(V_i))^{1/2}$. The time discretization is performed with constant time step Δt and $t_n = n\Delta t$.
Let us consider two adjacent control volumes V_i and V_j as in the Fig. 1. We can write the outward normal vector n_{ij} of V_i (orthogonal to Γ_{ij}) as follows:

$$n_{ij} = \frac{x_j^c - x_i^c}{\left\| x_j^c - x_i^c \right\|},$$

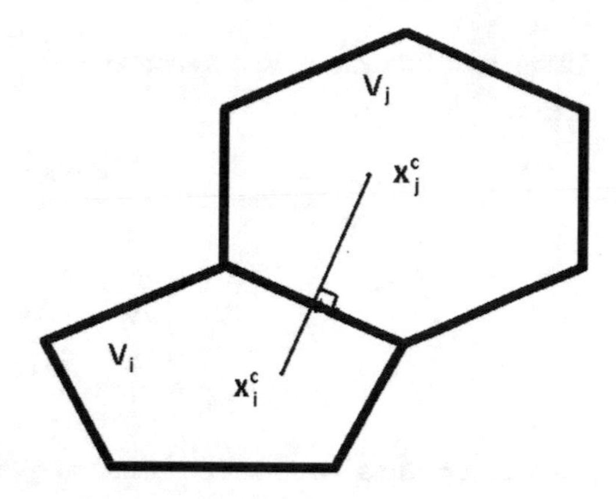

Fig. 1 Euclidian mesh: two adjacent control volumes V_i and V_j

2.1 Finite Volume Scheme for the Dimensionless Shallow Water Equation (SWE)

We perform in this subsection an explicit formulation of finite volume scheme for dimensionless SWE. For this we consider a control volume V_i, denote by

$$v_i^{\epsilon,n} = \frac{1}{|V_i|}\int_{V_i} v^{\epsilon}(t_n, x)dx, \tag{2.1}$$

$$v_{ij}^{\epsilon,n} = \frac{1}{|\Gamma_{ij}|}\int_{\Gamma_{ij}} v^{\epsilon}(t_n, \sigma)d\sigma, \tag{2.2}$$

the approximation of $v^{\epsilon,n}$ respectively in V_i and Γ_{ij}. Where $|\Gamma_{ij}|$ is the length of Γ_{ij} and $|V_i|$ is the area of V_i.

Integrating (1.2) on V_i we get:

$$\begin{aligned}\int_{V_i} \frac{\partial v^{\epsilon}}{\partial t}(t, x) + \frac{1}{\epsilon^4}\sum_{j=1}^{2}\int_{V_i} \frac{\partial}{\partial x_j} F_j(v^{\epsilon}(t, x)) = \\ \frac{1}{\epsilon}\int_{V_i} h(v^{\epsilon}(t, x)) - \frac{1}{\epsilon^4}\int_{V_i} P(v^{\epsilon}(t, x), z^{\epsilon}(t, x))\end{aligned} \tag{2.3}$$

Replacing t by t_n in (2.3) we have :

$$\begin{aligned}\int_{V_i} \frac{\partial v^{\epsilon}}{\partial t}(t_{n+1}, x) + \frac{1}{\epsilon^4}\sum_{j=1}^{2}\int_{V_i} \frac{\partial}{\partial x_j} F_j(v^{\epsilon}(t_n, x)) = \\ \frac{1}{\epsilon}\int_{V_i} h(v^{\epsilon}(t_n, x)) - \frac{1}{\epsilon^4}\int_{V_i} P(v^{\epsilon}(t_n, x), z^{\epsilon}(t_n, x))\end{aligned} \tag{2.4}$$

Using finite differences, we can approximate $\frac{\partial v^\epsilon}{\partial t}(t_n, x) \simeq \frac{v^\epsilon(t_{n+1}, x) - v^\epsilon(t_n, x)}{\Delta t}$.
Hence

$$\begin{aligned}\int_{V_i} \frac{\partial v^\epsilon}{\partial t}(t_n, x) &\simeq \int_{V_i} \frac{v^\epsilon(t_{n+1}, x) - v^\epsilon(t_n, x)}{\Delta t} \\ &\simeq \int_{V_i} \frac{v^\epsilon(t_{n+1}, x)}{\Delta t} - \int_{V_i} \frac{v^\epsilon(t_n, x)}{\Delta t} \\ &\simeq \frac{|V_i|}{\Delta t}(v_i^{\epsilon,n+1} - v_i^{\epsilon,n})\end{aligned}$$

For the approximation of the non linear transport term yields:

$$\begin{aligned}\frac{1}{\epsilon^4}\sum_{k=1}^{2}\int_{V_i}\frac{\partial}{\partial x_k}F_k(v^{\epsilon,n}) &= \frac{1}{\epsilon^4}\int_{\partial V_i}\sum_{k=1}^{2}F_k(v^{\epsilon,n})n_{ij}^k d\sigma \\ &= \frac{1}{\epsilon^4}\sum_{j\in J_i}\int_{\Gamma_{ij}}\sum_{k=1}^{2}F_k(v^{\epsilon,n})n_{ij}^k d\sigma\end{aligned}$$

Let's set $\Phi_{ij}^n = \Phi(v_i^{\epsilon,n}, v_j^{\epsilon,n}, n_{ij}) \simeq \frac{1}{|\Gamma_{ij}|}\int_{\Gamma_{ij}}\sum_{k=1}^{2}F_k(v^{\epsilon,n})n_{ij}d\sigma$ called numerical flux. Then (2.3) becomes:

$$\frac{|V_i|}{\Delta t}(v_i^{\epsilon,n+1} - v_i^{\epsilon,n}) + \frac{1}{\epsilon^4}\sum_{j\in J_i}|\Gamma_{ij}|\Phi_{ij}^n = \frac{1}{\epsilon}|V_i|h_i^n - \frac{1}{\epsilon^4}|V_i|(P(v^{\epsilon,n}, z^{\epsilon,n}))_i \tag{2.5}$$

where $h_i^n \simeq \frac{1}{|V_i|}\int_{V_i} h(v^\epsilon(t_n, x))$, and $(P(v^{\epsilon,n}, z^{\epsilon,n}))_i \simeq \int_{V_i} P(v^\epsilon(t_n, x), z^\epsilon(t_n, x))$
Multiplying by $\frac{\Delta t}{|V_i|}$ we get:

$$v_i^{\epsilon,n+1} - v_i^{\epsilon,n} = -\frac{1}{\epsilon^4}\frac{\Delta t}{|V_i|}\sum_{j\in J_i}|\Gamma_{ij}|\Phi_{ij}^n + \frac{1}{\epsilon}\Delta t\cdot h_i^n - \frac{1}{\epsilon^4}\Delta t\cdot(P(v^{\epsilon,n}, z^{\epsilon,n}))_i \tag{2.6}$$

Let's point out that we can fix following different methods, the value of Φ_{ij}^n. For this work we consider the Rusanov method:

$$\Phi(v_i^{\epsilon,n}, v_j^{\epsilon,n}, n_{ij}) = \sum_{k=1}^{2}\frac{F_k(v_i^{\epsilon,n}) + F_k(v_j^{\epsilon,n})}{2}n_{ij}^k - \frac{1}{2}S_{ij}(v_j^{\epsilon,n} - v_i^{\epsilon,n}),$$ where

$n_{ij} = (n_{ij}^1, n_{ij}^2)$, and $S_{ij} = S(v_{ij/2}^{\epsilon,n}, n_{ij})I$ is called numerical viscosity, where I is the identity matrix and S the maximum in absolute value of the eigenvalues of matrix

$$\mathcal{A}_{ij} = \mathcal{A}(v^{\epsilon,n}_{ij/2}, n_{ij}) = \sum_{k=1}^{2} A^k(v^{\epsilon,n}_{ij/2}) n^k_{ij}$$

S_{ij} is also known as the viscosity matrix or flux, A^k being the Jacobian matrices of F^k, $k = 1; 2$ and $v^{\epsilon,n}_{ij/2}$ is the intermediate Roe state. Hence with our choice of mesh we have:

$$\begin{aligned}\Phi(v^{\epsilon,n}_i, v^{\epsilon,n}_j, n_{ij}) &= \frac{1}{2}\mathcal{A}(v^{\epsilon,n}_{ij/2}, n_{ij})(v^{\epsilon,n}_j - v^{\epsilon,n}_i) \\ &\quad - \frac{1}{2} S_{ij}(v^{\epsilon,n}_{ij/2}, n_{ij})(v^{\epsilon,n}_j - v^{\epsilon,n}_i) \\ &= \frac{1}{2}\left(\mathcal{A}(v^{\epsilon,n}_{ij/2}, n_{ij}) - S_{ij}(v^{\epsilon,n}_{ij/2}, n_{ij})\right)(v^{\epsilon,n}_j - v^{\epsilon,n}_i)\end{aligned}$$

In order to approximate $\frac{1}{\epsilon^4}\int_{V_i} P(v^\epsilon(t_n, x), z^\epsilon(t_n, x))$, we introduce the following notation.

$\mathcal{L}(v^{\epsilon,n}) := d^1(m^{\epsilon,n} + b^1)$, then $P(v^{\epsilon,n}, z^{\epsilon,n}) = \begin{pmatrix} 0 \\ \mathcal{L}(v^{\epsilon,n})\nabla z^{\epsilon,n} \end{pmatrix}$.

$$\mathcal{L}(v^{\epsilon,n})_i = \frac{1}{|V_i|}\int_{V_i} \mathcal{L}(v^{\epsilon,n}) dx = \mathcal{L}(v^{\epsilon,n}_i).$$

$$(\nabla z^{\epsilon,n})_i = \frac{1}{|V_i|}\int_{V_i} \nabla z^{\epsilon,n} dx = \frac{1}{|V_i|}\sum_{j \in J_i} z^{\epsilon,n}_{ij} \left|\Gamma_{ij}\right|$$

$$\begin{aligned}\int_{V_i} \mathcal{L}(v^{\epsilon,n})\nabla z^{\epsilon,n} dx &= |Vi|\,(\mathcal{L}(v^{\epsilon,n})\nabla z^{\epsilon,n})_i \\ &\simeq |Vi|\,(\mathcal{L}(v^{\epsilon,n}))_i(\nabla z^{\epsilon,n})_i \\ &\simeq \mathcal{L}(v^{\epsilon,n}_i)\sum_{j \in J_i} z^{\epsilon,n}_{ij} \left|\Gamma_{ij}\right| n_{ij}.\end{aligned}$$

We consider $z^{\epsilon,n}_{ij} = (z^{\epsilon,n}_j - z^{\epsilon,n}_i)$ as the quantity of sand which pass through the edge Γ_{ij} in the direction of the normal vector n_{ij} of the edge (from the control volume V_i to the control volume V_j).

And finally we get:

$$\begin{aligned}v^{\epsilon,n+1}_i - v^{\epsilon,n}_i &= -\frac{1}{\epsilon^4}\frac{\Delta t}{2|V_i|}\sum_{j \in J_i} |\Gamma_{ij}|(\mathcal{A}_{ij} n_{ij} - S_{ij} n_{ij})(v^{\epsilon,n}_j - v^{\epsilon,n}_i) + \\ &\quad \frac{1}{\epsilon}\Delta t \cdot h^{\epsilon,n}_i - \frac{1}{\epsilon^4}\begin{pmatrix} 0 \\ \mathcal{L}(v^{\epsilon,n}_i)\sum_{j \in J_i}(z^{\epsilon,n}_j - z^{\epsilon,n}_i)\left|\Gamma_{ij}\right| n_{ij}\end{pmatrix}\end{aligned} \tag{2.7}$$

2.2 *Finite Volume Scheme for the Dimensionless Long Term Dynamic of Dunes of sand (LTDD)*

As in the previous section we take the Eq. (1.3) at $t = t_n$ to obtain:

$$\frac{\partial z^\epsilon}{\partial t}(t_n, x) - \frac{1}{\epsilon^2}\nabla \cdot [\mathcal{A}^\epsilon \nabla z^\epsilon(t_n, x)] = \frac{1}{\epsilon^2}\nabla \cdot \mathcal{C}^\epsilon(t_n, x)$$

Integrating the above equation on V_i we have:

$$\int_{V_i} \frac{\partial z^\epsilon}{\partial t}(t_n, x) - \frac{1}{\epsilon^2}\int_{V_i} \nabla \cdot [\mathcal{A}^\epsilon(v^\epsilon(t_n, x))\nabla z^\epsilon(t_n, x)] = \frac{1}{\epsilon^2}\int_{V_i} \nabla \cdot \mathcal{C}^\epsilon(t_n, x)$$

Now by using finite difference and integrating on control volume V_i, we get

$$\begin{aligned}\int_{V_i} \frac{\partial z^\epsilon}{\partial t}(t_n, x) &\simeq \int_{V_i} \frac{z^\epsilon(t_{n+1}, x) - z^\epsilon(t_n, x)}{\Delta t} \\ &\simeq \frac{|V_i|}{\Delta t}(z_i^{\epsilon,n+1} - z_i^{\epsilon,n})\end{aligned}$$

with $z_i^{\epsilon,n} = \frac{1}{|V_i|}\int_{V_i} z^\epsilon(t_n, x)$. Using Stokes formula the integral of the diffusion term becomes:

$$\begin{aligned}\frac{1}{\epsilon^2}\int_{V_i} \nabla \cdot [\mathcal{A}^\epsilon(v^\epsilon(t_n, x))\nabla z^\epsilon(t_n, x)] &= \frac{1}{\epsilon^2}\int_{\partial V_i} \mathcal{A}^\epsilon(v^{\epsilon,n}(x))\nabla z^{\epsilon,n}(x) d\sigma \\ &= \frac{1}{\epsilon^2}\sum_{j\in J_i}\int_{\Gamma_{ij}} \mathcal{A}^\epsilon(v^{\epsilon,n}(x))\nabla z^{\epsilon,n}(x) n_{ij} d\sigma,\end{aligned}$$

Let's set $G_{ij}(z^{\epsilon,n}, v^{\epsilon,n}) = \int_{\Gamma_{ij}} \mathcal{A}^\epsilon(v^{\epsilon,n}(x))\nabla z^{\epsilon,n}(x) n_{ij} d\sigma$. The next step is to have an approximation of $G_{ij}(z^{\epsilon,n}, v^{\epsilon,n})$. For this purpose, we need to apply the Taylor approximation as in [7] to $z^{\epsilon,n}$. First, let's do the expansion of $z^{\epsilon,n}$ in V_i, where $z_\sigma^{\epsilon,n}$ is the approximation of $z^{\epsilon,n}$ in the ridge $V_i|V_j$:

$$z_\sigma^{\epsilon,n} - z_i^{\epsilon,n} = \nabla z^{\epsilon,n}(x)(x_\sigma - x_i^c) + \int_{x_i^c}^{x_\sigma} (x_\sigma - x)\nabla^2 z^{\epsilon,n}(x)(x_\sigma - x)dx \tag{2.8}$$

We can write also the same formula for the control volume V_j and get:

$$z_\sigma^{\epsilon,n} - z_j^{\epsilon,n} = \nabla z^{\epsilon,n}(x)(x_\sigma - x_j^c) + \int_{x_j^c}^{x_\sigma} (x_\sigma - x)\nabla^2 z^{\epsilon,n}(x)(x_\sigma - x)dx \tag{2.9}$$

Doing the difference between (2.8) and (2.9), we have:

$$z_j^{\epsilon,n} - z_i^{\epsilon,n} = \nabla z^{\epsilon,n}(x)(x_j^c - x_i^c) + \int_{x_i^c}^{x_j^c} (x_\sigma - x)\nabla^2 z^{\epsilon,n}(x)(x_\sigma - x)dx \tag{2.10}$$

Setting $H_\sigma(x) = (x_\sigma - x)\nabla^2 z^{\epsilon,n}(x)(x_\sigma - x)$, we have:

$$\frac{z_j^{\epsilon,n} - z_i^{\epsilon,n}}{\left\|x_j^c - x_i^c\right\|} = \nabla z^{\epsilon,n}(x)n_{ij} + \frac{1}{\left\|x_j^c - x_i^c\right\|}\int_{x_i^c}^{x_j} H_\sigma(x)dx$$

$$\mathcal{A}^\epsilon(v^{\epsilon,n})\frac{z_j^{\epsilon,n} - z_i^{\epsilon,n}}{\left\|x_j^c - x_i^c\right\|} = \mathcal{A}^\epsilon(v^{\epsilon,n})\nabla z^{\epsilon,n}(x)n_{ij} + \frac{\mathcal{A}^\epsilon(v^{\epsilon,n})}{\left\|x_j^c - x_i^c\right\|}\int_{x_i^c}^{x_j} H_\sigma(x)dx$$

$$\int_{\Gamma_{ij}} \mathcal{A}^\epsilon(v^{\epsilon,n})\frac{z_j^{\epsilon,n} - z_i^{\epsilon,n}}{\left\|x_j^c - x_i^c\right\|}d\sigma = \int_{\Gamma_{ij}} \mathcal{A}^\epsilon(v^{\epsilon,n})\nabla z^{\epsilon,n}(x)n_{ij}d\sigma$$

$$+ \int_{\Gamma_{ij}} \frac{\mathcal{A}^\epsilon(v^{\epsilon,n})}{\left\|x_j^c - x_i^c\right\|}\int_{x_i^c}^{x_j} H_\sigma(x)dxd\sigma$$

We can prove that $\int_{\Gamma_{ij}} \frac{\mathcal{A}^\epsilon(v^{\epsilon,n})}{\left\|x_j^c - x_i^c\right\|}\int_{x_i^c}^{x_j} H_\sigma(x)dxd\sigma$ is a $o(h^2)$, and consequently the approximation of the flux diffusion is given by:

$$G_{ij}(z^{\epsilon,n}, v^{\epsilon,n}) = \int_{\Gamma_{ij}} \mathcal{A}^\epsilon(v^{\epsilon,n})\nabla z^{\epsilon,n}(x)n_{ij}d\sigma \approx \int_{\Gamma_{ij}} \mathcal{A}^\epsilon(v^{\epsilon,n})\frac{z_j^{\epsilon,n} - z_i^{\epsilon,n}}{\left\|x_j^c - x_i^c\right\|}d\sigma$$

Now let's approximate $\int_{\Gamma_{ij}} \mathcal{A}^\epsilon(v^{\epsilon,n})\frac{z_j^{\epsilon,n} - z_i^{\epsilon,n}}{\left\|x_j^c - x_i^c\right\|}d\sigma$. For this we consider (2.2) to have for all $x \in \Gamma_{ij}$:

$$\mathcal{A}^\epsilon(v^{\epsilon,n}) \approx \mathcal{A}^\epsilon(v_{ij}^{\epsilon,n}) \text{ and then}$$

$$\int_{\Gamma_{ij}} \mathcal{A}^\epsilon(v^{\epsilon,n})d\sigma \approx \int_{\Gamma_{ij}} \mathcal{A}^\epsilon(v_{ij}^{\epsilon,n})d\sigma = \mathcal{A}^\epsilon(v_{ij}^{\epsilon,n})|\Gamma_{ij}|$$

$$\frac{1}{|\Gamma_{ij}|}\int_{\Gamma_{ij}} \mathcal{A}^\epsilon(v^{\epsilon,n})d\sigma \approx \int_{\Gamma_{ij}} \mathcal{A}^\epsilon(v_{ij}^{\epsilon,n})d\sigma = \mathcal{A}^\epsilon(v_{ij}^{\epsilon,n})$$

Then we have:

$$\int_{\Gamma_{ij}} \mathcal{A}^{\epsilon}(v^{\epsilon,n})\frac{z_j^{\epsilon,n}-z_i^{\epsilon,n}}{\left\|x_j^c-x_i^c\right\|}d\sigma = \int_{\Gamma_{ij}}(\mathcal{A}^{\epsilon}(v^{\epsilon,n})d\sigma)\frac{z_j^{\epsilon,n}-z_i^{\epsilon,n}}{\left\|x_j^c-x_i^c\right\|}$$

$$= \mathcal{A}^{\epsilon}(v_{ij}^{\epsilon,n})|\Gamma_{ij}|\frac{z_j^{\epsilon,n}-z_i^{\epsilon,n}}{\left\|x_j^c-x_i^c\right\|}$$

Finally we have:

$$\int_{\Gamma_{ij}} \mathcal{A}^{\epsilon}(v^{\epsilon,n})\nabla z^{\epsilon,n}(x)n_{ij}d\sigma \approx \mathcal{A}^{\epsilon}(v_{ij}^{\epsilon,n})|\Gamma_{ij}|\frac{z_j^{\epsilon,n}-z_i^{\epsilon,n}}{\left\|x_j^c-x_i^c\right\|}$$

It remains to give an approximation of $\frac{1}{\epsilon^2}\int_{V_i}\nabla\cdot\mathcal{C}^{\epsilon}(v^{\epsilon}(t_n,x))dx$.

In fact, $\frac{1}{\epsilon^2}\int_{V_i}\nabla\cdot\mathcal{C}^{\epsilon}(v^{\epsilon,n}(x)) = \frac{1}{\epsilon^2}\sum_{j\in J_i}\int_{\Gamma_{ij}}\mathcal{C}^{\epsilon}(v^{\epsilon,n}(x))n_{ij}(x)d\sigma$.

Using the same deduction as for $\mathcal{A}^{\epsilon}$ we have:

$\frac{1}{|\Gamma_{ij}|}\int_{\Gamma_{ij}}\mathcal{C}^{\epsilon}(v^{\epsilon,n}(x))n_{ij}(x)d\sigma = \mathcal{C}^{\epsilon}(v_{ij}^{\epsilon,n})$.

And finally we get an explicit time formulation of (1.3):

$$\frac{|V_i|}{\Delta t}(z_i^{\epsilon,n+1}-z_i^{\epsilon,n}) = \frac{1}{\epsilon^2}\sum_{j\in J_i}\mathcal{A}^{\epsilon}(v_{ij}^{\epsilon,n})|\Gamma_{ij}|\frac{z_j^{\epsilon,n}-z_i^{\epsilon,n}}{\left\|x_j^c-x_i^c\right\|}+\frac{1}{\epsilon^2}\sum_{j\in J_i}|\Gamma_{ij}|\mathcal{C}^{\epsilon}(v_{ij}^{\epsilon,n}) \tag{2.11}$$

This implies

$$z_i^{\epsilon,n+1}-z_i^{\epsilon,n} = \frac{1}{\epsilon^2}\frac{\Delta t}{|V_i|}\sum_{j\in J_i}\mathcal{A}^{\epsilon}(v_{ij}^{\epsilon,n})|\Gamma_{ij}|\frac{z_j^{\epsilon,n}-z_i^{\epsilon,n}}{\left\|x_j^c-x_i^c\right\|}+\frac{1}{\epsilon^2}\frac{\Delta t}{|V_i|}\sum_{j\in J_i}|\Gamma_{ij}|\mathcal{C}^{\epsilon}(v_{ij}^{\epsilon,n}) \tag{2.12}$$

2.3 *Stability of the Scheme*

To study the stability of the global finite volume scheme of SW-LTDD, we shall first do it for the LTDD scheme followed by the SWE scheme. And then we deduce a Courant-Friedrichs-Lewy (CFL) which satisfies to the coupled scheme.

Let's start by defining the following norm:

$\|v^{\epsilon,n}\|_{\infty} = \sup_{1\le i\le N}\left\|v_i^{\epsilon,n}\right\|$, with $v^{\epsilon,n} = (v_i^{\epsilon,n})_{1\le i\le N}\in\mathbb{R}^3\times\mathbb{R}^N$, where N is the number of control volume. For a scalar as $z^{\epsilon,n}$, $\|z^{\epsilon,n}\|_{\infty} = \sup_{1\le i\le N}\left|z_i^{\epsilon,n}\right|$ and for a function $M(v^{\epsilon,n})$, $\|M(v^{\epsilon,n})\|_{\infty} = \sup_{1\le i\le N}\left\|M(v_i^{\epsilon,n})\right\|$.

2.3.1 Stability of the LTDD Scheme

We have the following theorem:

Theorem 2.1 *For $\epsilon > 0$ small enough and the following CFL condition:*

$$\Delta t \cdot \max_{ij} \frac{|\Gamma_{ij}|}{d_{ij}\,|V_i|} \leq \frac{\epsilon^2}{\|\mathcal{A}^\epsilon\|_\infty}$$

then we have the L^∞ stability of the LTDD scheme in the sense that

$$\left\|z^{\epsilon,n+1}\right\|_\infty \leq \|z^{\epsilon,n}\|_\infty + C \tag{2.13}$$

where $C = \dfrac{c}{a}\max_{ij}(d_{ij})$, c and a are given respectively in the expressions of $\mathcal{C}^\epsilon$ and $\mathcal{A}^\epsilon$.

Proof Considering (2.12) we have:

$$\begin{aligned}
z_i^{\epsilon,n+1} &= z_i^{\epsilon,n} + \frac{1}{\epsilon^2}\frac{\Delta t}{|V_i|}\sum_{j\in J_i}\mathcal{A}^\epsilon(v_{ij}^{\epsilon,n})|\Gamma_{ij}|\frac{z_j^{\epsilon,n} - z_i^{\epsilon,n}}{\left\|x_j^c - x_i^c\right\|} + \frac{1}{\epsilon^2}\frac{\Delta t}{|V_i|}\sum_{j\in J_i}|\Gamma_{ij}|\mathcal{C}^\epsilon(v_{ij}^{\epsilon,n}) \\
&= z_i^{\epsilon,n} + \frac{1}{\epsilon^2}\frac{\Delta t}{|V_i|}\sum_{j\in J_i}\mathcal{A}^\epsilon(v_{ij}^{\epsilon,n})\frac{|\Gamma_{ij}|}{d_{ij}}(z_j^{\epsilon,n} - z_i^{\epsilon,n}) + \frac{1}{\epsilon^2}\frac{\Delta t}{|V_i|}\sum_{j\in J_i}|\Gamma_{ij}|\mathcal{C}^\epsilon(v_{ij}^{\epsilon,n}) \\
&= \frac{1}{\epsilon^2}\frac{\Delta t}{|V_i|}\sum_{j\in J_i}\mathcal{A}^\epsilon(v_{ij}^{\epsilon,n})\frac{|\Gamma_{ij}|}{d_{ij}}z_j^{\epsilon,n} + (1 - \frac{1}{\epsilon^2}\frac{\Delta t}{|V_i|}\sum_{j\in J_i}\mathcal{A}^\epsilon(v_{ij}^{\epsilon,n})\frac{|\Gamma_{ij}|}{d_{ij}})z_i^{\epsilon,n} \\
&\quad + \frac{1}{\epsilon^2}\frac{\Delta t}{|V_i|}\sum_{j\in J_i}|\Gamma_{ij}|\mathcal{C}^\epsilon(v_{ij}^{\epsilon,n})
\end{aligned}$$

Hence

$$\begin{aligned}
|z_i^{\epsilon,n+1}| \leq{}& \frac{1}{\epsilon^2}\frac{\Delta t}{|V_i|}\sum_{j\in J_i}\left|\mathcal{A}^\epsilon(v_{ij}^{\epsilon,n})\right|\frac{|\Gamma_{ij}|}{d_{ij}}\left|z_j^{\epsilon,n}\right| + \left|1 - \frac{1}{\epsilon^2}\frac{\Delta t}{|V_i|}\sum_{j\in J_i}\mathcal{A}^\epsilon(v_{ij}^{\epsilon,n})\frac{|\Gamma_{ij}|}{d_{ij}}\right||z_i^{\epsilon,n}| \\
&+ \frac{1}{\epsilon^2}\frac{\Delta t}{|V_i|}\sum_{j\in J_i}|\Gamma_{ij}|\left|\mathcal{C}^\epsilon(v_{ij}^{\epsilon,n})\right|
\end{aligned}$$

and finally

$$\begin{aligned}
|z_i^{\epsilon,n+1}| \leq{}& (\frac{1}{\epsilon^2}\frac{\Delta t}{|V_i|}\sum_{j\in J_i}\left|\mathcal{A}^\epsilon(v_{ij}^{\epsilon,n})\right|\frac{|\Gamma_{ij}|}{d_{ij}} + \left|1 - \frac{1}{\epsilon^2}\frac{\Delta t}{|V_i|}\sum_{j\in J_i}\mathcal{A}^\epsilon(v_{ij}^{\epsilon,n})\frac{|\Gamma_{ij}|}{d_{ij}}\right|)\,\|z^{\epsilon,n}\|_\infty \\
&+ \frac{1}{\epsilon^2}\frac{\Delta t}{|V_i|}\sum_{j\in J_i}|\Gamma_{ij}|\left|\mathcal{C}^\epsilon(v_{ij}^{\epsilon,n})\right|
\end{aligned} \tag{2.14}$$

If $\Delta t \cdot \max\limits_{ij} \dfrac{|\Gamma_{ij}|}{d_{ij}\,|V_i|} \leq \dfrac{\epsilon^2}{\|\mathcal{A}^\epsilon\|_\infty}$ then $\dfrac{1}{\epsilon^2}\Delta t \cdot \max\limits_{ij} \dfrac{|\Gamma_{ij}|}{d_{ij}\,|V_i|} \left\|\mathcal{A}^\epsilon\right\|_\infty \leq 1$. So we can deduce that $\dfrac{1}{\epsilon^2}\dfrac{\Delta t}{|V_i|}\sum\limits_{j\in J_i}\left|\mathcal{A}^\epsilon(v_{ij}^{\epsilon,n})\right|\dfrac{|\Gamma_{ij}|}{d_{ij}} \leq 1$ since $\dfrac{1}{\epsilon^2}\dfrac{\Delta t}{|V_i|}\sum\limits_{j\in J_i}\left|\mathcal{A}^\epsilon(v_{ij}^{\epsilon,n})\right|\dfrac{|\Gamma_{ij}|}{d_{ij}} \leq \dfrac{1}{\epsilon^2}\Delta t \cdot \max\limits_{ij} \dfrac{|\Gamma_{ij}|}{d_{ij}\,|V_i|}\left\|\mathcal{A}^\epsilon\right\|_\infty$, with multiplicative constant due to the summation.

About the positiveness of $\mathcal{A}^\epsilon(v_{ij}^{\epsilon,n})$, we have $\mathcal{A}^\epsilon(v^\epsilon) = \dfrac{a(1-\epsilon bm)\,|q^\epsilon|^3}{\left|m^\epsilon + b^1\right|^3} < 0$ only if $m^\epsilon > \dfrac{1}{\epsilon b}$, or for ϵ sufficiently small $\dfrac{1}{\epsilon b}$ goes to infinite and so m^ϵ which is a dimensionless water variation that can't be infinite. Then we can consider for ϵ sufficiently small $m^\epsilon < \dfrac{1}{\epsilon b}$ this implies $\mathcal{A}^\epsilon(v^\epsilon) > 0$.

Now (2.14) yields

$$|z_i^{\epsilon,n+1}| \leq \|z^{\epsilon,n}\|_\infty + \frac{1}{\epsilon^2}\frac{\Delta t}{|V_i|}\sum_{j\in J_i}|\Gamma_{ij}|\left|\mathcal{C}^\epsilon(v_{ij}^{\epsilon,n})\right| \tag{2.15}$$

And finally

$$\left\|z^{\epsilon,n+1}\right\|_\infty \leq \|z^{\epsilon,n}\|_\infty + \frac{1}{\epsilon^2}\frac{\Delta t}{|V_i|}\sum_{j\in J_i}|\Gamma_{ij}|\,\|\mathcal{C}^\epsilon\|_\infty \tag{2.16}$$

Considering the second term in the second member of (2.16) we have

$$\begin{aligned}
\frac{1}{\epsilon^2}\frac{\Delta t}{|V_i|}\sum_{j\in J_i}|\Gamma_{ij}|\,\|\mathcal{C}^\epsilon\|_\infty &\leq \frac{1}{\epsilon^2}\Delta t \cdot \max_{ij}\frac{|\Gamma_{ij}|}{|V_i|}\,\|\mathcal{C}^\epsilon\|_\infty \\
&\leq \frac{1}{\epsilon^2}\Delta t \cdot \max_{ij}\frac{\max_{ij}(d_{ij})\,|\Gamma_{ij}|}{d_{ij}\,|V_i|}\,\|\mathcal{C}^\epsilon\|_\infty \\
&\leq \frac{1}{\epsilon^2}\Delta t \cdot \max_{ij}(d_{ij})\max_{ij}\frac{|\Gamma_{ij}|}{d_{ij}\,|V_i|}\,\|\mathcal{C}^\epsilon\|_\infty \\
&\leq \frac{1}{\epsilon^2}\max_{ij}(d_{ij})\frac{\epsilon^2}{\|\mathcal{A}^\epsilon\|_\infty}\,\|\mathcal{C}^\epsilon\|_\infty \\
&\leq \max_{ij}(d_{ij})\frac{\|\mathcal{C}^\epsilon\|_\infty}{\|\mathcal{A}^\epsilon\|_\infty}
\end{aligned}$$

In addition,

$$\left\|\mathcal{A}^{\epsilon}(v^{\epsilon,n})\right\|_{\infty} = \sup_{1\le i\le N}\left|\frac{a(1-\epsilon b m_i^{\epsilon,n})\left\|q_i^{\epsilon,n}\right\|^3}{\left|m_i^{\epsilon,n}+b^1\right|^3}\right|$$

$$= \sup_{1\le i\le N}\frac{a\left|1-\epsilon b m_i^{\epsilon,n}\right|\left\|q_i^{\epsilon,n}\right\|^3}{\left|m_i^{\epsilon,n}+b^1\right|^3}$$

$$\left\|\mathcal{C}^{\epsilon}(v^{\epsilon,n})\right\|_{\infty} = \sup_{1\le i\le N}\left\|\frac{c(1-\epsilon b m_i^{\epsilon,n})\left\|q_i^{\epsilon,n}\right\|^2 q_i^{\epsilon,n}}{(m_i^{\epsilon,n}+b^1)^3}\right\|$$

$$= \sup_{1\le i\le N}\frac{c\left|1-\epsilon b m_i^{\epsilon,n}\right|\left\|q_i^{\epsilon,n}\right\|^3}{\left|m_i^{\epsilon,n}+b^1\right|^3}$$

Then $\dfrac{\|\mathcal{C}^{\epsilon}\|_{\infty}}{\|\mathcal{A}^{\epsilon}\|_{\infty}} = \dfrac{c}{a}$. Then $C = \max_{ij}(d_{ij})\dfrac{c}{a}$.
Finally with (2.16) we get (2.13). □

Remark 2.2

- It is well known that in practice (see [5]) for the stability of (2.7) it suffice to satisfy the so called Courant-Friedrichs- Lewy condition (CFL condition) adapted to our case:

$$CFL := \frac{\Delta t}{2\epsilon^4}\max_k\{\max_{ij}\lambda_{ij}^k\frac{|\Gamma_{ij}|}{|V_i|}\} \le 1. \tag{2.17}$$

- Using a finite volume scheme technique for solving hyperbolic system or balance law, the condition (2.17) can be seen as the definition of a time step small enough so that the different Riemann problems at each inter-cell do not interact between each other, that is, the information of each Riemann problem does not cross more than one cell (see [2]).

Theorem 2.3 *Under the CFL condition (2.17) the coupled scheme (2.7)–(2.12) is stable.*

Proof If the CFL (2.17) is satisfied then the scheme (2.7) is stable. In addition if ϵ is small enough, we have:

$$\frac{1}{\epsilon^2}\Delta t\cdot\max_{ij}\frac{\left|\Gamma_{ij}\right|}{d_{ij}\left|V_i\right|}\left\|\mathcal{A}^{\epsilon}\right\|_{\infty} \le \frac{\Delta t}{2\epsilon^4}\max_k\{\max_{ij}\lambda_{ij}^k\frac{|\Gamma_{ij}|}{|V_i|}\} \le 1,$$

$$\text{d'où } \frac{1}{\epsilon^2}\Delta t\cdot\max_{ij}\frac{\left|\Gamma_{ij}\right|}{d_{ij}\left|V_i\right|}\left\|\mathcal{A}^{\epsilon}\right\|_{\infty} \le 1.$$

Hence the scheme (2.12) is stable by the Theorem 2.1.

Finally the coupled scheme (2.7)–(2.12) is stable since the CFL condition (2.17) is satisfied. □

3 Numerical Simulation

In this section we do some simulations with the coupled scheme.

Let us fix the following parameters: $\epsilon = 0.0052$, $a^1 = 0.333$, $c^1 = 16$, $b = 4$, $a = 1$, $c = 20$, $d^1 = 3.33$, $fc = 3.8$, $kf = 19$, $CFL = 0.8$. The code of the program is done with Freefem++ 3.42 http://www.freefem.org/. The figures are realized with gnuplot version 5.0 patchlevel 3 http://www.gnuplot.info.

We give here four tests. In each test we have initial condition and fixed parameters. For all the tests we have performed the initial condition and all the others parameters are dimensionless. The first test concern the steady state at rest in order to verify if the numerical scheme conserves the well balanced property. In the second test the water variation is sinusoidal and water discharge is equal to zero. In the third test we consider as initial condition, a constant water variation and a flow of water equal to zero. In the fourth test we choose also a constant water variation but the flow of water is not equal to zero. Some of these tests are adapted to our situation from the tests given in [6].

Let us recall what situation we intend to simulate. The level of the water is given by $b^1 + \xi$, where ξ is the free level of the water. The height of the water from dune of sand to the free level of the water is given by $h = \xi + b^1 - z$. See Fig. 2. The time step is calculated with the CFL condition given by 2.17. We have fixed the CFL close to 1. So the time step is very small. The algorithm runs as follows:

- Calculation of the time step.
- Calculation of the approximation of $z^{\epsilon,n}$ using the LTDD scheme.
- Calculation of the approximation of $v^{\epsilon,n}$ by introducing in the SWE scheme the approximation of $z^{\epsilon,n}$ we have calculated before. And so on.

Remark 3.1 As the time step is calculated by using the CFL condition 2.17, it is very small since ϵ is small.

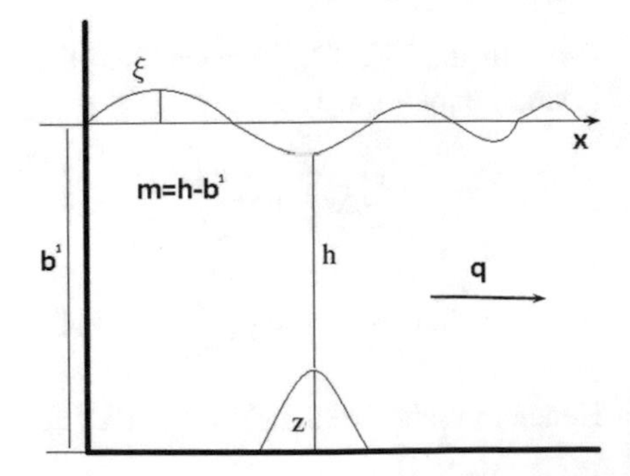

Fig. 2 Situation to simulate with unknowns: water flow variation, height of the dune, water discharge

3.1 First Test

Let us consider a square domain of $[0, 1] \times [0, 1]$, and the initial condition as follows (Fig. 3):

$$f(x_1, x_2) = \sin^2(\frac{\pi(x_1 - 0.4)}{0.2}) \sin^2(\frac{\pi(x_2 - 0.4)}{0.2})$$

$$z^0(x_1, x_2) = \begin{cases} 0.1 + f(x_1, x_2) & x \in [0.4, 0.6] \times [0.4, 0.6] \\ 0.1 & \text{otherwise} \end{cases}$$

$$m^0(x_1, x_2) = -z^0$$

$$q^0(x_1, x_2) = 0$$

The dimensionless mean height is $b^1 = 10$. In this situation the water level is constant, equal to $m^0 + z^0 + b^1 = b^1$, since $m^0 + z^0 = 0$ and the free level of the water $\xi^0 = 0$. We suppose that the dune can't be eroded out of the domain $[0.4, 0.6] \times [0.4, 0.6]$. In the domain $[0.4, 0.6] \times [0.4, 0.6]$ the minimal level of the dune is fixed to -0.1. We suppose also that $|m^{\epsilon,n} - m^0| < 1.5$.

Initially the water discharge $q = 0$ and the $m^0 + z^0 = 0$. If for all n $q^{\epsilon,n} = 0$ and the $m^{\epsilon,n} + z^{\epsilon,n} = 0$, then we say that the steady state at rest is preserved. The numerical scheme presented in the following section dœs not preserves immediately the steady state. It is necessary to add correction terms to the scheme to obtain a well balanced scheme.

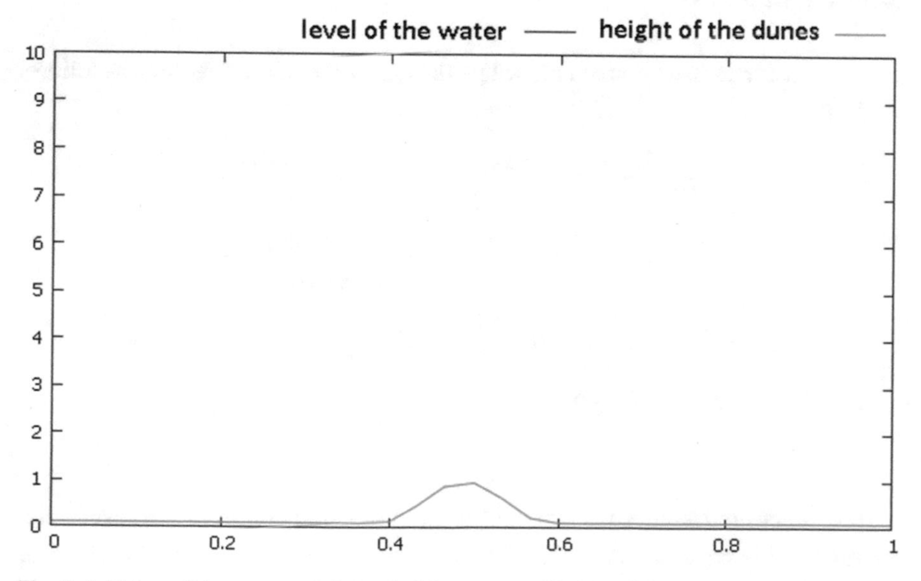

Fig. 3 Initial condition at $x_2 = 0.5$: level of the water and height of the dunes

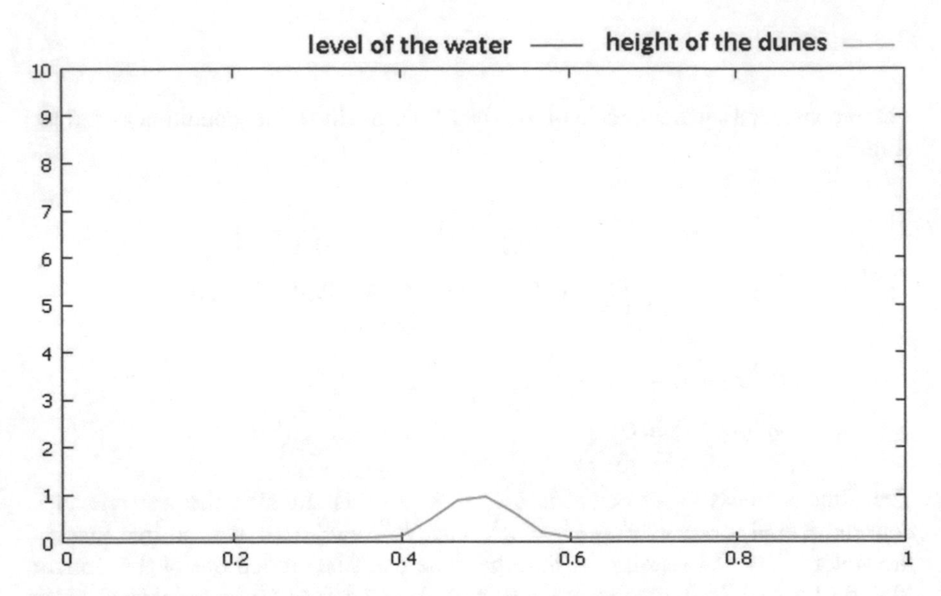

Fig. 4 Iteration 1500 at $x_2 = 0.5$: level of the water and height of the dunes

The result is given in the Fig. 4. We can see that the steady state at rest is preserved since we add correction terms in the scheme.

3.2 *Second Test*

Let us consider a square domain of $[0, 1] \times [0, 1]$, and the initial condition as follows (Fig. 5):

$$f(x_1, x_2) = \sin^2\left(\frac{\pi(x_1 - 4)}{2}\right)\sin^2\left(\frac{\pi(x_2 - 4)}{2}\right)$$

$$z^0(x_1, x_2) = \begin{cases} 0.1 + f(x_1, x_2) & x \in [4, 6] \times [4, 6] \\ 0.1 & \text{otherwise} \end{cases}$$

$$m^0(x_1, x_2) = 0.1\sin\left(\frac{\pi x_1}{2}\right) - z^0$$

$$q^0(x_1, x_2) = 0$$

The dimensionless mean height is $b^1 = 5$. In this situation the water level is not constant and is equal to $m^0 + z^0 + b^1 = 0.1\sin(\frac{\pi x_1}{2}) + b^1$. So the free level of

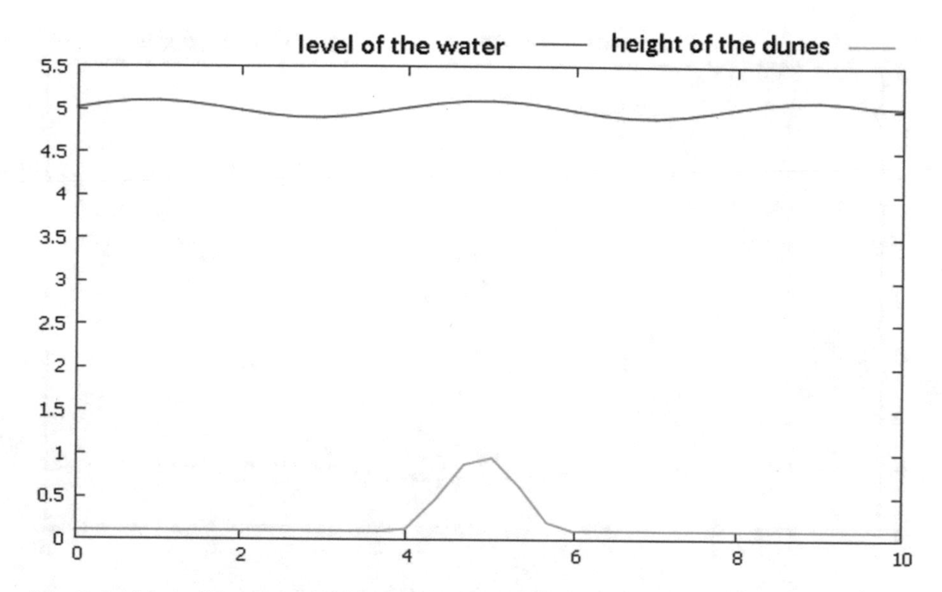

Fig. 5 Initial condition at $x_2 = 5$: level of the water and height of the dunes

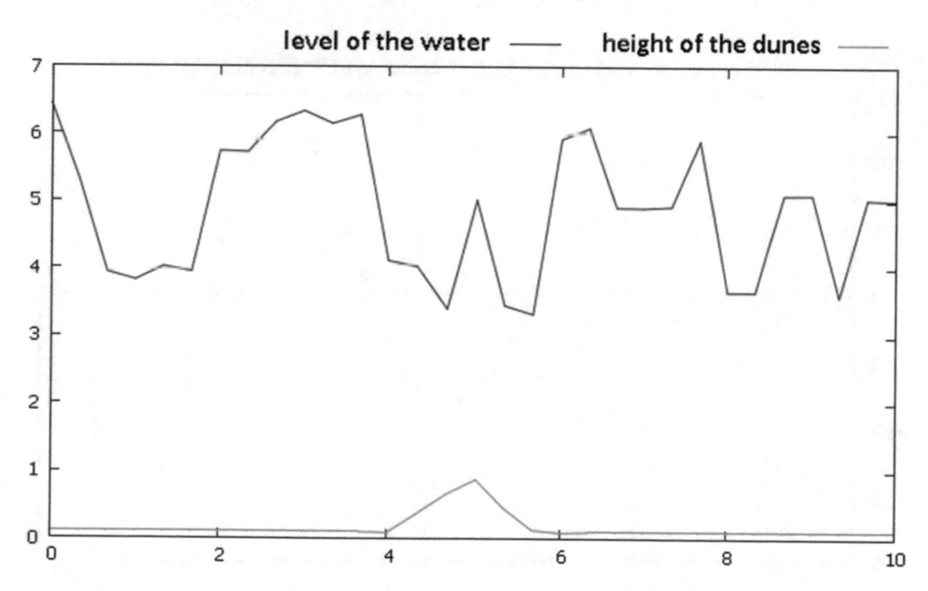

Fig. 6 Iteration 100 at $x_2 = 5$: level of the water and height of the dunes

the water $\xi^0 = 0.1 \sin(\frac{\pi x_1}{2})$. We suppose that the dune can't be eroded out of the domain $[4, 6] \times [4, 6]$. In the domain $[4, 6] \times [4, 6]$ the minimal level of the dune is fixed to -0.1. We suppose also that $|m^{\epsilon,n} - m^0| < 1.5$.

The Fig. 6 shows the evolution at 100 iterations and the Figs. 7, 8 and 9 at 1500 iterations.

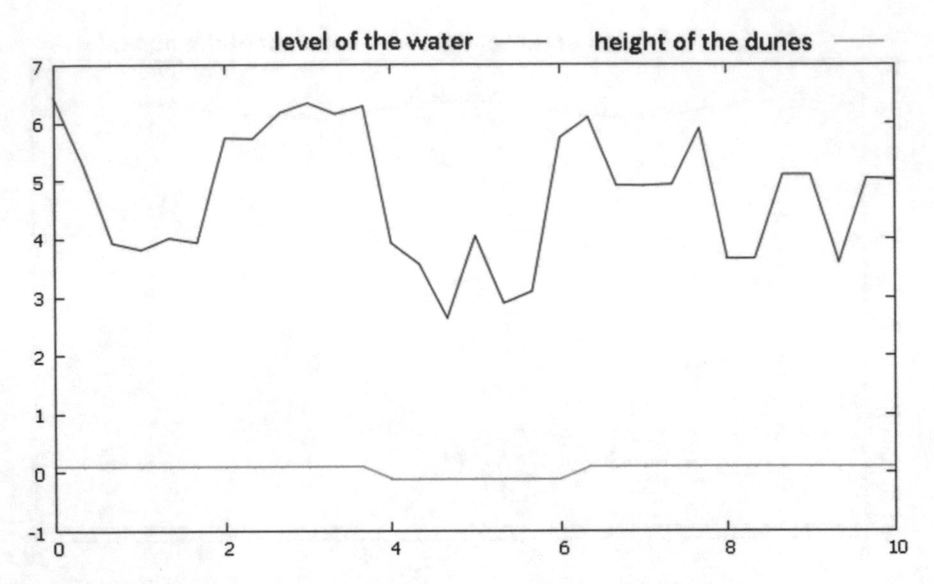

Fig. 7 Iteration 1500 at $x_2 = 5$: level of the water and height of the dunes

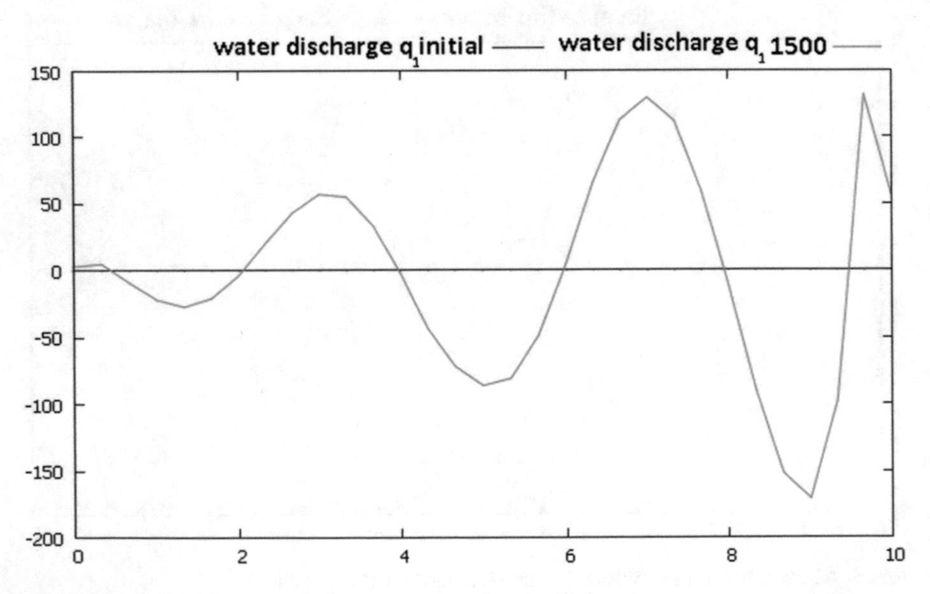

Fig. 8 Iteration 1500 at $x_2 = 5$: first component of water discharge q_1

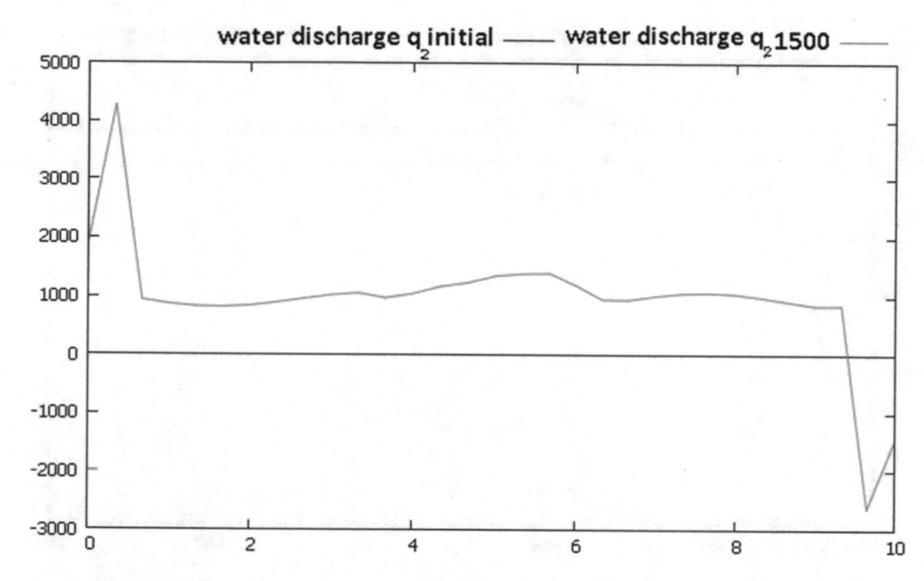

Fig. 9 Iteration 1500 at $x_2 = 5$: second component of water discharge q_2

3.3 *Third Test*

Let us consider now a square domain of 1000×1000, and the initial condition as follows (Fig. 10):

$$z^0(x_1, x_2) = \begin{cases} 0.1 + f(x_1, x_2) & x \in [300, 500] \times [400, 600] \\ 0.1 & \text{otherwise} \end{cases}$$

$$m^0(x_1, x_2) = 0$$

$$q_1^0(x_1, x_2) = 0 = q_2^0(x_1, x_2)$$

with $f(x_1, x_2) = \sin^2(\dfrac{\pi(x_1 - 300)}{200}) \sin^2(\dfrac{\pi(x_2 - 400)}{200})$.

We suppose that the dune can't be eroded out of the domain $[400, 600] \times [400, 600]$. In the domain $[400, 600] \times [400, 600]$ the minimal level of the dune is fixed to -0.1. We suppose also that $|m^{\epsilon,n} - m^0| < 1.5$.

The Fig. 11 shows the evolution at 100 iterations and the Figs. 12, 13 and 14 at 1500 iterations.

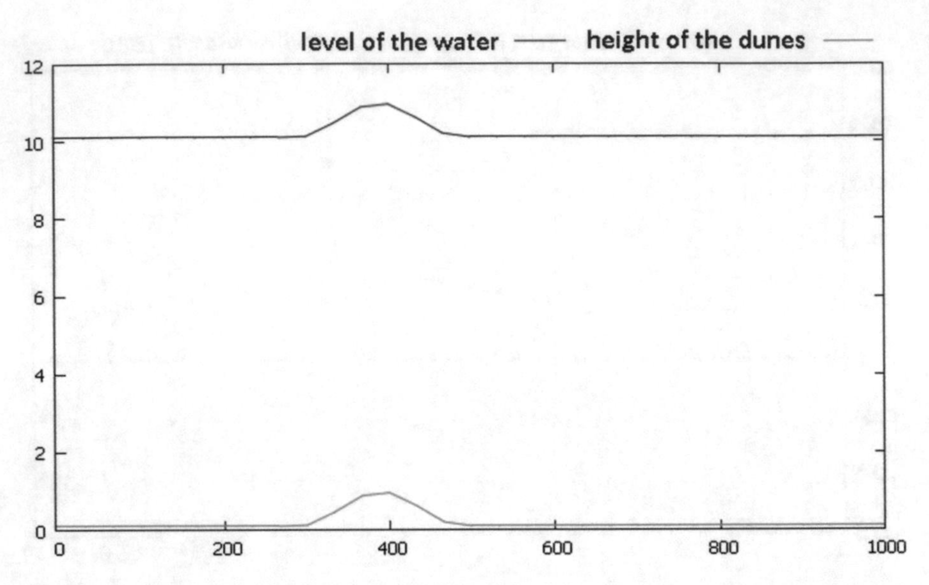

Fig. 10 Initial condition at $x_2 = 500$: level of the water and height of the dunes

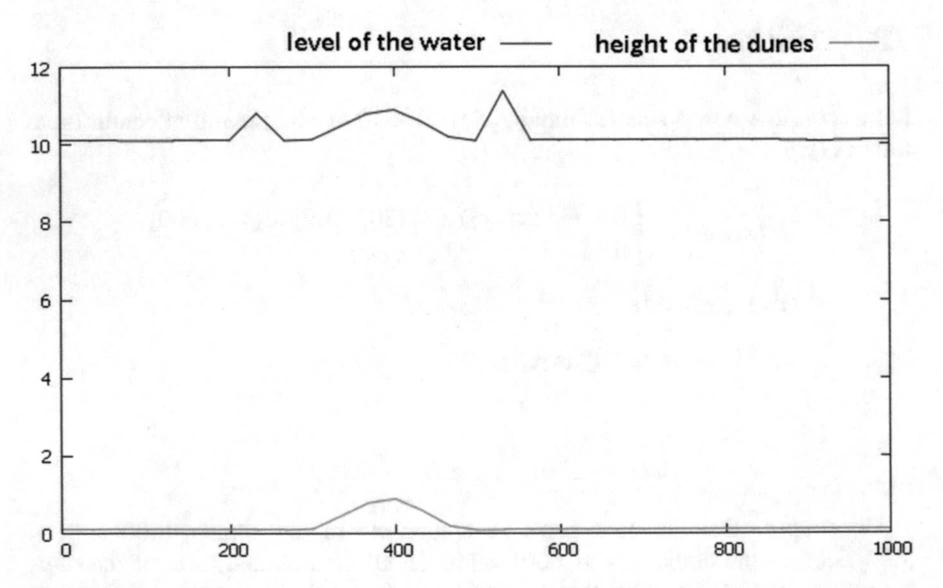

Fig. 11 Iteration 100 at $x_2 = 500$: level of the water and height of the dunes

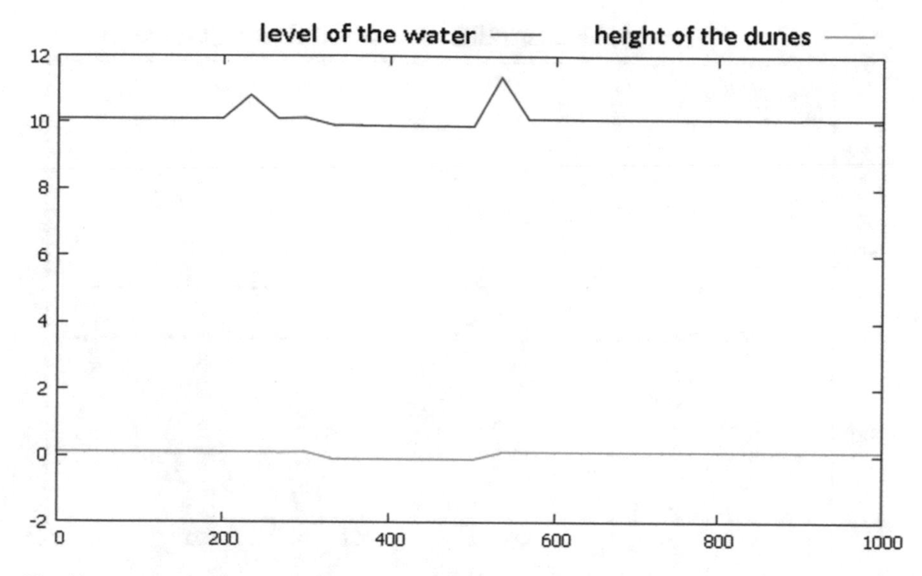

Fig. 12 Iteration 1500 at $x_2 = 500$: level of the water and height of the dunes

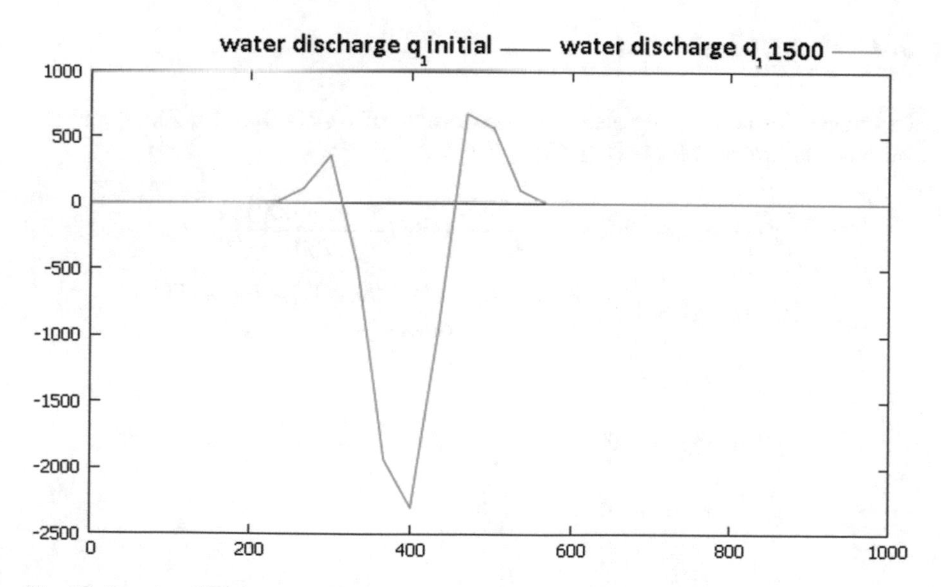

Fig. 13 Iteration 1500 at $x_2 = 500$: first component of water discharge q_1

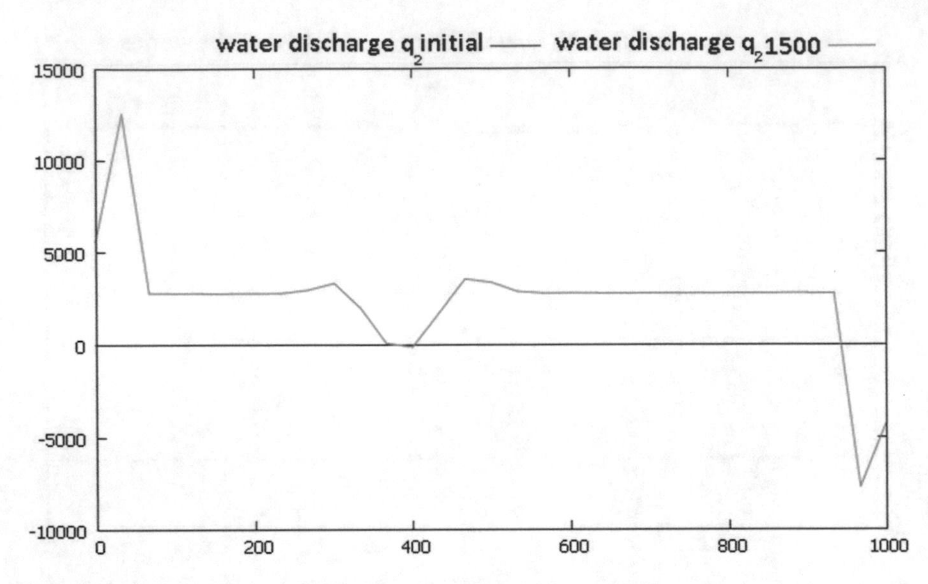

Fig. 14 Iteration 1500 at $x_2 = 500$: second component of water discharge q_2

3.4 *Fourth Test*

In this test let us consider also a square domain of $[0, 1000] \times [0, 1000]$, and the initial condition as follows (Fig. 15):

$$f(x_1, x_2) = \sin^2(\frac{\pi(x_1 - 300)}{200}) \sin^2(\frac{\pi(x_2 - 400)}{200})$$

$$z^0(x_1, x_2) = \begin{cases} 0.1 + f(x_1, x_2) & x \in [300, 500] \times [400, 600] \\ 0.1 & \text{otherwise} \end{cases}$$

$$m^0(x_1, x_2) = 0$$

$$q_1^0(x_1, x_2) = 1000 \sin(\frac{\pi x_2}{1000})$$

$$q_2^0(x_1, x_2) = -1000 \sin(\frac{\pi x_1}{1000})$$

We suppose that the dune can't be eroded out of the domain $[400, 600] \times [400, 600]$. In the domain $[400, 600] \times [400, 600]$ the minimal level of the dune is fixed to -0.1. We suppose also that $|m^{\epsilon,n} - m^0| < 1.5$.

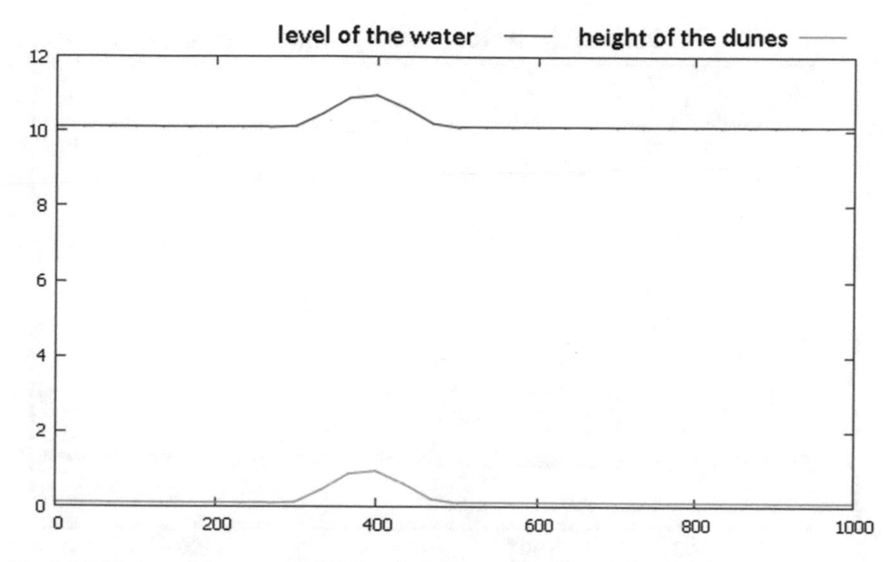

Fig. 15 Initial condition at $x_2 = 500$: level of the water and height of the dunes

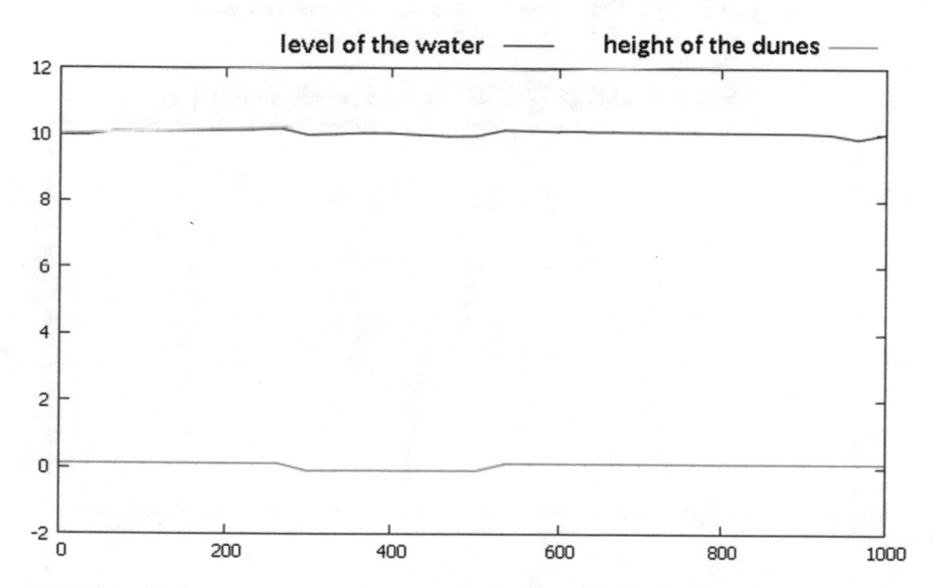

Fig. 16 Iteration 100 at $x_2 = 500$: level of the water and height of the dunes

The Fig. 16 shows the evolution at 100 iterations and the Figs. 17, 18 and 19 at 1500 iterations.

We can see that dunes and water level are hollow in areas with high water discharge.

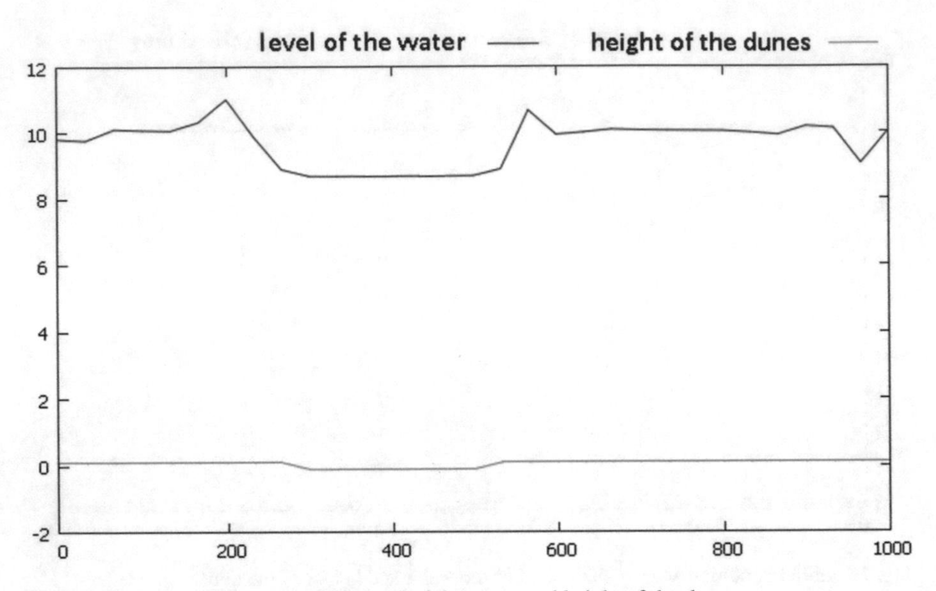

Fig. 17 Iteration 1500 at $x_2 = 500$: level of the water and height of the dunes

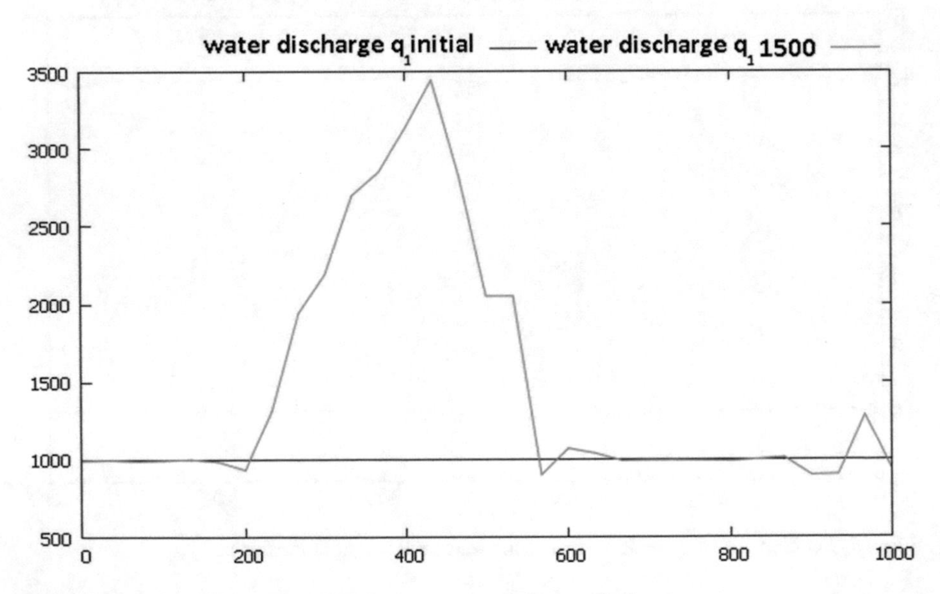

Fig. 18 Iteration 1500 at $x_2 = 500$: first component of water discharge q_1

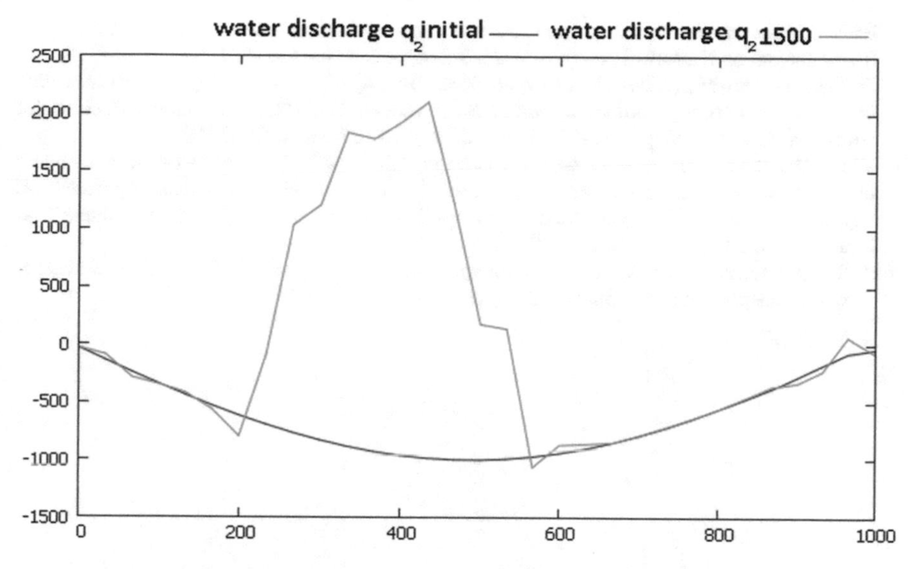

Fig. 19 Iteration 1500 at $x_2 = 500$: second component of water discharge q_2

4 Conclusion and Perspectives

In this paper we have done a numerical study of the dimensionless coupled system of shallow water equation and long term dynamic of dunes of sand equation. A numerical scheme has been performed and its stability has been shown. Some numerical tests have been performed and we have seen that the time step is too small since ϵ is small. The absolute value of the water discharge increase rapidly due to the small parameter ϵ. An other consequence of the small parameter ϵ is that to observe a short time of the tide, a long time of execution of the program may be necessary.

So in an upcoming work it would be interesting to simulate for different values of ϵ and compare the results. It would also be useful to do a numerical study of the dimensionless homogenization of the coupled system presented in [1] and to do a comparison of the result of the schemes of these two systems.

References

1. M.A.M.T. Baldé and D. Seck. *Coupling the Shallow Water Equation with a Long Term Dynamics of Sand Dunes*. *Discrete and Continuous Dynamical Systems Series S. Vol. 9, n°5, 2016, 1521–1551.*
2. S. Cordier, M. Le, T. M. de Luna. *Bedload transport in shallow water models: why splitting (may) fail, how hyperbolicity (can) help*. 13 pages. 2010. <hal-00536267v>.
3. I. Faye, E. Frénod and D. Seck. *Long term behavior of singularity perturbed parabolic degenerated equation.* Journal of Nonlinear Analysis and Application. Vol. 2016 n°2 (2016), 82–105.

4. Parés C., *Numerical methods for nonconservative hyperbolic systems: a theoretical framework*. SIAM Journal on Numerical Analysis, 2006, Vol. **44**, No. 1: pp. 300–321.
5. Parés C. et Castro M., (2004). *On the well-balance property of Roe's method for nonconservative hyperbolic systems. applications to shallow-water systems*. ESAIM: Mathematical Modelling et Numerical Analysis, **38**, pp 821–852. https://doi.org/10.1051/m2an:2004041.
6. Castro Díaz M.J., Fernández-Nieto E.D., Ferreiro A.M., Parés C., *Two-dimensional sediment transport models in shallow water equations. A second order finite volume approach on unstructured meshes*, Computer Methods in Applied Mechanics et Engineering, Volume **198**, Issues 33–36, 1 July 2009, Pages 2520–2538.
7. Z. Sheng et G. Yuan. *A finite volume scheme for diffusion equations on distorted quadrilateral meshes*. Transport Theory et Statistical Physics, **37**:171–207, 2008.

Coupling the Navier-Stokes Equations with a Short Term Dynamic of Sand Dunes

Ibrahima Faye, Mariama Ndiaye, and Diaraf Seck

Abstract Our paper deals about sand transport problem near the seabed. We consider model for short-term dynamics of dune (STDD) and megariple morphodynamics built in (Faye et al., Discrete and Continuous Dynamical Systems, **29**; N^o3 March 2011, 1001–1030), that we coupled with a Navier-Stokes equations. We study the evolution of the dunes and an existence and uniqueness results are established for coupled short-term model. In this framework we derive an asymptotic expansion with respect to the small parameter ϵ of its solution, and characterize the terms of the expansion.

Keywords Degenerate parabolic equation · Space-time periodic solutions · Homogenization · Asymptotic analysis · Short term behaviour · Dune · Navier-Stokes equations

Mathematics Subject Classification (2010). Primary 35K65, 35B25, 35B40, 35B10, 86A60

I. Faye (✉)
Université Alioune Diop de Bambey, UFR SATIC, Laboratoire de Mathématique de la Décision et d'Anlyse Numérique, Equipe de recherche: Analyse Non Linéaire et Géométrie, Bambey, Senegal
e-mail: ibrahima.faye@uadb.edu.sn

M. Ndiaye
Université Gaston Berger de Saint Louis, Laboratoire de Mathématique de la Décision et d'Analyse Numérique, St Louis, Senegal
e-mail: mariama-ndiaye.diakhaby@ugb.edu.sn

D. Seck
Université Cheikh Anta Diop de Dakar, Laboratoire de Mathématique de la Décision et d' Analyse Numérique, Dakar-Fann, Senegal
e-mail: diaraf.seck@ucad.edu.sn

V. Schulz, D. Seck (eds.), *Shape Optimization, Homogenization and Optimal Control*, International Series of Numerical Mathematics 169,
https://doi.org/10.1007/978-3-319-90469-6_10

1 Introduction and Modeling

Dunes and megaripples and dynamics, on the seabed, are the results of interaction between the seabed and water currents. The aim of this work is to study the dynamical of dunes in tidal area. We combine a short term dynamical of sand transportation equation with a Navier-Stokes equation. This couplage models the short term dynamics of dunes transportation with equations of the dynamical water in the ocean. Faye et al [4] studied the problem by considering that the water velocity near the sea bed is a given regular vector field in $L^\infty([0,T), H^1(\mathbb{T}^2))$. In the work, as there is a flow, we suppose that the fluid motion is described by the incompressible Navier-Stokes equation.

The coupled model that we shall consider consists of combining an equation modeling sand transport equation near the seabed with an equation of the dynamics of the oceans modeled by the Navier-Stokes equations. The equation modeling the sand transport [4, 6, 20] is the following

$$\frac{\partial z}{\partial t} - \frac{1}{1-p}\nabla \cdot q = 0. \tag{1.1}$$

In this system of equations the fields z depends on time $t \in [0, T)$, for $T > 0$, on the horizontal position $x = (x_1, x_2) \in \mathbb{T}^2$, where $\mathbb{T}^2$ being the two dimensional torus $\mathbb{R}^2/\mathbb{Z}^2$.The field $z = z(t, x)$ is the height of the seabed in position x and at time t and $q = q(x, t)$ is the sand volume flow in x and at t. The parameter $p \in [0, 1)$ is called sand porosity. The constant λ is the inverse value of the maximum slope of the sediment surface when the water velocity $\mathbf{u}$ is 0. In this work, we will restrict to the q given by Van Rijn [20] which consists in writing [4]

$$q = \alpha\chi\Big(D_G\rho\frac{|\mathbf{u}|^2 - u_c^2}{C^2}\Big)\Big(\frac{\mathbf{u}}{|\mathbf{u}|} - \lambda\nabla z\Big) \tag{1.2}$$

where ρ is the water density, C is a constant defined by $C = \ln(\frac{12d}{3D_G})$, d being the water height on the seabed and D_G, the sand speck diameter. The function $\chi : \mathbb{R} \to \mathbb{R}_+$ is defined by

$$\chi(\sigma) = \begin{cases} 0 \text{ if } \sigma < 0 \\ |\sigma|^{\frac{3}{2}} \text{ if } \sigma \geq 0. \end{cases} \tag{1.3}$$

Finally the model of sand transport is given by

$$\frac{\partial z}{\partial t} - \frac{\alpha}{1-p}\nabla \cdot \Big[\chi\Big(D_g\varphi\frac{|\mathbf{u}|^2 - u_c^2}{C^2}\Big)\Big(\frac{\mathbf{u}}{|\mathbf{u}|} - \lambda\nabla z\Big)\Big] = 0. \tag{1.4}$$

The sand transport flow given by (1.2) and the Exner's equation given by (1.4) depend on $\mathbf{u}$, the velocity of the water. We assume in the sequel that $\mathbf{u}$ is a solution of the incompressible Navier-Stokes equation.

$$\begin{cases} \dfrac{\partial \mathbf{u}}{\partial t} - \nu \Delta \mathbf{u} + (\mathbf{u} \cdot \nabla)\mathbf{u} + \nabla p = f \\ \nabla \cdot \mathbf{u} = 0. \end{cases} \tag{1.5}$$

where p is the pressure in the considered domain and ν is the viscosity.

Let us recall that the Navier-Stokes equations are nonlinear partial differential equations that describe the motion of Newtonian fluids. In this study, Eq. (1.4), has to be coupled with the two-dimensional Navier-Stokes equations given by (1.5) describing the flow of the water in the ocean. Thus, we obtain the following reference model

$$\begin{cases} \dfrac{\partial z}{\partial t} - \dfrac{\alpha}{1-p} \nabla \cdot \left[\chi\Big(D_g \varphi \dfrac{|\mathbf{u}|^2 - u_c^2}{C^2}\Big)\Big(\dfrac{\mathbf{u}}{|\mathbf{u}|} - \lambda \nabla z\Big)\right] = 0 \text{ in }]0, T[\times \mathbb{T}^2 \\ \frac{\partial \mathbf{u}}{\partial t} - \nu \Delta \mathbf{u} + (\mathbf{u} \cdot \nabla)\mathbf{u} + \nabla p = f \quad \text{in }]0, T[\times \mathbb{T}^2 \\ \nabla \cdot \mathbf{u} = 0 \text{ in }]0, T[\times \mathbb{T}^2 \end{cases} \tag{1.6}$$

The system (1.6) need to be provided with initial conditions

$$\begin{cases} z(0, x) = z_0(x) \text{ in } \mathbb{T}^2 \\ u(0, x) = u_0(x) \text{ in } \mathbb{T}^2, \end{cases} \tag{1.7}$$

where z_0 belongs to $H^1(\mathbb{T}^2)$ and $u_0 \in H^1(\mathbb{T}^2)$.

2 Scaling of the Reference Model and Parametrized Model

In this section, we shall use the scaling in [4]. Let us introduce the characteristic values: $\bar{t}$, $\bar{L}$, $\bar{z}$ and the fields velocity $\bar{u}$ and height $\bar{m}$ are the same as in [4]. We define also the dimensionless variable t' and x', the seabed height z' and velocity $\mathbf{u}'$.

So let $x = \bar{L}x'$, $\quad t = \bar{t}t'$, $\quad u = \bar{u}\mathbf{u}'$,

$\mathbf{u}(\bar{t}t', \bar{L}x') = \bar{u}\mathbf{u}'(t', x')$ and $z(\bar{t}t', \bar{L}x') = \bar{z}z'(t', x')$.

Following the idea developed in [4], the dimensionless model of the sand transport equation is given by

$$\frac{\partial z'}{\partial t'} - \frac{\lambda}{1-p} \alpha \frac{\bar{t}\bar{u}^3 (\rho D_G)^{3/2}}{\left(\ln(\frac{4H}{D_G})\right)^3 \bar{L}^2} \nabla' \cdot \left(\left(1 - 3\frac{\bar{M}}{H \ln(\frac{4H}{D_G})}\mathbf{m}'\right) \chi\Big(|\mathbf{u}'|^2 - \frac{{u_c}^2}{\bar{u}^2}\Big) \nabla' z' \right)$$

$$= \frac{1}{1-p} \alpha \frac{\bar{t}\bar{u}^3 (\rho D_G)^{3/2}}{\left(\ln(\frac{4H}{D_G})\right)^3 \bar{L}\bar{z}} \nabla' \cdot \left(\left(1 - 3\frac{\bar{M}}{H \ln(\frac{4H}{D_G})}\mathbf{m}'\right) \chi\Big(|\mathbf{u}'|^2 - \frac{{u_c}^2}{\bar{u}^2}\Big) \frac{\mathbf{u}'}{|\mathbf{u}'|} \right). \tag{2.1}$$

In the same way, we easily obtain the dimensionless Navier-Stokes equations. Then, we can write

$$\frac{\partial \mathbf{u}}{\partial t} = \bar{u}\frac{\partial \mathbf{u}'}{\partial t'}\frac{\partial t'}{\partial t} = \frac{\bar{u}}{\bar{t}}\frac{\partial \mathbf{u}'}{\partial t'}$$

$$(\mathbf{u}\cdot\nabla)\mathbf{u} = (\bar{u}\mathbf{u}'\cdot\nabla)(\bar{u}\mathbf{u}') = \bar{u}^2(\mathbf{u}'\cdot\nabla)\mathbf{u}'\frac{1}{\bar{L}} = \frac{u^2}{\bar{L}}(\mathbf{u}'\cdot\nabla)\mathbf{u}'$$

$$-\nu\Delta\mathbf{u} = -\nu\bar{u}\Delta\mathbf{u}' = -\frac{\nu\bar{u}}{\bar{L}^2}\Delta\mathbf{u}'$$

$$\nabla p = \frac{\bar{u}^2}{\bar{L}}\nabla p'$$

$$f = \frac{\bar{u}^2}{\bar{L}}f'.$$

We get from the following equation for $\mathbf{u}'$

$$\begin{cases} \dfrac{\partial \mathbf{u}'}{\partial t} - \dfrac{\nu\bar{t}}{\bar{L}^2}\Delta\mathbf{u}' - \dfrac{\bar{u}\bar{t}}{\bar{L}}(\mathbf{u}'\cdot\nabla)\mathbf{u}' + \dfrac{\bar{u}\bar{t}}{\bar{L}}\nabla p' = \dfrac{\bar{u}\bar{t}}{\bar{L}}f', \\ \nabla\cdot\mathbf{u}' = 0. \end{cases} \tag{2.2}$$

Finally, the dimensionless model of the coupled system is given by

$$\begin{cases} \dfrac{\partial z'}{\partial t'} - \dfrac{\lambda}{1-p}\alpha\dfrac{\bar{t}\bar{u}^3(\rho D_G)^{3/2}}{\left(\ln(\frac{4H}{D_G})\right)^3\bar{L}^2}\nabla'\cdot\left(\left(1-3\dfrac{\bar{M}}{H\ln(\frac{4H}{D_G})}\mathbf{m}'\right)\chi\left(|\mathbf{u}'|^2-\dfrac{u_c{}^2}{\bar{u}^2}\right)\nabla' z'\right) \\ = \dfrac{1}{1-p}\alpha\dfrac{\bar{t}\bar{u}^3(\rho D_G)^{3/2}}{\left(\ln(\frac{4H}{D_G})\right)^3\bar{L}\bar{z}}\nabla'\cdot\left(\left(1-3\dfrac{\bar{M}}{H\ln(\frac{4H}{D_G})}\mathbf{m}'\right)\chi\left(|\mathbf{u}'|^2-\dfrac{u_c{}^2}{\bar{u}^2}\right)\dfrac{\mathbf{u}'}{|\mathbf{u}'|}\right) \\ \dfrac{\partial \mathbf{u}'}{\partial t} - \dfrac{\nu\bar{t}}{\bar{L}^2}\Delta\mathbf{u}' - \dfrac{\bar{u}\bar{t}}{\bar{L}}(\mathbf{u}'\cdot\nabla)\mathbf{u}' + \dfrac{\bar{u}\bar{t}}{\bar{L}}\nabla p' = \dfrac{\bar{u}\bar{t}}{\bar{L}}f' \\ \nabla\cdot\mathbf{u}' = 0. \end{cases} \tag{2.3}$$

Having this dimensionless model on hand, we shall now consider the characteristic values for short term dynamics of dunes evolutions and for small specks given in[4]. Let us recall them here:

$$\bar{u} = 1\text{m/s}, \quad \bar{L} = 300\,\text{m},$$

$$\bar{t} = 100\,\text{days} = 100\times 24\times 3600\,\text{s} = 864.104$$

From these data, ν is given by $\nu = 10^{-6}$.

Let $\bar{\omega}$ be the main tide frequency, then $\bar{t}$ is compared with the main tide period $\frac{1}{\omega} = 13\,\text{h} \sim 4.710^4\,\text{s}$. This makes it possible to define the small parameter

$$\epsilon = \frac{1/\bar{\omega}}{\bar{t}} = \frac{46810^2}{86410^4} = 5.42 \times 10^{-3} \sim 0.5 \times 10^{-2} \sim \frac{1}{200}.$$

Concerning the coefficients of the Navier-Stokes equations, we have

$$\frac{\frac{\nu\bar{t}}{L^2}}{\epsilon} = \frac{9.610^{-5}}{1/200} = 0.0192 \simeq 0.02 \simeq = \frac{2}{100} \simeq = \frac{4}{200} = 4\epsilon,$$

So $\frac{\nu\bar{t}}{L^2} \simeq 4\epsilon^2$

$$\frac{\bar{u}\bar{t}}{\bar{L}} \times \epsilon = 288 \times 10^2 \times \frac{1}{200} = 144,$$
$$\frac{\bar{u}\bar{t}}{\bar{L}} \times \epsilon^2 = 144 \times \frac{1}{200} = 0.72 \simeq \frac{3}{4},$$

and then we have $\frac{\bar{u}\bar{t}}{\bar{L}} \simeq \frac{3}{4\epsilon^2}$.

In the short-term dynamics of dunes with small sand specks, by removing the' in order to lighten the writings, we obtain the following system:

$$\begin{cases} \dfrac{\partial z}{\partial t} - \dfrac{1}{2\epsilon}\nabla\cdot\left((1-3\epsilon\mathbf{m})\chi(|\mathbf{u}^\epsilon|^2 - \dfrac{1}{2})\nabla z\right) \\ \quad = \frac{5}{\epsilon}\nabla\cdot\left((1-3\epsilon\mathbf{m})\chi(|\mathbf{u}^\epsilon|^2 - \frac{1}{2})\frac{\mathbf{u}^\epsilon}{|\mathbf{u}^\epsilon|}\right) \\ z^\epsilon(x,0) = z_0(x) \\ \dfrac{\partial \mathbf{u}^\epsilon}{\partial t'} - 4\epsilon^2\Delta\mathbf{u}^\epsilon + \dfrac{3}{4\epsilon^2}[(\mathbf{u}^\epsilon\cdot\nabla)\mathbf{u}^\epsilon + \nabla p^\epsilon] = \dfrac{3}{4\epsilon^2}f \\ \nabla\cdot\mathbf{u}^\epsilon = 0 \\ \mathbf{u}^\epsilon(x,0) = u_0(x), \end{cases} \tag{2.4}$$

where $u_0 \in H^1(\mathbb{T}^2)$ and $f \in L^2(0,T;H^{-1})$. In the following, , we assume that the source term f given in (2.4) is negligible i.e $f = 0$. Since we want to study the asymptotic behavior of $(z^\epsilon, \mathbf{u}^\epsilon)$ as ϵ goes to zero, we need solutions of (2.4) and estimates which do not depend on ϵ. Existence of z^ϵ solution to the first equation of (2.4) is a consequence of a result of Ladyzenskaja et al [7] or Lions [12]. Existence of $\mathbf{u}^\epsilon$ solution to the Navier-Stokes equations

$$\begin{cases} \dfrac{\partial \mathbf{u}^\epsilon}{\partial t} - 4\epsilon^2\Delta\mathbf{u}^\epsilon + \dfrac{3}{4\epsilon^2}\left((\mathbf{u}^\epsilon\cdot\nabla)\mathbf{u}^\epsilon + \nabla p^\epsilon\right) = 0 \\ \nabla\cdot\mathbf{u}^\epsilon = 0 \\ \mathbf{u}^\epsilon(x,0) = u_0(x) \end{cases} \tag{2.5}$$

shall be studied in the following section.

3 On the Incompressible Navier-Stokes Equation

We look first for existence and uniqueness of solutions $\mathbf{u}^\epsilon$ to (2.5) for a given p^ϵ and secondly estimates of $\mathbf{u}^\epsilon$ which do not depends on ϵ.

The mathematical study of incompressible Navier-Stokes equations goes back to the pioneering work of J. Leray [9–11] in a series of papers emanating from his doctoral studies. Leray proved in these papers the existence of a global in time weak. Many other results concerning the Navier-Stokes equations are obtained in mathematics, see for example [2, 3, 8, 12–14, 18, 19]. Existence of a weak and strong solutions to (2.5) for a given ϵ is also known, see for example [13, 14, 16, 17]. We have the following theorem

Theorem 3.1 *Let $\epsilon > 0$ and $u_0 \in H = \{\mathbf{u} \in L^2(\mathbb{T}^2),\ div\mathbf{u} \in L^2(\mathbb{T}^2)\}$ such that $u_0 \in H^1(\mathbb{T}^2)$.There exists a sequence of solutions $\mathbf{u}^\epsilon \in L^\infty([0,T), L^2(\mathbb{T}^2))$ to (2.5). Moreover, $\mathbf{u}^\epsilon$ satisfies the following hypotheses:*

$$\left\|\mathbf{u}^\epsilon\right\|_{L^\infty([0,T),L^2(\mathbb{T}^2))} \leq C_0, \tag{3.1}$$

$$\left\|\epsilon\nabla\mathbf{u}^\epsilon\right\|_{L^2([0,T),L^2(\mathbb{T}^2))} \leq C_0', \tag{3.2}$$

$$\left\|\epsilon\Delta\mathbf{u}^\epsilon\right\|_{L^2([0,T),L^2(\mathbb{T}^2))} \leq \tilde{C}_0, \tag{3.3}$$

where $C_0,\ C_0',\ \tilde{C}_0'$ depend only on u_0.

Before giving the proof of this theorem we shall need the following theorem which ensures the existence of a solution for any fixed ϵ. Let's consider now a sequence $(\mathbf{u}_n^\epsilon)_n$ that satisfying: $\forall n \in \mathbb{N}$, $\forall\epsilon$

$$\begin{cases} \dfrac{\partial \mathbf{u}_n^\epsilon}{\partial t} - 4\epsilon^2\Delta\mathbf{u}_n^\epsilon + \dfrac{3}{4\epsilon^2}\Big((\mathbf{u}_n^\epsilon\cdot\nabla)\mathbf{u}_n^\epsilon + \nabla p^\epsilon\Big) = 0 \\ \nabla\cdot\mathbf{u}_n^\epsilon = 0 \\ \mathbf{u}_n^\epsilon(x,0) = u_0(x). \end{cases} \tag{3.4}$$

Then we have:

Theorem 3.2 *Let $(\mathbf{u}_n^\epsilon)$ a sequence of $L^2(H)$, solution to the Navier-Stokes (2.5) such that $\frac{\partial \mathbf{u}_n^\epsilon}{\partial t} \in L^2(H')$, then, up to a subsequence, there exists a subsequence denoted $(\mathbf{u}_n^\epsilon)$ which weakly converges to $\mathbf{u}^\epsilon$ in $L^2(H)$, solution of (2.5) and satisfying the energy estimates:*

$$\frac{1}{2}\int_{\mathbb{T}^2} \mid \mathbf{u}^\epsilon(t,x)dx \mid^2 - \frac{1}{2}\int_{\mathbb{T}^2} \mid \mathbf{u}_0^\epsilon(x)dx \mid^2 dx$$

$$+4\epsilon^2\int_0^t\int_{\mathbb{T}^2} \mid \nabla\mathbf{u}^\epsilon \mid^2 dx\,dt = 0\ \forall t \in]0,T[. \tag{3.5}$$

Proof Let $\mathbf{u}_n^\epsilon \in L^2(H)$ such that $\frac{\partial \mathbf{u}_n^\epsilon}{\partial t} \in L^2(H')$ and solution to (3.4), then multiplying (3.4) by $\mathbf{u}_n^\epsilon$ and integrating over $\mathbb{T}^2$ we get

$$\frac{1}{2}\frac{d}{dt}\int_{\mathbb{T}^2} \mid \mathbf{u}_n^\epsilon \mid^2 dx + 4\epsilon^2 \int_{\mathbb{T}^2} \mid \nabla \mathbf{u}_n^\epsilon \mid^2 dx = 0.$$

Integrating the above identity on $[0, T[$, we have

$$\frac{1}{2}\int_{\mathbb{T}^2} \mid \mathbf{u}_n^\epsilon(t,x) \mid^2 dx + 4\epsilon^2 \int_0^t \int_{\mathbb{T}^2} \mid \nabla \mathbf{u}_n^\epsilon \mid^2 dxdt = \frac{1}{2}\int_{\mathbb{T}^2} u_0(x)^2 dx \tag{3.6}$$

Since the two terms in the left hand side of equality (3.6) are positive, we have the two following inequalities

$$\int_{\mathbb{T}^2} \mid \mathbf{u}_n^\epsilon(t,x) \mid^2 \leq \int_{\mathbb{T}^2} u_0(x)^2 dx \tag{3.7}$$

and

$$4\epsilon^2 \int_0^t \int_{\mathbb{T}^2} \mid \nabla \mathbf{u}_n^\epsilon \mid^2 \leq \frac{1}{2}\int_{\mathbb{T}^2} u_0(x)^2 dx \tag{3.8}$$

Then,

$$\sup_{t\in[0,T)} \|\mathbf{u}_n^\epsilon\|_{L^2(\mathbb{T}^2)} \leq \|u_0\|_{L^2(\mathbb{T}^2)} \tag{3.9}$$

and

$$\|\epsilon \nabla \mathbf{u}_n^\epsilon\|_{L^2((0,T),L^2(\mathbb{T}^2))} \leq \frac{1}{8}\|u_0\|_{L^2(\Omega)} \tag{3.10}$$

Thus, there is $C_\epsilon = (\frac{1}{8\epsilon} + \sqrt{T})$ such that

$$\|\mathbf{u}_n^\epsilon\|_{L^2([0,T),H^1(\Omega))} \leq C_\epsilon \|u_0\|_{L^2(\Omega)} \tag{3.11}$$

then the sequence $\mathbf{u}_n^\epsilon$ is bounded in $L^2([0,T), H^1(\mathbb{T}^2)) \cap L^\infty([0,T), L^2(\mathbb{T}^2))$. Since the space H is reflexive, we can extract from the sequence $\mathbf{u}_n^\epsilon$ a subsequence which converges weakly on $\mathbf{u}^\epsilon$ in $L^2(H)$.

Following the idea developed by Temam [13, 14, 17], one can prove that, for all ϵ, the sequence $\mathbf{u}_n^\epsilon$ satisfies the following inequalities

$$\|\mathbf{u}_n^\epsilon\|_{L^{8/3}((0,T),L^4(\mathbb{T}^2))} \leq \|\mathbf{u}_n^\epsilon\|^{\frac{1}{2}}_{L^\infty((0,T),L^2(\mathbb{T}^2))} \|\mathbf{u}_n^\epsilon\|^{\frac{3}{4}}_{L^2((0,T),L^6(\mathbb{T}^2))},$$

then $(\mathbf{u}_n^\epsilon)$ is bounded in $L^{8/3}([0,T), L^4(\mathbb{T}^2))$.

Multiplying the first equation of (3.4) by φ, a regular valued function with compact support, and integrating over $\mathbb{T}^2 \times [0, T)$ we have

$$\int_0^T \int_{\mathbb{T}^2} \frac{\partial \mathbf{u}_n^\epsilon}{\partial t}\varphi + 4\epsilon^2 \int_0^T \int_{\mathbb{T}^2} \nabla \mathbf{u}_n^\epsilon \nabla \varphi + \frac{3}{4\epsilon^2} \int_0^T \int_{\mathbb{T}^2} \Big((\mathbf{u}_n^\epsilon \cdot \nabla)\mathbf{u}_n^\epsilon + \nabla p^\epsilon)\varphi = 0 \tag{3.12}$$

The previous equality allow us to pass to the limit at each term and to show the limit $\mathbf{u}^\epsilon$ of $\mathbf{u}_n^\epsilon$ satisfies the following equality

$$\int_0^T \int_{\mathbb{T}^2} \frac{\partial \mathbf{u}^\epsilon}{\partial t}\varphi + 4\epsilon^2 \int_0^T \int_{\mathbb{T}^2} \nabla \mathbf{u}^\epsilon \nabla \varphi + \frac{3}{4\epsilon^2} \int_0^T \int_{\mathbb{T}^2} \Big((\mathbf{u}^\epsilon \cdot \nabla)\mathbf{u}^\epsilon + \nabla p^\epsilon\Big)\varphi = 0. \tag{3.13}$$

which is a weak formulation of (2.5). Replacing φ par $\mathbf{u}^\epsilon$ we get

$$\int_{\mathbb{T}^2} \frac{\partial \mathbf{u}^\epsilon}{\partial t}\mathbf{u}^\epsilon - 4\epsilon^2 \int_{\mathbb{T}^2} \Delta \mathbf{u}^\epsilon\, \mathbf{u}^\epsilon + \underbrace{\frac{3}{4\epsilon^2} \int_{\mathbb{T}^2} \Big((\mathbf{u}^\epsilon \cdot \nabla)\mathbf{u}^\epsilon + \nabla p^\epsilon\Big)\mathbf{u}^\epsilon}_{=0} = 0,$$

which gives the result (3.5) by using Green's formula in the second term and integrating over $[0, T)$. □

Proof (of Theorem 3.1) Let us consider a test function $\varphi \in \mathcal{C}^\infty(\mathbb{T}^2 \times [0, T])$ with compact support and satisfying the condition $\nabla \cdot \varphi = 0$ in $\mathbb{T}^2 \times [0, T]$.

Multiplying the first equation of (2.5) par φ and integrating in $\mathbb{T}^2 \times [0, T]$ we have

$$\int_{\mathbb{T}^2} u_0 \varphi dx + 4\epsilon^2 \int_0^T \int_{\mathbb{T}^2} \nabla \mathbf{u} \nabla \varphi dxdt - \frac{3}{4\epsilon^2} \int_0^T \int_{\mathbb{T}^2} \sum_{i,j} u_i u_j \frac{\partial \varphi_j}{\partial x_i} dxdt -$$
$$\int_0^T \int_{\mathbb{T}^2} \mathbf{u} \frac{\partial \varphi}{\partial t} dxdt = 0. \tag{3.14}$$

Multiplying (2.5) by $\mathbf{u}^\epsilon$ and integrating over $\mathbb{T}^2$ we get

$$\int_{\mathbb{T}^2} \frac{\partial \mathbf{u}^\epsilon}{\partial t}\mathbf{u}^\epsilon - 4\epsilon^2 \int_{\mathbb{T}^2} \Delta \mathbf{u}^\epsilon\, \mathbf{u}^\epsilon + \frac{3}{4\epsilon^2} \int_{\mathbb{T}^2} \Big((\mathbf{u}^\epsilon \cdot \nabla)\mathbf{u}^\epsilon + \nabla p^\epsilon\Big)\mathbf{u}^\epsilon = 0.$$

But since $\nabla \cdot \mathbf{u}^\epsilon = 0$, we have

$$\int_{\mathbb{T}^2} \nabla p^\epsilon . \mathbf{u}^\epsilon = \int_{\mathbb{T}^2} p^\epsilon \nabla \cdot \mathbf{u}^\epsilon = 0 \tag{3.15}$$

and

$$\int_{\mathbb{T}^2} (\mathbf{u}\cdot\nabla)\mathbf{u}^{\epsilon}\mathbf{u}^{\epsilon} = \int_{\mathbb{T}^2} \nabla\cdot(\mathbf{u}^{\epsilon}\,\mathbf{u}^{\epsilon})\cdot\mathbf{u}^{\epsilon} = -\sum_{i,j}\int_{\mathbb{T}^2} u^j u^i \partial_j u^i =$$

$$-\int_{\mathbb{T}^2}\sum_{i,j} u^j \frac{1}{2}\partial_j (u^i)^2 = \frac{1}{2}\sum_{i,j}\int_{\mathbb{T}^2} \partial_j (u^j)(u^i)^2 dx = 0. \tag{3.16}$$

Using the above two last equalities, we get

$$\frac{1}{2}\frac{d}{dt}\int_{\mathbb{T}^2} |\, \mathbf{u}^{\epsilon}\,|^2 + 4\epsilon^2 \int_{\mathbb{T}^2} |\,\nabla\mathbf{u}^{\epsilon}\,|^2 = 0.$$

Integrating this equality in $(0, T)$, we have:

$$\frac{1}{2}\int_{\mathbb{T}^2} |\, \mathbf{u}^{\epsilon}(t,x)dx\,|^2 - \frac{1}{2}\int_{\mathbb{T}^2} |\, \mathbf{u}_0^{\epsilon}(x)dx\,|^2\, dx + 4\epsilon^2 \int_0^t\int_{\mathbb{T}^2} |\,\nabla\mathbf{u}^{\epsilon}\,|^2\, dx\, dt = 0 \tag{3.17}$$

Or $\forall t \in (0, T),\ \forall\epsilon$,

$$\frac{1}{2}\int_{\mathbb{T}^2} |\, \mathbf{u}^{\epsilon}(t,x)dx\,|^2 + \underbrace{4\epsilon^2 \int_0^t\int_{\mathbb{T}^2} |\,\nabla\mathbf{u}^{\epsilon}\,|^2\, dx\, dt}_{\geq 0} = \frac{1}{2}\int_{\mathbb{T}^2} |\, \mathbf{u}_0^{\epsilon}(x)dx\,|^2\, dx.$$

So, we can deduce that, from the non negativity of the second term in the left hand side , we

$$\frac{1}{2}\int_{\mathbb{T}^2} |\, \mathbf{u}^{\epsilon}(t,x)dx\,|^2 \leq \frac{1}{2}\int_{\mathbb{T}^2} |\, \mathbf{u}_0^{\epsilon}(x)dx\,|^2\, dx.$$

This implies:

$$\sup_{0\leq t\leq T}\left(\int_{\mathbb{T}^2} (\mathbf{u}^{\epsilon}(x,t))^2\right) \leq \int_{\mathbb{T}^2} |\, u_0(x)dx\,|^2\, dx. \tag{3.18}$$

From (3.19) we can deduce

$$\sup_{0\leq t\leq T}\left(\int_{\mathbb{T}^2} (\mathbf{u}^{\epsilon}(x,t))^2\right) \leq C_0, \tag{3.19}$$

where C_0 depends only on the choice of u_0.

Also we have:

$$4\int_0^t\int_{\mathbb{T}^2} \mid \epsilon\,\nabla \mathbf{u}^\epsilon \mid^2 \, dx\, dt \leq \frac{1}{2}\int_{\mathbb{T}^2} \mid \mathbf{u}_0^\epsilon(x)dx \mid^2 dx.$$

Thus,

$$\int_0^t\int_{\mathbb{T}^2} \mid \epsilon\,\nabla \mathbf{u}^\epsilon \mid^2 \, dx\, dt \leq \frac{1}{8}\underbrace{2\int_{\mathbb{T}^2} \mid u_0(x)dx \mid^2 dx}_{<+\infty} \tag{3.20}$$

then $\epsilon\nabla\mathbf{u}^\epsilon \in L^2([0,T), L^2(\mathbb{T}^2))$.
Multiplying (2.5) by $-\Delta\mathbf{u}^\epsilon$ and integrating over $\mathbb{T}^2$ we get

$$\int_{\mathbb{T}^2}\frac{\partial \mathbf{u}^\epsilon}{\partial t}\Delta\mathbf{u}^\epsilon + 4\epsilon^2\int_{\mathbb{T}^2}\left|\Delta\mathbf{u}^\epsilon\right|^2 - \frac{3}{4\epsilon^2}\int_{\mathbb{T}^2}\Big((\mathbf{u}^\epsilon\cdot\nabla)\mathbf{u}^\epsilon + \nabla p^\epsilon\Big)\Delta\mathbf{u}^\epsilon = 0$$

Then we have,

$$\int_{\mathbb{T}^2}\nabla(\frac{\partial \mathbf{u}^\epsilon}{\partial t})\nabla\mathbf{u}^\epsilon + 4\epsilon^2\int_{\mathbb{T}^2}\left|\Delta\mathbf{u}^\epsilon\right|^2 + \frac{3}{4\epsilon^2}b(\mathbf{u}^\epsilon,\mathbf{u}^\epsilon,\Delta\mathbf{u}^\epsilon) = 0$$

where

$$b(u,v,w) = \sum_{i,j=1}\int_{\mathbb{T}^2} u_i\frac{\partial v_j}{\partial x_i}w_j dx, \quad \forall u,\ v\, w \in H.$$

Following the idea used in [13, 14, 17], we have $b(\mathbf{u}^\epsilon,\mathbf{u}^\epsilon,\Delta\mathbf{u}^\epsilon,) = 0,\ \forall\mathbf{u}^\epsilon \in H$.
Integrating from 0 to t, we have,

$$\frac{1}{2}\int_{\mathbb{T}^2}\left|\nabla\mathbf{u}^\epsilon(t,\cdot)\right|^2 dx + 4\left\|\epsilon\Delta\mathbf{u}^\epsilon\right\|_{L^2([0,T),L^2(\mathbb{T}^2))} \leq \frac{1}{2}\int_{\mathbb{T}^2}\left|\nabla u_0\right|^2 dx \tag{3.21}$$

Because of the positivity of the first term, we that

$$\left\|\epsilon\Delta\mathbf{u}^\epsilon\right\|_{L^2([0,T),L^2(\mathbb{T}^2))} \leq \tilde{C}_0, \tag{3.22}$$

where $\tilde{C}_0$ depends only on u_0. □

4 Asymptotics Analysis of the Navier-Stokes Equation

Our aim in this section is to see how it is possible to get an expansion of $\mathbf{u}^\epsilon$ with respect to the parameter ϵ. $\mathbf{u}^\epsilon$ is proposed as follow:

$$\mathbf{u}^\epsilon(t,x) = \sum_{i=0}^{+\infty} \epsilon^i U^i(t, \frac{t}{\epsilon}, x), \tag{4.1}$$

where the function $U^k(t,\theta,x)$, $k = 1,2,\dots$ are periodic in θ of period 1. The objective of this section is to intend to characterize the equation satisfied by the function U^k, $k = 0,1,2,\dots$, if the expression (4.1) is valid. And mainly, we aim to prove the following estimates:

$$\begin{aligned} &\left\| \mathbf{u}^\epsilon(t,x) - U^0(t,\frac{t}{\epsilon},x) \right\|_{L^\infty([0,T),L^2(\mathbb{T}^2))} \leq c_1\epsilon \quad \text{and} \\ &\left\| \mathbf{u}^\epsilon(t,x) - U^0(t,\frac{t}{\epsilon},x) - \epsilon U^1(t,\frac{t}{\epsilon},x) \right\|_{L^\infty([0,T),L^2(\mathbb{T}^2))} \leq c_2\epsilon, \end{aligned} \tag{4.2}$$

where c_1, c_2 are constants.

Let's see formally how by simple identification, it is possible to derive candidates U^0 and U^1 with PDE verified by these functions.

On the one hand, plugging expansion (4.1) into (2.5) we obtain

$$\begin{aligned} &\frac{\partial}{\partial t}\Big(\sum_{i=0}^{+\infty} \epsilon^i U^i(t,\frac{t}{\epsilon},x)\Big) - 4\epsilon^2 \Delta(\sum_{i=0}^{+\infty} \epsilon^i U^i(t,\frac{t}{\epsilon},x)) + \\ &\frac{3}{4\epsilon^2}\Big((\sum_{i=0}^{+\infty} \epsilon^i U^i(t,\frac{t}{\epsilon},x)\cdot\nabla)\sum_{i=0}^{+\infty} \epsilon^i U^i(t,\frac{t}{\epsilon},x) + \nabla P^\epsilon\Big) = 0. \end{aligned} \tag{4.3}$$

On the other hand the initial condition gives:

$$U^0(0,0,x) = u_0(x), \;\; \nabla\cdot U^0 = 0 \tag{4.4}$$

and for $i \geq 1$,

$$U^i(0,0,x) = 0, \;\; \nabla\cdot U^i = 0 \tag{4.5}$$

By computations, it is easy to see that

$$\frac{\partial \mathbf{u}^\epsilon}{\partial t} = \sum_{i=0}^{+\infty} \epsilon^i \frac{\partial U^i}{\partial t} + \epsilon^{i-1}\frac{\partial U^i}{\partial \theta}, \qquad \Delta\mathbf{u}^\epsilon = \sum_{i=0}^{+\infty} \epsilon^i \Delta U^i \tag{4.6}$$

and

$$\left((\mathbf{u}^{\epsilon}\cdot\nabla)\mathbf{u}^{\epsilon}\right)_i=\sum_{l=0}^{+\infty}\sum_{k=0}^{+\infty}\epsilon^{l+k}\Big(\sum_{j=1}^{n}U_j^l\frac{\partial U_i^k}{\partial x_j}\Big) \tag{4.7}$$

Let's also assume that we can write $P^{\epsilon}(t,x)$ as follows

$$P^{\epsilon}(t,x)=\sum_{i=0}^{\infty}\epsilon^i P^i(t,\frac{t}{\epsilon},x) \tag{4.8}$$

in a way that

$$\nabla P^{\epsilon}(t,x)=\sum_{i=0}^{\infty}\epsilon^i\nabla P^i(t,\frac{t}{\epsilon},x). \tag{4.9}$$

Then the first equation of the Navier-Stokes equation becomes

$$\sum_{j=0}^{\infty}\epsilon^j\frac{\partial U_i^j}{\partial t}+\epsilon^{j-1}\frac{\partial U_i^j}{\partial\theta}-4\sum_{j=0}^{+\infty}\epsilon^{i+2}\Delta U_i^j+\frac{3}{4}\sum_{l=0}^{+\infty}\sum_{k=0}^{+\infty}\epsilon^{l+k-2}\Big(\sum_{j=1}^{n}U_j^l\frac{\partial U_i^k}{\partial x_j}\Big)$$
$$=-\frac{3}{4}\sum_{j=0}^{\infty}\epsilon^{i-2}\frac{\partial P^j}{\partial x_i}. \tag{4.10}$$

Taking the power of order less than or equal to 2, we get

$$\frac{\partial U_i^0}{\partial t}+\epsilon\frac{\partial U_i^1}{\partial t}+\frac{1}{\epsilon}\frac{\partial U_i^0}{\partial\theta}+\frac{\partial U_i^1}{\partial\theta}+\epsilon\frac{\partial U_i^2}{\partial\theta}+\frac{3}{4\epsilon^2}\sum_{j=1}^{n}U_j^0\frac{\partial U_i^0}{\partial x_j}$$
$$+\frac{3}{4\epsilon}\sum_{j=1}^{n}U_j^0\frac{\partial U_i^1}{\partial x_j}+\frac{3}{4\epsilon}\sum_{j=1}^{n}U_j^1\frac{\partial U_i^0}{\partial x_j}+\frac{3}{4}\sum_{j=1}^{n}U_j^1\frac{\partial U_i^1}{\partial x_j}$$
$$=-\frac{3}{4\epsilon^2}\frac{\partial P^0}{\partial x_i}-\frac{3}{4\epsilon}\frac{\partial P^1}{\partial x_i}-\frac{3}{4}\frac{\partial P^2}{\partial x_i}-\epsilon\frac{3}{4}\frac{\partial P^3}{\partial x_i}. \tag{4.11}$$

And finally, this yields:

$$\epsilon^{-2}\Big(\frac{3}{4}\sum_{j=1}^{n}U_j^0\frac{\partial U_i^0}{\partial x_j}+\frac{3}{4}(\frac{\partial P^0}{\partial x_i})\Big)=0 \tag{4.12}$$

$$\epsilon^{-1}\Big(\frac{\partial U_i^0}{\partial\theta}+\sum_{j=1}^{n}U_j^0\frac{\partial U_i^1}{\partial x_j}+\frac{3}{4}\sum_{j=1}^{n}U_j^1\frac{\partial U_i^0}{\partial x_j}\Big)=-\frac{3}{4\epsilon}\frac{\partial P^1}{\partial x_i} \tag{4.13}$$

$$\frac{\partial U_i^0}{\partial t} + \frac{\partial U_i^1}{\partial \theta} + \frac{3}{4}\sum_{j=1}^{n} U_j^1 \frac{\partial U_i^1}{\partial x_j} = -\frac{3}{4}\frac{\partial P^2}{\partial x_i} \tag{4.14}$$

And the proposed system is expressed as follows:

The functions $U^0 = (U_1^0, U_2^0)$, $U^1 = (U_1^1, U_2^1)$ satisfy for $i = 1, 2$ the following equation

$$(U^0 \cdot \nabla)U^0 = -\nabla P^0 \tag{4.15}$$

$$\frac{\partial U_i^0}{\partial \theta} + \frac{3}{4}\Big(U^0 \nabla U_i^1 + U^1 \cdot \nabla U_i^0\Big) = -\frac{3}{4}\frac{\partial P^1}{\partial x_i}$$

$$\frac{\partial U_i^0}{\partial t} + \frac{\partial U_i^1}{\partial \theta} + \frac{3}{4}(U^1 \cdot \nabla U_i^1) = -\frac{3}{4}\frac{\partial P^2}{\partial x_i} \tag{4.16}$$

Theorem 4.1 *Let u_0, $\mathbf{u}^\epsilon \in H$ and $\epsilon > 0$. For any $T > 0$, if $\mathbf{u}^\epsilon$ is solution to (2.5), then $\mathbf{u}^\epsilon$ stays bounded in $L^\infty((0,T); L^2(\mathbb{R}^2))$ and two-scale converges to the function $U^0 \in L^\infty([0,T), L^\infty_\#(\mathbb{R}, L^2(\mathbb{R}^2)))$ solution to (4.15).*
Moreover, $\nabla \cdot U^0 = 0$ and we have

$$\lim_{\epsilon \to 0} \int_0^T \int_{\mathbb{T}^2} \left| \mathbf{u}^\epsilon(t,x) - U^0(t, \frac{t}{\epsilon}, x) \right|^2 dxdt = 0. \tag{4.17}$$

Proof Let $\Psi = \Psi(t, \theta, x)$ be a function of class $\mathcal{C}^1$, 1-periodic in θ such that $\Psi(\cdot, t, \cdot) \in \mathcal{C}_c^1(\mathbb{R}^2 \times [0,T))$. We define $\Psi_\epsilon(t,x) = \Psi(t, \frac{t}{\epsilon}, x)$, so that $\psi_\epsilon \in \mathcal{C}_c^1(\mathbb{R}^2 \times [0,T))$. Multiplying (2.5) by ψ_ϵ and integrating over $\mathbb{T}^2 \times [0,T)$ we get

$$\int_0^T \int_{\mathbb{T}^2} \frac{\partial \mathbf{u}^\epsilon}{\partial t}\psi_\epsilon dxdt - 4\epsilon^2 \int_0^T \int_{\mathbb{T}^2} \Delta \mathbf{u}^\epsilon \psi_\epsilon dxdt +$$
$$\frac{3}{4\epsilon^2}\int_0^T \int_{\mathbb{T}^2} \Big((\mathbf{u}^\epsilon \cdot \nabla)\mathbf{u}^\epsilon + \nabla p^\epsilon\Big)\psi_\epsilon dxdt = 0$$

We obtain

$$\int_0^T \int_{\mathbb{T}^2} \frac{\partial \psi_\epsilon}{\partial t}\mathbf{u}^\epsilon dxdt + 4\epsilon^2 \int_0^T \int_{\mathbb{T}^2} \nabla \mathbf{u}^\epsilon \nabla \psi_\epsilon dxdt +$$
$$\frac{3}{4\epsilon^2}\int_0^T \int_{\mathbb{T}^2} \Big((\mathbf{u}^\epsilon \cdot \nabla)\mathbf{u}^\epsilon + \nabla p^\epsilon\Big)\psi_\epsilon dxdt = \int_{\mathbb{T}^2} u_0(x)\psi(0,0,x)dx$$

We have

$$\frac{\partial \psi_\epsilon}{\partial t}(t,x) = \frac{\partial \psi}{\partial t}(t, \frac{t}{\epsilon}, x) + \frac{1}{\epsilon}\frac{\partial \psi}{\partial \theta}(t, \frac{t}{\epsilon}, x).$$

Multiplying by ϵ^2, we get

$$\int_0^T \int_{\mathbb{T}^2} \Big(\epsilon(\frac{\partial \psi}{\partial \theta}) + \epsilon^2(\frac{\partial \psi}{\partial t}\Big) dxdt + 4\epsilon^4 \int_0^T \int_{\mathbb{T}^2} \nabla \mathbf{u}^\epsilon \nabla \psi_\epsilon dxdt +$$

$$\frac{3}{4}\int_0^T \int_{\mathbb{T}^2} \Big((\mathbf{u}^\epsilon \cdot \nabla)\mathbf{u}^\epsilon + \nabla p^\epsilon\Big)\psi_\epsilon dxdt = \epsilon^2 \int_{\mathbb{T}^2} u_0(x)\psi(0,0,x)dx$$

As $\mathbf{u}^\epsilon$ in bounded in $L^\infty([0,T), L^\infty_\#(\mathbb{R}, L^\infty(\mathbb{T}^2)))$ then $\mathbf{u}^\epsilon$ two-scale converges to U^0, P^ϵ two-scale converge to P^0 and as ϵ tends to zero we get

$$\lim_{\epsilon\to 0} \frac{3}{4}\int_0^T \int_{\mathbb{T}^2} \Big((\mathbf{u}^\epsilon \cdot \nabla)\mathbf{u}^\epsilon + \nabla p^\epsilon\Big)\psi_\epsilon dxdt = 0$$

which give the weak formulation of

$$(U^0 \cdot \nabla)U^0 + \nabla P^0 = 0 \tag{4.18}$$

□

Since $\mathbf{u}^\epsilon$ two-scale converges to U^0, we are interested first, by existence and uniqueness of z^ϵ solution to (2.4). Since we want to study the asymptotic behavior of z^ϵ as ϵ goes to 0, we need estimate of z^ϵ which do not depends on ϵ.

As in the previous paragraph, we have shown that the solution $\mathbf{u}^\epsilon$ of the Navier-Stokes equation (2.5) is bounded as well as its derivatives, we will proceed in the same way as in [4], to show that the sequence of solutions z^ϵ to (2.4) admits a unique solution. Before proceeding further, let's show that the coefficients of the Eq. (2.4) are bounded. Let

$$\mathcal{A}^\epsilon(t,x) = (1 - 3\epsilon\mathbf{m})\chi(|\mathbf{u}^\epsilon|^2 - \frac{1}{2}) \tag{4.19}$$

and

$$\mathcal{C}^\epsilon(t,x) = (1 - 3\epsilon\mathbf{m})\chi(|\mathbf{u}^\epsilon|^2 - \frac{1}{2})\frac{\mathbf{u}^\epsilon}{|\mathbf{u}^\epsilon|} \tag{4.20}$$

where $\mathbf{m} : \mathbb{T}^2 \to \mathbb{R}$ is a regular function defined as the variation of water and the function $\chi : \mathbb{R} \to \mathbb{R}_+$ by

$$\chi(\sigma) = \begin{cases} 0 \text{ if } \sigma < 0 \\ |\sigma|^{\frac{3}{2}} \text{ if } \sigma \geq 0 \end{cases} \tag{4.21}$$

The first equation of (2.4) can be set in the form

$$\begin{cases} \dfrac{\partial z^\epsilon}{\partial t} - \dfrac{1}{\epsilon}\nabla \cdot \Big(\mathcal{A}^\epsilon \nabla z^\epsilon\Big) = \dfrac{1}{\epsilon}\nabla \cdot \mathcal{C}^\epsilon \\ z^\epsilon(0,x) = z_0(x) \end{cases} \tag{4.22}$$

We have the following result:

Proposition 4.2 *Let $\epsilon > 0$ and $T > 0$, then the coefficients $\mathcal{A}^\epsilon$ and $\mathcal{C}^\epsilon$ of Eq. (4.22) given by (4.19) and (4.20) belong to $L^\infty((0,T), L^2(\mathbb{T}^2))$.*

Proof As $\mathbf{m}$ is regular and χ is continuous, we have

$$\int_{\mathbb{T}^2} \left|\mathcal{A}^\epsilon(t,x)\right|^2 \leq \int_{\mathbb{T}^2} \left|(1-3\epsilon\mathbf{m})(|\mathbf{u}^\epsilon|^2 - \frac{1}{2})^{\frac{3}{2}}\right|^2 \leq \int_{\mathbb{T}^2} |1+3\epsilon|\mathbf{m}||^2|\mathbf{u}^\epsilon|^3$$
$$\leq (3\gamma+1)^2 \int_{\mathbb{T}^2} |\mathbf{u}^\epsilon|^3$$

Taking the supremum for every t belonging to [0, T]; we have

$$\|\mathcal{A}^\epsilon\|_{L^\infty[0,T),L^2(\mathbb{T}^2)} \leq (\gamma+1)^2 C_0$$

Following the same idea, we have also

$$\|\mathcal{C}^\epsilon\|_{L^\infty[0,T),H^1(\mathbb{T}^2)} \leq (\gamma+1)C_0.$$

□

Let's assume in the sequel that the water variation can be also set in the form

$$\mathbf{m}(t,x) = \mathcal{M}(t, \frac{t}{\epsilon}, x), \tag{4.23}$$

where $\theta \to \mathcal{M}(\cdot,\theta,\cdot)$ is periodic of period 1.

Theorem 4.3 *Let $\epsilon > 0$ and $T > 0$, under hypothesis (4.23), if $\mathbf{u}^\epsilon$ is solution (2.5), and $\mathcal{A}^\epsilon$ and $\mathcal{C}^\epsilon$ are given by (4.19), then there exists a unique sequence of functions $z^\epsilon \in L^\infty([0,T), L^2(\mathbb{T}^2))$, solution to (4.22), Moreover, for any $t \in [0,T]$, z^ϵ satisfy*

$$\|z^\epsilon\|_{L^\infty([0,T),L^2(\mathbb{T}^2))} \leq \gamma_1 \tag{4.24}$$

where γ_1 is a constant not depending on ϵ and

$$\frac{d\Big(\int_{\mathbb{T}^2} z^\epsilon(t,x)dx\Big)}{dt} = 0 \tag{4.25}$$

Proof The proof of this theorem is done in few steps

Integrating the first equation of (2.4) with respect to $x \in \mathbb{T}^2$ we get directly equality (4.25).

Because of Proposition 4.2, the coefficients $\mathcal{A}^\epsilon$ and $\mathcal{C}^\epsilon$ are bounded. We also recall that the solution $\mathbf{u}^\epsilon$ of Eq. (2.5) is also bounded and because of the two-scale results, we have proved that the sequence $\mathbf{u}^\epsilon$ satisfy the following asymptotic results

$$\mathbf{u}^\epsilon(t, x) \sim U^0(t, \frac{t}{\epsilon}, x) \tag{4.26}$$

when $\epsilon \to 0$ and where $\theta \to U^0(\cdot, \theta, \cdot)$ is periodic of period 1.

Under conditions (4.26) and (4.23), the coefficients $\mathcal{A}^\epsilon$ and $\mathcal{C}^\epsilon$ writes

$$\mathcal{A}^\epsilon(t, x) = \widetilde{\mathcal{A}}(t, \frac{t}{\epsilon}, x) \text{ and } \mathcal{C}^\epsilon(t, x) = \widetilde{\mathcal{C}}(t, \frac{t}{\epsilon}, x) \tag{4.27}$$

where $\theta \mapsto (\widetilde{\mathcal{A}}, \widetilde{\mathcal{M}})$ is periodic of period 1. Under hypotheses (4.27) and (4.23), the coefficients $\widetilde{\mathcal{A}}$ and $\widetilde{\mathcal{C}}$ and its derivatives are also bounded and satisfy the same hypotheses as [4]. Then there exists a constant γ not depending on ϵ such that

$$\begin{gathered} |\widetilde{\mathcal{A}}| \leq \gamma, \ |\widetilde{\mathcal{C}}| \leq \gamma, \ \left|\frac{\partial \widetilde{\mathcal{A}}}{\partial t}\right| \leq \gamma \ \left|\frac{\partial \widetilde{\mathcal{C}}}{\partial t}\right| \leq \gamma \\ \left|\frac{\partial \widetilde{\mathcal{A}}}{\partial \theta}\right| \leq \gamma, \ \left|\frac{\partial \widetilde{\mathcal{C}}}{\partial \theta}\right| \leq \gamma, \ |\nabla \widetilde{\mathcal{A}}| \leq \gamma, \ |\nabla \cdot \widetilde{\mathcal{C}}| \leq \gamma, \\ \left|\frac{\partial \nabla \widetilde{\mathcal{A}}}{\partial t}\right| \leq \gamma, \left|\frac{\partial \nabla \widetilde{\mathcal{C}}}{\partial t}\right| \leq \gamma. \end{gathered} \tag{4.28}$$

Under assumptions (4.26), (4.23),(4.27) and (4.28), following the idea of Faye et al [4], based on periodic solutions to parabolic equations, there exists a regular function $\mathcal{S}$ periodic of period 1 with respect to θ solutions to

$$\frac{\partial \mathcal{S}}{\partial \theta} - \nabla \cdot ((\widetilde{\mathcal{A}}(t, \cdot, \cdot))\nabla \mathcal{S}) = \nabla \cdot \widetilde{\mathcal{C}}(t, \cdot, \cdot). \tag{4.29}$$

Considering $\mathcal{S}$, we consider Z^ϵ as follows: $Z^\epsilon(t, x) = \mathcal{S}(t, \frac{t}{\epsilon}, x)$ where $\mathcal{S}$ is solution to (4.29) and using variational formula, we proof that

$$\|z^\epsilon(t, x) - Z^\epsilon(t, x)\|_2 \leq 2\|z_0 - \mathcal{S}(0, 0, \cdot)\|_2 \tilde{\psi} \tag{4.30}$$

where $\tilde{\psi}$ is a constant not depending on ϵ. From this last inequality, we get the desired estimate.

To prove uniqueness, we consider z_1^ϵ and z_2^ϵ two solutions of (4.22). Then $z_1^\epsilon - z_2^\epsilon$ is solution to

$$\begin{cases} \dfrac{\partial(z_1^\epsilon - z_2^\epsilon)}{\partial t} - \dfrac{1}{\epsilon}\nabla\cdot\left(\widetilde{\mathcal{A}}\nabla(z_1^\epsilon - z_2^\epsilon)\right) = 0, \\ \left(z_1^\epsilon - z_2^\epsilon\right)_{|t=0} = 0. \end{cases} \tag{4.31}$$

Multiplying the first equation of (4.22) by $(z_1^\epsilon - z_2^\epsilon)$ and integrating in $\mathbb{T}^2$, we get

$$\frac{d\left(\|z_1^\epsilon - z_2^\epsilon\|_2^2\right)}{dt} \leq 0. \tag{4.32}$$

From this last equality, we have

$$\|z_1^\epsilon(t) - z_2^\epsilon(t)\| = 0 \quad \text{for any } t, \tag{4.33}$$

and giving uniqueness.

□

In the following, we are interested by the behaviour of z^ϵ as $\epsilon \to 0$. The sequence z^ϵ of solution to (4.22) is bounded in $L^\infty([0,T), L^2(\mathbb{T}^2)$. Using a result of two scales convergence due to Nguetseng [15], Allaire [1], we proof that z^ϵ two scale converge to $Z \in L^\infty(\mathbb{R}, L^\infty_\#([0,T), L^2(\mathbb{T}^2))$. Our aim is to characterize the equation satisfied by the two scales limit Z.

We recall that the solution $\mathbf{u}^\epsilon$ to (2.5) two-scale converges to U^0 solution to (4.15). Then we have

$$\mathbf{u}^\epsilon \text{ two scales converges to } U^0(t,\theta,x) \in \left(L^\infty(\mathbb{R}, L^\infty_\#([0,T), L^2(\mathbb{T}^2))\right)^2. \tag{4.34}$$

Using (4.34), we obtain that $\mathcal{A}^\epsilon$ and $\mathcal{C}^\epsilon$ given by (4.19) and (4.20) satisfy:

$$\mathcal{A}^\epsilon(t,x) \text{ two scale converges to } \widetilde{\mathcal{A}}(t,\theta,x) \in L^\infty(\mathbb{R}, L^\infty_\#([0,T), L^2(\mathbb{T}^2)) \tag{4.35}$$

and

$$\mathcal{C}^\epsilon(t,x) \text{ two scale converges to } \widetilde{\mathcal{C}}(t,\theta,x) \in \left(L^\infty(\mathbb{R}, L^\infty_\#([0,T), L^2(\mathbb{T}^2))\right)^2 \tag{4.36}$$

where

$$\widetilde{\mathcal{A}}(t,\theta,x) = \chi(|U^0(t,\theta,x)|^2 - \frac{1}{2}) \text{ and } \widetilde{\mathcal{C}}(t,\theta,x) = \chi(|U^0(t,\theta,x)|^2 - \frac{1}{2}) \frac{U^0(t,\theta,x)}{|U^0(t,\theta,x)|}. \tag{4.37}$$

We have the following theorem

Theorem 4.4 *Under assumptions (4.26) and (4.23), for any T, not depending on ϵ, the sequence of solutions $(\mathbf{u}^\epsilon, z^\epsilon)$ to (2.4) two-scale converges to $(U^0, Z) \in \left(L^\infty(\mathbb{R}, L^\infty_\#([0,T), L^2(\mathbb{T}^2))\right)^2 \times L^\infty(\mathbb{R}, L^\infty_\#([0,T), L^2(\mathbb{T}^2))$ solution to*

$$\frac{\partial Z}{\partial \theta} - \nabla \cdot (\widetilde{\mathcal{A}} \nabla Z) = \nabla \cdot \widetilde{\mathcal{C}}, \tag{4.38}$$

where $\widetilde{\mathcal{A}}$ and $\widetilde{\mathcal{C}}$ are given by (4.37).

Proof The proof of this theorem is similar to the proof Defining test function $\psi^\epsilon(t,x) = \psi(t, \frac{t}{\epsilon}, x)$ for any $\psi(t,\theta,x)$, regular with compact support in $[0,T) \times \mathbb{T}^2$ and periodic in θ with period 1, multiplying (2.4) by ψ^ϵ and integrating in $[0,T) \times \mathbb{T}^2$ gives

$$\int_{\mathbb{T}^2} \int_0^T \frac{\partial z^\epsilon}{\partial t} \psi^\epsilon dt dx - \frac{1}{\epsilon} \int_{\mathbb{T}^2} \int_0^T \nabla \cdot (\mathcal{A}^\epsilon \nabla z^\epsilon) \psi^\epsilon dt dx = \frac{1}{\epsilon} \int_{\mathbb{T}^2} \int_0^T \nabla \cdot \mathcal{C}^\epsilon \psi^\epsilon dt dx. \tag{4.39}$$

Then integrating by parts in the first integral over $[0,T)$ and using the Green formula in $\mathbb{T}^2$ in the second integral we have

$$\begin{aligned} &- \int_{\mathbb{T}^2} z_0(x) \psi(0,0,x) dx - \int_{\mathbb{T}^2} \int_0^T \frac{\partial \psi^\epsilon}{\partial t} z^\epsilon dt dx \\ &+ \frac{1}{\epsilon} \int_{\mathbb{T}^2} \int_0^T \mathcal{A}^\epsilon \nabla z^\epsilon \nabla \psi^\epsilon dt dx = \frac{1}{\epsilon} \int_{\mathbb{T}^2} \int_0^T \nabla \cdot \mathcal{C}^\epsilon \psi^\epsilon dt dx. \end{aligned} \tag{4.40}$$

Again using the green formula in the third integral we obtain

$$\begin{aligned} &- \int_{\mathbb{T}^2} z_0(x) \psi(0,0,x)\, dx - \int_{\mathbb{T}^2} \int_0^T \frac{\partial \psi^\epsilon}{\partial t} z^\epsilon dt dx \\ &- \frac{1}{\epsilon} \int_{\mathbb{T}^2} \int_0^T z^\epsilon\, \nabla \cdot (\mathcal{A}^\epsilon \nabla \psi^\epsilon)\, dt dx = \frac{1}{\epsilon} \int_{\mathbb{T}^2} \int_0^T \nabla \cdot \mathcal{C}^\epsilon \psi^\epsilon dt dx. \end{aligned} \tag{4.41}$$

But

$$\frac{\partial \psi^\epsilon}{\partial t} = \left(\frac{\partial \psi}{\partial t}\right)^\epsilon + \frac{1}{\epsilon} \left(\frac{\partial \psi}{\partial \theta}\right)^\epsilon, \tag{4.42}$$

where

$$\left(\frac{\partial \psi}{\partial t}\right)^{\epsilon}(t,x) = \frac{\partial \psi}{\partial t}(t,\frac{t}{\epsilon},x) \text{ and } \left(\frac{\partial \psi}{\partial \theta}\right)^{\epsilon}(t,x) = \frac{\partial \psi}{\partial \theta}(t,\frac{t}{\epsilon},x), \tag{4.43}$$

then we have

$$\int_{\mathbb{T}^2}\int_0^T z^{\epsilon}\left(\left(\frac{\partial \psi}{\partial t}\right)^{\epsilon} + \frac{1}{\epsilon}\left(\frac{\partial \psi}{\partial \theta}\right)^{\epsilon} + \frac{1}{\epsilon}\nabla\cdot(\mathcal{A}^{\epsilon}\nabla\psi^{\epsilon})\right)dxdt$$
$$+\frac{1}{\epsilon}\int_{\mathbb{T}^2}\int_0^T \nabla\cdot\mathcal{C}^{\epsilon}\psi^{\epsilon}dtdx = -\int_{\mathbb{T}^2} z_0(x)\psi(0,0,x)\,dx. \tag{4.44}$$

Using the two-scale convergence due to Nguetseng [15] and Allaire [1] (see also Frénod Raviart and Sonnendrücker [5], Faye et al [4]), if a sequence z^{ϵ} is bounded in $L^{\infty}(0,T,L^2(\mathbb{T}^2))$, then there exists a profile $Z(t,\theta,x)$, periodic of period 1 with respect to θ, such that for all $\psi(t,\theta,x)$, regular with compact support with respect to (t,x) and periodic of period 1 with respect to θ, we have

$$\int_{\mathbb{T}^2}\int_0^T z^{\epsilon}\psi^{\epsilon}dtdx \longrightarrow \int_{\mathbb{T}^2}\int_0^T\int_0^1 Z\psi\,d\theta dtdx, \tag{4.45}$$

for a subsequence extracted from (z^{ϵ}).

Multiplying (4.44) by ϵ and passing to the limit as $\epsilon \to 0$ and using (4.45) we have

$$\int_{\mathbb{T}^2}\int_0^T\int_0^1 Z\frac{\partial \psi}{\partial \theta}\,d\theta dtdx + \lim_{\epsilon\to 0}\int_{\mathbb{T}^2}\int_0^T z^{\epsilon}\nabla\cdot(\mathcal{A}^{\epsilon}\nabla\psi^{\epsilon})\,dtdx$$
$$= \lim_{\epsilon\to 0}\int_{\mathbb{T}^2}\int_0^T \mathcal{C}^{\epsilon}\cdot\nabla\psi^{\epsilon}dtdx, \tag{4.46}$$

for an extracted subsequence. As $\mathcal{A}^{\epsilon}$ and $\mathcal{C}^{\epsilon}$ are bounded (see Theorem 4.2) and ψ^{ϵ} is a regular function, $\mathcal{A}^{\epsilon}\nabla\psi^{\epsilon}$ and $\nabla\psi^{\epsilon}$ can be considered as test functions. Using (4.35) and (4.36) we have

$$\int_{\mathbb{T}^2}\int_0^T z^{\epsilon}\,\nabla\cdot(\mathcal{A}^{\epsilon}\nabla\psi^{\epsilon})dtdx \longrightarrow \int_{\mathbb{T}^2}\int_0^T\int_0^1 Z\nabla\cdot(\widetilde{\mathcal{A}}\nabla\psi)\,d\theta dtdx, \tag{4.47}$$

and

$$\int_{\mathbb{T}^2}\int_0^T \mathcal{C}^{\epsilon}\cdot\nabla\psi^{\epsilon}dtdx \text{ two scale converges to } \int_{\mathbb{T}^2}\int_0^T\int_0^1 \widetilde{\mathcal{C}}\cdot\nabla\psi\,d\theta dtdx. \tag{4.48}$$

From this we obtain from (4.46) the equation satisfied by Z:

$$\frac{\partial Z}{\partial \theta} - \nabla \cdot (\widetilde{\mathcal{A}} \nabla Z) = \nabla \cdot \widetilde{\mathcal{C}}. \tag{4.49}$$

□

Discussion From (4.15) and Theorem 4.4, the sequence $(\mathbf{u}^\epsilon, z^\epsilon)$ of solutions to (2.4) two-scale converges to (U^0, Z) solution to

$$\begin{cases} \frac{\partial Z}{\partial \theta} - \nabla \cdot (\widetilde{\mathcal{A}} \nabla Z) = \nabla \cdot \widetilde{\mathcal{C}}, \\ (U^0 \cdot \nabla) U^0 = \nabla P^0 \\ \nabla \cdot U^0 = 0, \end{cases} \tag{4.50}$$

where $\widetilde{\mathcal{A}}$ and $\widetilde{\mathcal{C}}$ are given by (4.37).

But in this work we could not characterize the function U^1 given by (4.2). Does it possible to get a corrector result for U^1?

One of the challenges would be to find an equation satisfied by U^1 and to get a valid expansion of $\mathbf{u}^\epsilon$ in the form

$$\mathbf{u}^\epsilon(t, x) = U^0(t, \frac{t}{\epsilon}, x) + \epsilon U^1(t, \frac{t}{\epsilon}, x) + o(\epsilon). \tag{4.51}$$

Acknowledgement This work is supported by NLAGA project (Non Linear Analysis, Geometry and Applications).

References

1. G. Allaire, *Homogenization and Two-Scale convergence*, SIAM J. Math. Anal. **23** (1992), 1482–1518.
2. O. Besson and M. R. Laidi, *Some estimates for the anisotropic navier-stokes equations and for the hydrostatic approximation,* M2AN - Mod. Math. Ana. Num. **26** (1992), 855–865.
3. L. Caffarelli, R. Kohn, and L. Nirenberg, *Partial regularity of suitable weak solutions of the Navier-Stokes equations.* Comm. Pure and Appl. Math. **35** (1982), 771–831.
4. I. Faye, E. Frénod, D. Seck, *Singularly perturbed degenerated parabolic equations and application to seabed morphodynamics in tided environment,*Discrete and Continuous Dynamical Systems, **29**; N^o3 March 2011, 1001–1030.
5. E. Frénod, Raviart P. A., and E. Sonnendrücker, *Asymptotic expansion of the Vlasov equation in a large external magnetic field*, J. Math. Pures et Appl. **80** (2001), 815–843.
6. D. Idier, *Dunes et bancs de sables du plateau continental: observations in-situ et modélisation numérique.* PhD thesis, 2002.
7. O. A. Ladyzenskaja, V. A. Solonnikov, and N. N. Ural'ceva. *Linear and Quasi-linear Equations of Parabolic Type.* **23**. AMS, Translation of Mathematical Monographs.
8. O. A. Ladyzhenskaya, *The mathematical theory of viscous incompressible flow.* Second English edition, revised and enlarged. Gordon and Breach, Science Publishers, New York-London-Paris 1969.

9. J. Leray, *Sur le système d'équations aux dérivées partielles qui régit l'écoulement permanent des fuides visqueux.* C. R. Acad. Sci. Paris,192 (1931), 1180–1182.
10. J. Leray,*Etude de diverses équations intégrales non linéaires et de quelques problèmes que pose l'hydrodynamique.* J. Math. Pures Appl., **12**, 1933, 1–82.
11. J. Leray, *Sur le mouvement d'un liquide visqueux emplissant l'espace.* Acta Math. **63** (1934), no. 1, 193–248.
12. J.-L. Lions, *Remarques sur les équations différentielles ordinaires.* Osaka Math. J. **15** (1963), 131–142.
13. P. L. Lions, *Mathematical topics in fuid mechanics.* **1**, Oxford, Science Publications, 1996.
14. P. L. Lions, *Mathematical topics in fluid mechanics.* **2**, Compressible models. The Clarendon Press, Oxford University Press, New York, 1998.
15. G. Nguetseng, A general convergence result for a functional related to the theory of homogenization, SIAM J. Math. Anal. **20** (1989), 608–623.
16. J. Simon, *Ecoulement d'un fluide non homogène avec une densité initiale s'annulant.* C. R. Acad. Sci. Paris, **287** (1978), 1009–1012.
17. R. Temam, *Navier-Stokes Equations, Theory and Numerical Analysis.* North Holland Publishing Company-Amsterdam. New York Oxford, 1977.
18. R. Temam, *Navier-Stokes Equations and Nonlinear Functional Analysis, Second Edition.* Copyright 1983, 1985 by the Society ro industrial and Applied Mathematics.
19. R. Temam, *Navier-Stokes equations. Theory and numerical analysis.* North-Holland Publishing Co., Amsterdam-New York-Oxford, 1977.
20. L. C. Van Rijn, *Handbook on sediment transport by current and waves*, Tech. Report H461:12.1–12.27, Delft Hydraulics, 1989.

Branching Structures in Elastic Shape Optimization

Nora Lüthen, Martin Rumpf, Sascha Tölkes, and Orestis Vantzos

Abstract Fine scale elastic structures are widespread in nature, for instances in plants or bones, whenever stiffness and low weight are required. These patterns frequently refine towards a Dirichlet boundary to ensure an effective load transfer. The paper discusses the optimization of such supporting structures in a specific class of domain patterns in 2D, which composes of periodic and branching period transitions on subdomain facets. These investigations can be considered as a case study to display examples of optimal branching domain patterns.

In explicit, a rectangular domain is decomposed into rectangular subdomains, which share facets with neighbouring subdomains or with facets which split on one side into equally sized facets of two different subdomains. On each subdomain one considers an elastic material phase with stiff elasticity coefficients and an approximate void phase with orders of magnitude softer material. For given load on the outer domain boundary, which is distributed on a prescribed fine scale pattern representing the contact area of the shape, the interior elastic phase is optimized with respect to the compliance cost. The elastic stress is supposed to be continuous on the domain and a stress based finite volume discretization is used for the optimization. If in one direction equally sized subdomains with equal adjacent subdomain topology line up, these subdomains are consider as equal copies including the enforced boundary conditions for the stress and form a locally periodic substructure.

An alternating descent algorithm is employed for a discrete characteristic function describing the stiff elastic subset on the subdomains and the solution of the elastic state equation. Numerical experiments are shown for compression and shear load on the boundary of a quadratic domain.

N. Lüthen
Departement Maschinenbau und Verfahrenstechnik, ETH Zürich, Zürich, Switzerland

M. Rumpf (✉) · S. Tölkes
Institut für Numerische Simulation, Universität Bonn, Bonn, Germany
e-mail: martin.rumpf@uni-bonn.de

O. Vantzos
Department of Mathematics, Israel Institute of Technology, Haifa, Israel

V. Schulz, D. Seck (eds.), *Shape Optimization, Homogenization and Optimal Control*, International Series of Numerical Mathematics 169,
https://doi.org/10.1007/978-3-319-90469-6_11

Keywords Shape optimization · Elasticity · Branching patterns

1 Introduction

The formation of microstructures is a common phenomenon in elastic shape optimization. We refer to [2] and [1] for an overview about these topics. Depending on the geometry of the computational domain and the loads applied to it different type of microstructures appear. In fact, besides locally periodic structures one also observes branching type patterns when optimizing with respect to a compliance cost functional, a volume cost, and the perimeter of the structure. These patterns refine towards a Dirichlet boundary of the configuration [17, Figure 13]. Such branching patterns can also been observed in nature, for instance in the spongiosa of bones [14]. For the basic load configurations of uniaxial load and shear load Kohn and Wirth [9, 10] considered scaling laws for the cost functional and for the weight in front of the perimeter tending to zero.

Besides the question how locally periodic or branching periodic patterns might look like, a central challenge is also to identify optimal decompositions of elastic material devices or objects into such spatially varying patterns. This paper should be regarded as a case study in this direction. It is intended as a first step towards a truly multiscale modeling of optimized elastic objects involving periodic and branching periodic patterns. Such a multiscale model would enable to apply techniques from *homogenization*, an important concept to upscale the microscopic properties to a macroscale. This methodology has been described for instance in [4, 12] and in detail in the context of elastic shape optimization in [1] and in the context of engineering applications for instance in [15] and [16]. Two-scale materials can be numerically computed by the Heterogeneous Multiscale Method (HMM) [5–8], which explicitly simulates periodic microstructures at positions in the macroscopic domain.

Here we consider domain patterns which explicitly prescribe locally periodic or branching periodic structures on a fine scale instead of taking into account a truly multiscale model. To this end we study computational domains that can be decomposed into a set of rectangular subdomains with compatibility conditions on the facets of these domains. The actual elastic structure will be a subset of the computational domain, which we aim to identify via elastic shape optimization. This elastic structure correspondingly splits into components on the subdomains. Each subdomain will be a copy of one rectangular reference cell on which the shape of a hard phase will be optimized. Then these elastic structures in the subdomains assemble to a locally periodic or a locally branching type structures. The discretization discussed in this article is an extension of the approach suggested by one of the authors, O. Vantzos and also used in [11], now going beyond purely branching periodic ensembles of cells.

The paper is organized as follows: in Sect. 2 we describe the admissible subdivision of the computational domain into cells. Then, Sect. 3 discusses the underlying elasticity model and the optimization problem. Section 4 presents the spatial discretization and the numerical solution of the state equation, whereas in Sect. 5 the alternating descent scheme for the optimization of the discrete characteristic function is investigated. Finally, results for two different load configurations are depicted in Sect. 6.

2 Composite Structures

Let $D \in \mathbb{R}^2$ be a computational domain consisting of several subdomains D_i, $i = 1, \dots, M$ for some $M \in \mathbb{N}$, such that $\cup_{i=1}^M \overline{D_i} = D$ and $\cap_{i=1}^M D_i = \emptyset$. Each subdomain is supposed to be a rectangle $(a_1, a_2) \times (b_1, b_2)$. Each facet of a subdomain

(i) is either also a facet of an adjacent subdomain (e.g. there is a subdomain $(a_2, a_3) \times (b_1, b_2)$ sharing the facet $\{a_2\} \times (b_1, b_2)$ with the subdomain $(a_1, a_2) \times (b_1, b_2)$),
(ii) or splits into two facets of two adjacent subdomains (e.g. there are subdomains $(a_2, a_3) \times (b_1, \frac{b_1+b_2}{2})$ and $(a_2, a_3) \times (\frac{b_1+b_2}{2}, b_2)$ whose facets $\{a_2\} \times (b_1, \frac{b_1+b_2}{2})$ and $\{a_2\} \times (\frac{b_1+b_2}{2}, b_2)$ results from a splitting of the facet $\{a2\} \times (b_1, b_2)$),
(iii) or is on the facets resulting from such a splitting of a facet of an adjacent subdomain,
(iv) or is a boundary facet.

We always assume that the splitting of facets is in two halves of equal length. The subdomain configurations at a single facet are shows in Fig. 1. The four facets of a single rectangular subdomain can be of different type.

Let us assume that each subdomain D_i contains an inscribed subpart of the actual elastic object, the shape of which will be optimized. Let us denote by χ_i the associated characteristic function. for which we consider the continuous

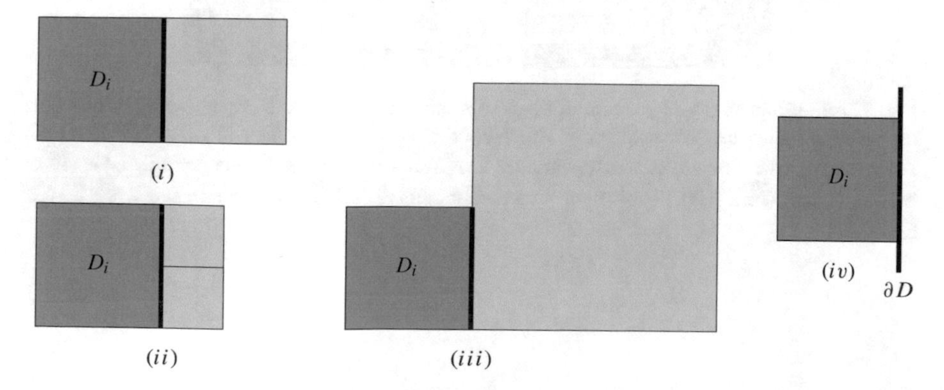

Fig. 1 The different local subdomain configurations at a single facet

extension in BV onto the closure of the subdomain. Thus, $\sum_{i=1}^{M} \chi_i$ as function on $\overline{D}$ is the characteristic function of the elastic objects we are investigating. Each (geometric) subdomain D_i will have a reference domain assigned and several geometric subdomains can have the same reference domain. The referenced domain is mapped onto a geometric subdomain via a translation and a rotation by a multiple of $\frac{\pi}{2}$. The characteristic function χ_i and later also the force distribution on $[\chi_i = 1]$ is handled and updated on the associated reference domain. Thus, the computational complexities scales with the number of reference domains. If adjacent subdomains share the same reference domain and the same rotation, then they are building blocks of local, period elastic structures. Figure 2 illustrates such a domain decomposition into rectangular boxes.

As a consequence, the domain sketched in Fig. 2 can later be numerically optimized by a shape optimization on 13 (coupled) reference domains only.

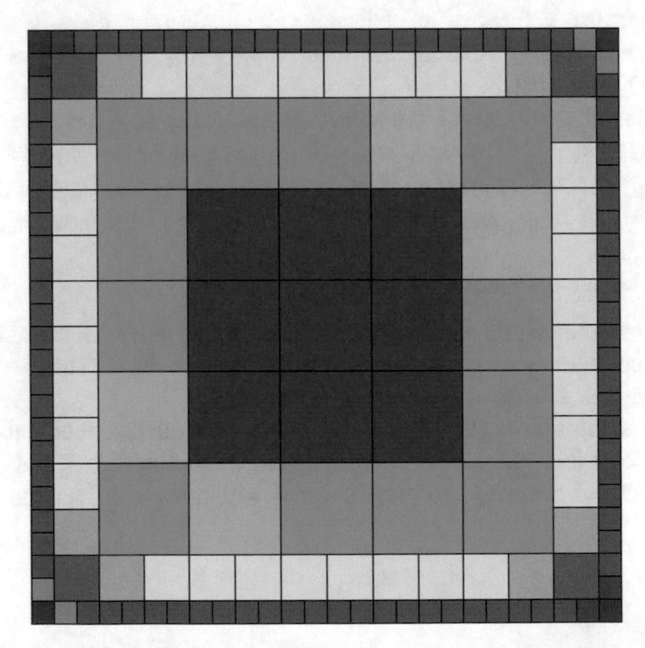

Fig. 2 An exemplary decomposition of the computational domain is displayed with 13 different reference domains that are mapped to the subdomains of the decomposition, where identical colors classify the different reference domains. The different types are locally periodic cells (■), branching cells (■■■), double branching corner cells (■■■) and coupling cells (■■■ ■■■)

3 Elastic State Equation and Compliance Optimization

To solve the elastic state equation in the context of the shape optimization we use the quadratic energy of a stress based formulation of linearized elasticity. On the same basis we evaluate the compliance-type target functional. On each reference domain the shape will be modelled by a phase field $v : D_i \to \mathbb{R}$ where v is assumed to approximate the characteristic function of the elastic object χ. Thus, $\{x \in D_i \mid v(x) \approx 1\}$ corresponds to the actual elastic domain and $\{x \in D_i \mid v(x) \approx 0\}$ to the void (or in our calculation very soft) phase.

Let us consider elastic stresses $\tau : D_i \to \mathbb{R}^2$ as extensions of the elastic stresses on the elastic object $[\chi = 1]$. Let us recall that stresses act on normals of infinitesimal area elements and represent the force density acting on this area element. We apply solely boundary forces and no volume forces. Hence, elastic stresses are divergence free, which constitutes together with the boundary condition the state equation of our optimization problem. Given the elasticity tensor $C(v)$, which depends on the phase (hard or soft) the stored elastic energy elastic on D_i is defined as

$$\mathcal{E}[v,\tau] = \int_{D_i} \mathcal{C}^{-1}(v)\tau : \tau \,\mathrm{d}x \,. \tag{3.1}$$

Here $\mathcal{C}(v)$ is a fourth order tensor satisfying $\mathcal{C}_{ijkl}(v) = \mathcal{C}_{jikl}(v) = \mathcal{C}_{ijlk}(v) = \mathcal{C}_{klij}(v)$. In the context of this paper, we define

$$\mathcal{C}(v) = v\mathcal{C}_{NL}$$

where $\mathcal{C}_{NL}$ is the elasticity tensor of the linearized Navier-Lamé elasticity model and $\delta > 0$. We define the inverse $\mathcal{C}^{-1}$ (needed in (3.1)) using *Young's modulus E* and the Poisson ratio ν (in Voigt notation)

$$\mathcal{C}_{NL}^{-1} = \frac{1}{E}\begin{pmatrix} 1 & -\nu & 0 \\ -\nu & 1 & 0 \\ 0 & 0 & 2+2\nu \end{pmatrix}.$$

For the phase field function v we consider a Modica Mortola type functional (cf. [13])

$$\mathcal{L}^{\varepsilon}[v] = \int_{D_i} \frac{1}{\varepsilon} W(v) + \frac{\varepsilon}{2}\,|\nabla v|^2 \,\mathrm{d}x$$

which approximates the length of the interface between hard and soft material. Here,

$$W(v) = \begin{cases} \frac{32}{\pi^2}(1-v)(v-\delta) & v \in [\delta, 1] \\ \infty & \text{else} \end{cases}$$

denotes a double-well potential, i. e. a positive function that is attending its only two minima at the pure phases $v = \delta$ and $v = 1$. We assume that $v|_{\partial D}$ is prescribed and describes the imposed fine scale structure on the domain boundary. The parameter ϵ is proportional to the width of the diffused interface. Furthermore, given the phase field v we can easily compute an approximation of the area of the elastic object $\mathcal{V}^\varepsilon[v] = \int_{D_i} v(x)\,\mathrm{d}x$.

Combining the energies above, we obtain the objective functional

$$\mathcal{J}[v] = \min_{\tau \in \Sigma_{\text{ad}}} \mathcal{E}[v, \tau] + \beta \mathcal{V}^\varepsilon[v] + \eta \mathcal{L}^\varepsilon[v]\,, \tag{3.2}$$

of our constraint optimization problem, where the stress τ is minimized over the set of admissible stresses

$$\Sigma_{\text{ad}} = \{\sigma : D_i \to \mathbb{R}^{2,2} \,|\, \sigma = \sigma^T, \operatorname{div}\sigma = 0, \text{b.c.}\}\,. \tag{3.3}$$

Here, the boundary condition (b.c.) differs for facets on ∂D and for interior facets, where the stresses are continuous across the facet. For facets on ∂D we prescribe forces f with $\tau(x) \cdot n(x) = f(x)$ if $v(x) = 1$. Here, $n(x)$ denotes the outer normal at points $x \in \partial D$.

As in [9, 10], we consider the case of a vanishing Poisson ratio $\nu = \frac{\lambda}{2(\lambda+\mu)} = 0$. As expressed by Kohn and Wirth in [9], this restriction is not expected to have a strong influence because for truss-like structures the lateral contraction is less relevant. The tensor $\mathcal{C}^{-1}$ then reduces to

$$\mathcal{C}_{NL}^{-1} = \frac{1}{E}\begin{pmatrix} 1 & 0 & 0 \\ 0 & 1 & 0 \\ 0 & 0 & 2 \end{pmatrix}.$$

Thus, we obtain for the stored elastic energy on the subdomain D_i

$$\mathcal{E}[v, \tau] = \int_{D_i} \frac{1}{v\,E}(\tau_{11}^2 + \tau_{22}^2) + \frac{2}{v\,E}\tau_{12}^2\,\mathrm{d}x = \int_{D_i} \frac{1}{v\,E}|\tau|^2\,\mathrm{d}x\,.$$

In what follows we choose $E = 1$.

4 Discretization

In this section, we present a finite volume discretization of the state equation and discuss the optimization algorithm. For details we refer to [11]. To this end, the conditions for σ prescribed in the set of admissible stress fields Σ_{ad} in (3.3) have to be transcribed into a linear system of equations resulting from the finite volume discretization. For a subdomain D_i we consider a finite decomposition of D_i into

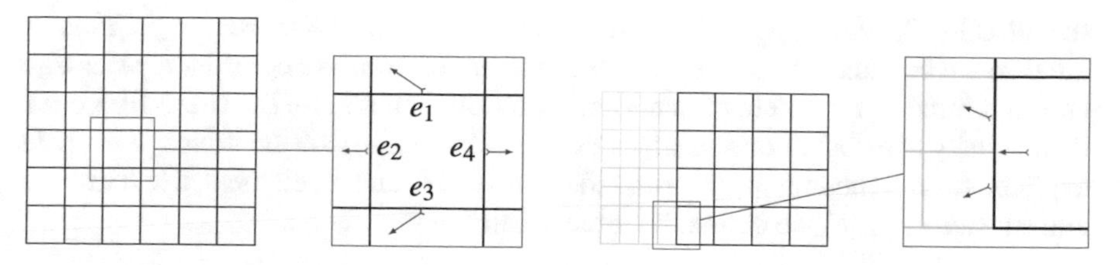

Fig. 3 Discretization of forces as vector valued degrees of freedom on the interior cells of the finite volume grid (left) and on a facet with prescribed branching-type boundary conditions (right). The numbering of the edges is displayed on the left

$N \times N$ rectangular cells $\mathbf{C}$ of equal size. The discrete forces are defined as constant vectors in $\mathbb{R}^2$ on edges and the discrete phase field is assumed to be constant on cells. As degrees of freedom we consider average forces across edges f_j which approximate $\frac{1}{h_j}\int_{\mathbf{e}_j} \sigma \cdot n_j \, dl$ of a volume cell $\mathbf{C}$ with edge length h_j. (cf. Fig. 3 for a sketch). The symmetry of the stress tensor is transformed into a conservation of torque:

$$0 = \int_{\partial\mathbf{C}} x \times (n \cdot \sigma) \, \mathrm{d}l = \int_{\partial\mathbf{C}} n \cdot (\sigma x^\perp) \, \mathrm{d}l = \int_{\mathbf{C}} \mathrm{div}(\sigma x^\perp) \, \mathrm{d}x$$
$$= \int_{\mathbf{C}} \mathrm{div}\sigma \, x^\perp + \sigma^T : \nabla x^\perp \, \mathrm{d}x = \int_{\mathbf{C}} \sigma_{21} - \sigma_{12} \, \mathrm{d}x \, ,$$

with $x = (x_1, x_2)$ and $x^\perp = (-x_2, x_1)$ for $\mathrm{div}\,\sigma = 0$. The conservation law $\mathrm{div}\,\tau = 0$ in the continuous set up translates to a balance relation on cells, i.e. one obtains

$$0 = \int_{\mathbf{C}} \mathrm{div}\tau \, \mathrm{d}x = \int_{\partial\mathbf{C}} n \cdot \tau \, \mathrm{d}l = \sum_{j=1}^{4} \int_{\mathbf{e}_j} n \cdot \tau \, \mathrm{d}l \approx \sum_{i=1}^{4} h_i f_i$$

where the discrete forces f_i for $i = 1, \ldots, 4$ are associated with the four edges $e_1, \ldots, e_4$ (numbered counter clock wise, starting with the upper edge, cf. Fig. 3) of the cell $\mathbf{C}$ and h_i are the corresponding edge lengths. Thus, the discrete balance of forces reads as

$$0 = \sum_{i=1}^{4} h_i f_i \tag{4.1}$$

for all cells $\mathbf{C}$. Given the above numbering of the edges, the balance of torques turns into the equation

$$0 = (f_{1,1} + f_{4,1}) - (f_{2,2} + f_{3,2}) \tag{4.2}$$

for all cells $\mathbf{C}$, where $f_{i,j}$ is the jth component of the force vector f_i. Finally, the discrete boundary conditions are encoded as follows. We consider a piecewise constant force f for facets of subdomains on the boundary ∂D. These forces are then equally distributed on the edges of cells $\mathbf{C}$ touching ∂D on which $v = 1$. In explicit, given a subdomain D_i with a facet F on ∂D and an element E with $v = 1$ and an edge e_j on F, we define the force density

$$f_j = \frac{\int_F v \, \mathrm{d}l}{\int_F \mathrm{d}l} f \tag{4.3}$$

All the discrete counterparts of the conditions in (3.3) are assembled in a linear system $\mathbf{Af} = \mathbf{b}$ for the vector of forces $\mathbf{f}$ on all edges. The matrix $\mathbf{A}$ and the right hand side $\mathbf{b}$ have the following block structure

$$\mathbf{A} = \begin{bmatrix} \mathbf{A}_f \\ \mathbf{A}_t \\ \mathbf{A}_{bc} \end{bmatrix}, \quad \mathbf{b} = \begin{bmatrix} 0 \\ 0 \\ \mathbf{b}_{bc} \end{bmatrix}.$$

Here, the index f refers to the force balance, the index t to the conservation of torque, and bc to the boundary condition. Now, solving the state equation coincides with minimizing the stored elastic energy

$$\mathbf{E} = \frac{1}{2} \sum_{\mathbf{C}} h^2 \sum_{i=1}^{4} \sum_{j=1}^{2} \frac{(f_{i,j}(\mathbf{C}))^2}{E}$$

with $f_i(\mathbf{C})$ denoting the force vectors on the edges of the cell $\mathbf{C}$, subject to the constraint $\mathbf{Af} = \mathbf{b}$. This can be rephrased in a Lagrangian formulation for the Lagrangian

$$\mathbf{L}(\mathbf{f}, \boldsymbol{\lambda}) = \frac{1}{2} \mathbf{f}^T \mathbf{M} \mathbf{f} + \mathbf{f} \boldsymbol{\lambda}^T (\mathbf{b} - \mathbf{Af}) .$$

Due to the invertibility of the matrix $\mathbf{M}$ we obtain the equations

$$\mathbf{f} = \mathbf{M}^{-1} \mathbf{A}^T \boldsymbol{\lambda} \quad \text{and}$$

$$\mathbf{A} \mathbf{M}^{-1} \mathbf{A}^T \boldsymbol{\lambda} = \mathbf{b} . \tag{4.4}$$

as the necessary conditions for a saddle point. They have to be solved first for the dual solution $\boldsymbol{\lambda}$ and then for the force vector $\mathbf{f}$. If $\ker \mathbf{A}^T = \{0\}$, then $\mathbf{Z} = \mathbf{A}\mathbf{M}^{-1}\mathbf{A}^T$ is positive definite and thus invertible. In general $\mathbf{A}^T$ is underdetermined with nontrivial kernel. Thus, we have to eliminate rows of $\mathbf{A}$ to reduce the kernel. For domains consisting of a single subdomain D_i with either periodic, branching periodic or non-periodic boundary conditions, the linear dependencies can be reduced to a small number of cases (cf. [11]). For composite domains consisting of several subdomains with different coupling and boundary conditions, the number, type and complexity of linear dependencies increases. Let us suppose that the number of boundary conditions (e.g. the number of edges on the boundary with prescribed forces is large enough to ensure that $\mathbf{A}^T$ has more rows than columns. Then, we apply a QR-decomposition of $\mathbf{A}^T$ to find a basis of its kernel and use this to reduce the linear system $\mathbf{Af} = \mathbf{b}$ by the elimination of redundant equations. This in turn successively reduces $\mathbf{A}$ and $\mathbf{b}$ until A has full (column-)rank and thus $\ker \mathbf{A}^T = \{0\}$. For the results presented here the `CHOLMOD` [3] as a tool in the `Suitesparse` package was used, which comes along with an efficient parallel implementation.

5 Shape Optimization

Optimization of the elastic shape coincides in the discrete set up with an optimization of the discrete phase field and thus a minimization of the objective functional (3.2) added up over all subdomains. To this end, we apply an alternating solution strategy, i.e. we alternatingly solve for the forces $\mathbf{f}$ for fixed discrete phase field $\mathbf{v}$ and improve the phase field v for given forces $\mathbf{f}$. A threshold for the difference between two consecutive phase fields in L^2 is taken into account as a stopping criterium for this descent scheme. To improve the discrete phase field given a force vector $\mathbf{f}$ we apply a Gauss-Seidel type iteration that optimizes the values of $\mathbf{v}$ on single cells of the finite volume mesh. Let us consider the discretized version of $\mathcal{J}$ given in (3.2)

$$\mathbf{J}[\mathbf{v}] := \min_{\tau \in \Sigma_{\mathrm{ad}}} \sum_{\mathbf{C}} \mathbf{J}_{\mathbf{C}}[\mathbf{v}] \text{ with}$$

$$\mathbf{J}_{\mathbf{C}}[\mathbf{v}] = \frac{|\sigma(\mathbf{C})|^2}{\mathbf{v}(\mathbf{C})} + \beta \mathbf{v}(\mathbf{C}) + \frac{\eta}{\epsilon} \frac{32}{\pi^2} (\mathbf{v}(\mathbf{C}) - \delta)(1 - \mathbf{v}(\mathbf{C}))$$
$$+ \frac{\eta \epsilon}{4} \sum_{i=1}^{4} \frac{(\mathbf{v}(\mathbf{C}) - \mathbf{v}(\mathbf{C}^{(i)}))^2}{h^2}$$

Here, $\mathbf{C}^{(i)}$ denotes the cell adjacent to $\mathbf{C}$ across the edge $\mathbf{e}_i$. To find a minimum of $\mathbf{J}[\mathbf{v}]$ for $\mathbf{v}(\mathbf{C})$ for all $\mathbf{C}$ and for fixed $\mathbf{v}(\mathbf{C}^{(i)})$, we we apply Newton's method to compute the minimum of the rescaled function $\mathbf{v}(\mathbf{C}) \mapsto \mathbf{J}_{\mathbf{C}}(e^{-\mathbf{v}})$. The rescaling turns out to be an appropriate reformulation to overcome difficulties due to the singularity at $v(\mathbf{C}) = 0$.

To ensure a good performance of Newton's method, it is important to choose a suitable initialization of v. To this end we consider the local terms of the cost functional dropping the term involving the discrete gradient and define: $\tilde{\mathbf{J}}_{\mathbf{C}}(\mathbf{v}) = \frac{|\sigma|^2}{\mathbf{v}} + \beta \mathbf{v} + \frac{\eta}{\epsilon} \frac{32}{\pi^2} (\mathbf{v} - \delta)(1 - \mathbf{v})$. It is easy to check that this function has two minima and we initialize $\mathbf{v}$ with the minimal value of these two minima, again using in the algorithm the rescaled function $\mathbf{v}(\mathbf{C}) \mapsto \tilde{\mathbf{J}}_{\mathbf{C}}(e^{-\mathbf{v}})$ (or an approximation of it) as long as this value is smaller or equal 1. Otherwise, we initialize with the value 1.

This led to an initialization $\mathbf{v}$ depending on the current stress value on $\mathbf{C}$. Starting from the second iteration of the alternating descent algorithm, the phase field could also be initialized using the result of the last phase field optimization. However, this approach is prone to becoming stuck in local minima.

6 Results

We applied our method for two different loads on a quadratic domain $D \subset [0, 1]^2$, subdivided into subdomain as displayed in Fig. 2. In particular, the central region of the domain is filled with subdomains with inscribed locally periodic elastic structure and periodic boundary conditions for the forces, where as in the vicinity of the boundary branching periodic subdomains are taken into account. The two load scenarios are compression and shear as depicted in Fig. 4.

For all domain types, the subdomains were discretized using $N \times N$ cells with $N = 200$. On each cell of the outer layer, forces were applied on the edge intervals $[\frac{2}{6}N, \frac{3}{6}N]$, $[\frac{4}{6}N, \frac{5}{6}N]$ on all horizontal and vertical boundaries. These forces are depicted in Fig. 4 for both scenario.

Compression Load Pillar-like structures support the load on the boundary, branching structures transfer load to a mesh-like structure in the center region of the domain as depicted in Fig. 5. The elastic structures on corner and coupling cells connect the branching periodic pillars. Let us point to a small artifact in the optimal shape shown in one of the magnifications, where the phase field could obviously not been fully optimized locally.

Shear Load Figure 6 shows the optimal elastic shape in case of the shear load. On the resulting compound, optimized elastic structure forces are transferred via branching type structures to the center.

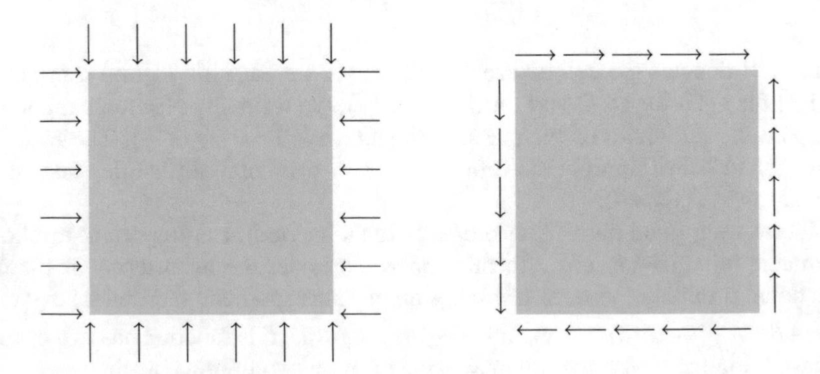

Fig. 4 Sketch of the applied loads: compression (left) and shear (right)

Fig. 5 The optimal shape for the compression load case and a subdomain structure with 13 cell types is depicted (top). We show in addition a color coding of the von Mises stresses using the colorbar on the set $[\mathbf{v} > 0.5]$ (bottom)

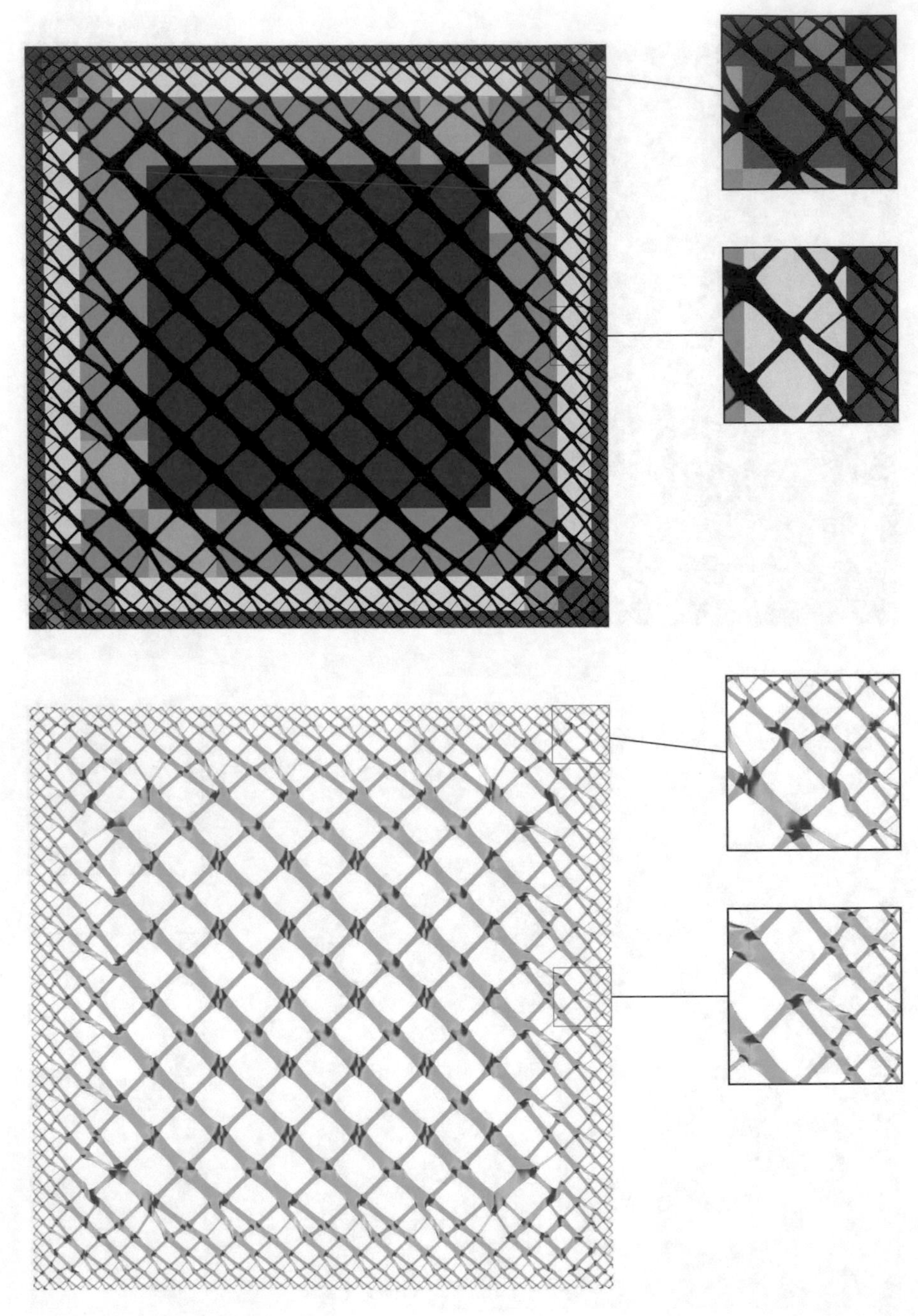

Fig. 6 The optimal shape for the shear load case and the same subdomain structure as in Fig. 5 is displayed. with two regions being magnified on the right. Again together with the optimal shape (top) the associated von Mises stresses are rendered (bottom)

References

1. ALLAIRE, G. *Shape optimization by the homogenization method*, vol. 146 of *Applied Mathematical Sciences*. Springer-Verlag, New York, 2002.
2. BENDSØE, M. P. *Optimization of structural topology, shape, and material*. Springer-Verlag, Berlin, 1995.
3. CHEN, Y., DAVIS, T. A., HAGER, W. W., AND RAJAMANICKAM, S. Algorithm 887: Cholmod, supernodal sparse cholesky factorization and update/downdate. *ACM Trans. Math. Softw. 35*, 3 (Oct. 2008), 22:1–22:14.
4. CIORANESCU, D., AND DONATO, P. *An Introduction to Homogenization*. Oxford University Press, Oxford, 1999.
5. E, W., AND ENGQUIST, B. The heterogeneous multiscale methods. *Commun. Math. Sci. 1*, 1 (2003), 87–132.
6. E, W., AND ENGQUIST, B. The heterogeneous multi-scale method for homogenization problems. In *Multiscale Methods in Science and Engineering*, vol. 44 of *Lecture Notes in Computational Science and Engineering*. Springer Berlin Heidelberg, 2005, pp. 89–110.
7. E, W., ENGQUIST, B., AND HUANG, Z. Heterogeneous multiscale method: A general methodology for multiscale modeling. *Physical Review B 67*, 9 (March 2003), 1–4.
8. E, W., MING, P., AND ZHANG, P. Analysis of the heterogeneous multiscale method for elliptic homogenization problems. *J. Amer. Math. Soc. 18*, 1 (2005), 121–156.
9. KOHN, R. V., AND WIRTH, B. Optimal fine-scale structures in compliance minimization for a uniaxial load. *Proceedings of the Royal Society of London A: Mathematical, Physical and Engineering Sciences 470*, 2170 (2014).
10. KOHN, R. V., AND WIRTH, B. Optimal fine-scale structures in compliance minimization for a shear load. *Communications in Pure and Applied Mathematics* (2015). to appear.
11. LÜTHEN, N. Numerical shape optimization of branching-periodic elastic structures. Master thesis, University of Bonn, 2016.
12. MILTON, G. W. *The Theory of Composites*. Cambridge University Press, 2002.
13. MODICA, L., AND MORTOLA, S. Un esempio di Γ^--convergenza. *Boll. Un. Mat. Ital. B (5) 14*, 1 (1977), 285–299.
14. MÜLLER, R. Hierarchical microimaging of bone structure and function. *Nat. Rev. Rheumatol. 5*, 7 (2009), 373–381.
15. MÜLLER, V., BRYLKA, B., DILLENBERGER, F., GLÖCKNER, R., AND BÖHLKE, T. Homogenization of elastic properties of short-fiber reinforced composites based on microstructure data. *J. Compos. Mater. 50*, 3 (mar 2015), 297–312.
16. NEMAT-NASSER, S., WILLS, J., SRIVASTAVA, A., AND AMIRKHIZI, A. Homogenization of periodic elastic composites and locally resonant materials. *Phys. Rev. B 83* (Mar 2011), 104103.
17. PENZLER, P., RUMPF, M., AND WIRTH, B. A phase-field model for compliance shape optimization in nonlinear elasticity. *ESAIM: Control, Optimisation and Calculus of Variations 18*, 1 (2012), 229–258.

Shape and Topological Derivatives via One Sided Differentiation of the Minimax of Lagrangian Functionals

Michel C. Delfour

Abstract A standard approach to the *minimization* of an *objective function* in the presence of *equality constraints* in Mathematical Programming or of a *state equation* in Control Theory is the introduction of Lagrange multipliers or an *adjoint state*, that is, a linear penalization of the equality constraints or the state equation. The initial minimization problem is equivalent to the minimax of the associated Lagrangian. This approach can also be used to compute the one-sided directional derivative with respect to the control or the shape or topology of a family of sets. It is sufficient to consider a Lagrangian parametrized by a positive parameter $t > 0$ as t goes to 0. This involves the introduction of some forms of equation for the *multiplier* or the *adjoint state* at $t > 0$. The simplest one is the *standard adjoint*, but other forms of closely related adjoint such as the *quasi adjoint* in Serovaiskii (Mat Zametki 54(2):85–95, 159, 1993; translation in Math Notes 54(1–2):825–832, 1994) and Kogut et al. (Z Anal Anwend (J Anal Appl) 34(2):199–219, 2015) or the new *averaged adjoint* in Sturm (On shape optimization with non-linear partial differential equations, Doctoral thesis, Technische Universiltät of Berlin, 2014; SIAM J Control Optim 53(4):2017–2039, 2015) can be considered. In general, the various forms of adjoints coincide at $t = 0$. For instance, by using the *averaged adjoint*, the minimax problem need not be related to a saddle point as in Correa-Seeger (Nonlinear Anal Theory Methods Appl 9:13–22, 1985) and the so-called *dual problem* need not make sense.

In this paper, we sharpen the results of Sturm (On shape optimization with non-linear partial differential equations, Doctoral thesis, Technische Universiltät of Berlin, 2014; SIAM J Control Optim 53(4):2017–2039, 2015) which have been recently extended in Delfour and Sturm (J Convex Anal 24(4):1117–1142, 2017; Minimax differentiability via the averaged adjoint for control/shape sensitivity. In: Proceedings of the 2nd IFAC Workshop on Control of Systems Governed by

---

M. C. Delfour (✉)
Centre de recherches mathématiques and Département de mathématiques et de statistique, Université de Montréal, Montréal, QC, Canada
e-mail: delfour@crm.umontreal.ca

V. Schulz, D. Seck (eds.), *Shape Optimization, Homogenization and Optimal Control*, International Series of Numerical Mathematics 169,
https://doi.org/10.1007/978-3-319-90469-6_12

Partial Differential Equations, IFAC-PaperOnLine 49-8, pp 142–149, 2016) from the single valued case to the case where the solutions of the state/averaged adjoint state equations are not unique. In such a case, a non-differentiability can occur and only a one-sided directional derivative is expected even if the functions at hand are infinitely differentiable as was illustrated in the seminal paper of Danskin (SIAM J Appl Math 14(4):641–644) in 1966. Some examples for control and for shape and topological derivatives will be given.

Keywords Minimax · One-sided differentiability · Averaged adjoint · Control · Shape · Topological derivatives · Rectifiability · Mathematical programming

1 Introduction

The generic notions of *shape* and *topological derivatives* have proven to be both pertinent and useful from the theoretical and numerical points of view. The *shape derivative* is a differential (see [5, 12, 20]) while the *topological derivative* of [23] obtained by the *method* of *matched* and *compound expansions* (for instance, see the recent book of Novotny and J. Sokołowski [21]) is only a *semidifferential* (one sided directional derivative). This arises from the fact that the tangent space to the underlying metric spaces of "geometries" is only a cone. By using the notions of *d-dimensional Minkowski content* [17] and *d-rectifiable set* [16], the definition of a topological derivative was extended in [6, 7] to perturbations obtained by creating holes around curves, surfaces, and, potentially, microstructures. In that context, the *Hadamard semidifferential* that retains the advantages of the standard differential calculus including the *chain rule* and the fact that semiconvex functions are Hadamard semidifferentiable (see [4, 18]) is a *natural notion* to study the semidifferentiability of objective functions with respect to the sets/geometries that belong to complete non-linear non-convex metric spaces.

An important advantage for *state constrained objective functions* is that theorems on the one-sided differentiation of minimax of Lagrangians can be used to get the semidifferential (see [2, 10–12]). For instance, a standard approach to the *minimization* of a state constrained *objective function* in Control/Shape Optimization problems is to consider the minimax of the associated Lagrangian. By using the new notion of *averaged adjoint* introduced by Sturm [24, 25], the minimax problem need not be related to a saddle point: non-convex objective functions and non-linear state equations can be directly considered. The same adjoint system operator will occur for both shape and topological derivatives. Recently, Delfour and Sturm [9] provided a new simpler and more general version of the original condition of [24, 25]. Finally, in [8] the new condition is extended from the single valued case to the case where the solutions of the state/averaged adjoint state equations are not unique in which case a non-differentiability can occur as predicted in the pioneering work of Danskin [3] followed by Dem'janov [13–15] from 1968 to 1974.

Note that other forms of adjoints have been used in the literature such as, for instance, the *quasi adjoint* in Serovaiskii [22] and Kogut et al. [19].

2 Illustrative Examples from PDE Control

Let $\Omega \subset \mathbb{R}^N$, $N \geq 1$, bounded open and $a \in L^2(\Omega)$ be the *control variable* to which is associated the *state* $u = u(a) \in H_0^1(\Omega)$ solution of the variational *state equation*

$$\int_\Omega \nabla u(a) \cdot \nabla \psi - a\,\psi\, dx = 0, \quad \forall \psi \in H_0^1(\Omega), \tag{2.1}$$

where $x \cdot y$ denotes the inner product of x and y in $\mathbb{R}^N$. Given a target function $g \in L^2(\Omega)$, associate with $u(a)$ the *objective function*

$$f(a) \stackrel{\text{def}}{=} \int_\Omega \frac{1}{2}|u(a) - g|^2\, dx. \tag{2.2}$$

By introducing the Lagrangian, we get an unconstrained minimax formulation

$$G(a, \varphi, \psi) \stackrel{\text{def}}{=} \int_\Omega \frac{1}{2}|\varphi - g|^2\, dx + \int_\Omega \nabla\varphi \cdot \nabla\psi - a\,\psi\, dx$$

$$\boxed{f(a) = \inf_{\varphi \subset H^1(\Omega)} \sup_{\psi \in H^1(\Omega)} G(a, \varphi, \psi).}$$

If we are only interested in a descent method, we can obtain the semidifferential of $f(a)$ by a similar minimax formulation. Given the direction $b \in L^2(\Omega)$, to compute

$$df(a; b) \stackrel{\text{def}}{=} \lim_{t \searrow 0} \frac{f(a + tb) - f(a)}{t},$$

where the *state* $u^t \in H_0^1(\Omega)$ at $t > 0$ is solution of

$$\int_\Omega \nabla u^t \cdot \nabla \psi - (a + tb)\,\psi\, dx = 0, \quad \forall \psi \in H_0^1(\Omega). \tag{2.3}$$

The *associated Lagrangian* is

$$L(t, \varphi, \psi) \stackrel{\text{def}}{=} \int_\Omega \frac{1}{2}|\varphi - g|^2\, dx + \int_\Omega \nabla\varphi \cdot \nabla\psi - (a + tb)\,\psi\, dx.$$

It is readily seen that

$$\boxed{\begin{gathered} g(t) \stackrel{\text{def}}{=} \inf_{\varphi \in H_0^1(\Omega)} \sup_{\psi \in H_0^1(\Omega)} L(t, \varphi, \psi) = f(a + tb) \\ dg(0) \stackrel{\text{def}}{=} \lim_{t \searrow 0} \frac{g(t) - g(0)}{t} = df(a; b). \end{gathered}}$$

The following more substantial example of a state constrained objective function is detailed in [24, p. 64]. Let Ω be a bounded open domain in $\mathbb{R}^N$, $N \geq 1$, and $a \in L^2(\Omega)$ be the *control variable* to which is associated the *state* $u = u(a)$ solution of the partial differential equation

$$-\Delta u(a) + \rho(u(a)) = a \text{ in } \Omega, \quad u(a) = 0 \text{ on } \partial\Omega,$$

where ρ is a continuously differentiable, bounded, and non-decreasing function. In weak form the *state* $u = u(a) \in H_0^1(\Omega)$ is solution of the variational *state equation*

$$\int_\Omega \nabla u \cdot \nabla \psi + \rho(u)\, \psi - a\, \psi\, dx = 0, \quad \forall \psi \in H_0^1(\Omega), \tag{2.4}$$

where $x \cdot y$ denotes the inner product of x and y in $\mathbb{R}^N$. Given a target function $g \in L^2(\Omega)$, associate with $u(a)$ the *state constrained objective function*

$$f(a) \stackrel{\text{def}}{=} \int_\Omega |u(a) - g|^2 + \alpha\, |a|^2\, dx, \quad \alpha > 0,$$

to be minimized over $L^2(\Omega)$. By introducing the Lagrangian, we get an unconstrained minimax expression of $f(a)$:

$$G(a, \varphi, \psi) \stackrel{\text{def}}{=} \int_\Omega |\varphi - g|^2 + \alpha\, |a|^2 + \nabla\varphi \cdot \nabla\psi + \rho(\varphi)\, \psi - a\, \psi\, dx,$$
$$f(a) = \inf_{\varphi \in H^1(\Omega)} \sup_{\psi \in H^1(\Omega)} G(a, \varphi, \psi).$$

The semidifferential of $f(a)$ is obtained by a similar minimax formulation. Given the direction $b \in L^2(\Omega)$, we want to compute $df(a; b) = \lim_{t \searrow 0}(f(a + tb) - f(a))/t$. The *state* $u^t \in H_0^1(\Omega)$ at $t \geq 0$ is solution of

$$\int_\Omega \nabla u^t \cdot \nabla\psi + \rho(u^t)\, \psi - (a + tb)\, \psi\, dx = 0, \quad \forall \psi \in H_0^1(\Omega), \tag{2.5}$$

and the *associated Lagrangian* is

$$L(t, \varphi, \psi) \stackrel{\text{def}}{=} \int_\Omega |\varphi - g|^2 + \alpha\, |a + tb|^2 + \nabla\varphi \cdot \nabla\psi + \rho(\varphi)\, \psi - (a + tb)\, \psi\, dx.$$

It is readily seen that

$$g(t) \stackrel{\text{def}}{=} f(a + tb) = \inf_{\varphi \in H_0^1(\Omega)} \sup_{\psi \in H_0^1(\Omega)} L(t, \varphi, \psi),$$

$$dg(0) \stackrel{\text{def}}{=} \lim_{t \searrow 0}(g(t) - g(0))/t = df(a; b).$$

The objective is to find general conditions under which the right-hand side t-derivative of a parametrized minimax at $t = 0$ exists and to give its expression. This is a first step toward the computation of a (semi) differential with respect to the control variable a in a direction b.

3 Shape Derivative via the Velocity Method

It is now well-established that the *Velocity Method* introduced by Zolésio [26] in 1979 is naturally associated with the construction of the *Courant metrics* on special groups of diffeomorphisms by Micheletti [20] in 1972. We first provide a simple example and then explain the relationship.

3.1 Simple Illustrative Example

Consider the state equation (2.4) and the associated objective function (2.2). In the *Velocity Method*, the domain Ω is perturbed by a family of diffeomorphisms T_t generated by a sufficiently smooth velocity field $V(t)$:

$$\frac{dx}{dt}(t; X) = V(t, x(t; X)), \; x(0; X) = X, \quad T_t(X) \overset{\text{def}}{=} x(t; X), \; t \geq 0.$$

Denote by $\Omega_t \overset{\text{def}}{=} T_t(\Omega)$ the perturbed domain. The state equation and objective function at $t > 0$ become

$$\int_{\Omega_t} \nabla u_t \cdot \nabla\psi - a\,\psi\,dx = 0, \quad \forall\psi \in H_0^1(\Omega_t), \quad g(t) \overset{\text{def}}{=} \int_{\Omega_t} |u_t - g|^2\,dx. \tag{3.1}$$

Introducing the composition $u^t = u_t \circ T_t$ to work in the fixed space $H_0^1(\Omega)$:

$$\int_\Omega \left[A(t)\nabla u^t \cdot \nabla\psi - a\,\psi\right] j(t)\,dx = 0, \quad \forall\psi \in H_0^1(\Omega), \tag{3.2}$$

$$A(t) = DT_t^{-1}\,(DT_t^{-1})^*, \; j(t) = \det DT_t, \quad DT_t \text{ is the Jacobian matrix,}$$

$$\Rightarrow g(t) = \int_{\Omega_t} |u_t - g|^2\,dx = \int_\Omega |u^t - g \circ T_t|^2\,j(t)\,dx, \tag{3.3}$$

$$\text{Lagrangian}: L(t, \varphi, \psi) \overset{\text{def}}{=} \int_\Omega \left[\frac{1}{2}|\varphi - g \circ T_t|^2 + A(t)\nabla\varphi \cdot \nabla\psi - a\,\psi\right] j(t)\,dx.$$

$$\Rightarrow g(t) = \inf_{\varphi \in H_0^1(\Omega)} \sup_{\psi \in H_0^1(\Omega)} L(t, \varphi, \psi),$$

$$dg(0) = \lim_{t \searrow 0} (g(t) - g(0))/t = df(\Omega; V(0)).$$

3.2 Generic Construction of Micheletti: Metric Group, Courant Metric, and Its Tangent Space

We sketch Micheletti's [20] generic construction of complete groups of diffeomorphisms and her *Courant metrics*. Associate with a real vector space (usually a Banach space) Θ of mappings $\theta : \mathbb{R}^N \to \mathbb{R}^N$ (Micheletti used the space $\Theta = C_0^k(\mathbb{R}^N, \mathbb{R}^N)$, $k \geq 1$), the following space of transformations (endomorphisms) of $\mathbb{R}^N$:

$$\boxed{\mathcal{F}(\Theta) \stackrel{\text{def}}{=} \left\{ F : \mathbb{R}^N \to \mathbb{R}^N \text{ bijective } : F - I \in \Theta, \text{ and } F^{-1} - I \in \Theta \right\},} \tag{3.4}$$

where $x \mapsto I(x) \stackrel{\text{def}}{=} x : \mathbb{R}^N \to \mathbb{R}^N$ is the identity mapping.

Given a fixed set $\Omega_0 \subset \mathbb{R}^N$ (Micheletti used a bounded open set of class C^k, hence *crack-free*), consider the set of images

$$\mathcal{X}(\Omega_0) \stackrel{\text{def}}{=} \{F(\Omega_0) : \forall F \in \mathcal{F}(\Theta)\} \tag{3.5}$$

of Ω_0 by the elements of $\mathcal{F}(\Theta)$ and the subgroup

$$\mathcal{G}(\Omega_0) \stackrel{\text{def}}{=} \{F \in \mathcal{F}(\Theta) : F(\Omega_0) = \Omega_0\}.$$

So there is a bijection between the set of images of Ω_0 and the quotient space

$$\boxed{\mathcal{X}(\Omega_0) \longleftrightarrow \mathcal{F}(\Theta)/\mathcal{G}(\Omega_0).}$$

The objective is to construct a metric on $\mathcal{F}(\Theta)/\mathcal{G}(\Omega_0)$ that will serve as a distance between two mages $F_1(\Omega_0)$ and $F_2(\Omega_0)$. Associate with $F \in \mathcal{F}(\Theta)$ the following candidate for a metric

$$\boxed{d_0(I, F) \stackrel{\text{def}}{=} \|F - I\|_\Theta + \|F^{-1} - I\|_\Theta, \quad d_0(F, G) \stackrel{\text{def}}{=} d_0(I, G \circ F^{-1}).} \tag{3.6}$$

Unfortunately, d_0 is only a semi-metric that will not satisfy the triangle inequality. Consider the following second candidate

$$\boxed{d(I, F) \stackrel{\text{def}}{=} \inf_{\substack{F=F_1\circ\cdots\circ F_n \\ F_i\in\mathcal{F}(\Theta)}} \sum_{i=1}^{n} \|F_i - I\|_\Theta + \|F_i^{-1} - I\|_\Theta,} \tag{3.7}$$

where the infimum is taken over all *finite factorizations* of F

$$F = F_1 \circ \cdots \circ F_n, \quad F_i \in \mathcal{F}(\Theta).$$

Extend this function to all F and G in $\mathcal{F}(\Theta)$:

$$\boxed{d(F, G) \stackrel{\text{def}}{=} d(I, G \circ F^{-1}).} \tag{3.8}$$

By definition, d is right-invariant since for all F, G and H in $\mathcal{F}(\Theta)$

$$d(F, G) = d(F \circ H, G \circ H).$$

This metric was called the *Courant metric* by Micheletti [20].

3.3 Examples of Θ and Continuity with Respect to the Courant Metric

$(\mathcal{F}(\Theta), d)$ is complete for Θ equal to the Banach spaces

$$C_0^k(\mathbb{R}^N, \mathbb{R}^N),\ C^k(\overline{\mathbb{R}^N}, \mathbb{R}^N) \subset \mathcal{B}^k(\mathbb{R}^N, \mathbb{R}^N) \text{ and } C^{k,1}(\overline{\mathbb{R}^N}, \mathbb{R}^N), \quad k \geq 0,$$

and, through special constructions, for the Fréchet spaces

$$C_0^\infty(\mathbb{R}^N, \mathbb{R}^N) \subset \mathcal{B}(\mathbb{R}^N, \mathbb{R}^N) = \cap_{k\geq 0}\mathcal{B}^k(\mathbb{R}^N, \mathbb{R}^N).$$

For any Banach or Fréchet space $\Theta \subset C^{0,1}(\overline{\mathbb{R}^N}, \mathbb{R}^N)$, $\mathcal{F}(\Theta)$ is an open subset of $I + \Theta$

- the tangent space is Θ at each point $F \in \mathcal{F}(\Theta)$
- and the associated smooth structure is trivial.

The analogue would be the general linear group GL(n) of invertible lincar functions from $\mathbb{R}^N$ to $\mathbb{R}^N$ which is an open subset of $\mathcal{L}(\mathbb{R}^N, \mathbb{R}^N)$. So, the tangent space in each point of GL(n) is the whole space $\mathcal{L}(\mathbb{R}^N, \mathbb{R}^N)$.

The continuity of a function $J : \mathcal{X}(\Omega_0) \to \mathbb{R}$ with respect to the Courant metric can be expressed in a rather nice way. For instance, choose $\Theta = C_0^k(\mathbb{R}^N, \mathbb{R}^N)$, $k \geq 1$, $\mathcal{F}(\Theta)$, and the set $\mathcal{X}(\Omega_0)$ of the images of an *open crack free set* $\Omega_0 \subset \mathbb{R}^N$.

Theorem 3.1 *Let $\Omega = F(\Omega_0) \in \mathcal{X}(\Omega_0)$ for some $F \in \mathcal{F}(\Omega_0)$. Then J is continuous at Ω for the Courant metric if and only if*

$$\lim_{t \searrow 0} J(T_t(\Omega)) = J(\Omega), \quad \frac{dT_t}{dt} = V(t) \circ T_t, \quad T_0 = F,$$

for all velocity fields $V \in C^0([0, \tau]; C_0^k(\mathbb{R}^N, \mathbb{R}^N))$.

3.4 Hadamard Semidifferentiability

As was shown in Theorem 3.1, the Velocity Method is naturally associated with the characterization of the continuity of a function on the metric space $(\mathcal{X}(\Omega_0), d)$. It is natural to introduce a slightly relaxed form of Hadamard semidifferentiability/differentiability.

Definition 3.1 Let $\Theta = C_0^k(\mathbb{R}^N, \mathbb{R}^N)$. The function $J : \mathcal{X}(\Omega_0) = \{F(\Omega_0) : F \in \mathcal{F}(\Theta)\} \to \mathbb{R}$ is *Hadamard semidifferentiable* at $F(\Omega_0)$, $F \in \mathcal{F}(\Theta)$, if

(i) for all $V \in C^0([0, \tau]; \mathcal{D}(\mathbb{R}^N, \mathbb{R}^N))$

$$dJ(F(\Omega_0); V) \stackrel{\text{def}}{=} \lim_{t \searrow 0} \frac{J(T_t(V)(F(\Omega_0)) - J(F(\Omega_0)}{t} \quad \text{exists},$$

$$\frac{dT_t}{dt} = V(t) \circ T_t, \; T_0 = F,$$

(ii) and there exists a function $dJ(F(\Omega_0)) : \mathcal{D}(\mathbb{R}^N, \mathbb{R}^N) \to \mathbb{R}$ such that for all $V \in C^0([0, \tau]; \mathcal{D}(\mathbb{R}^N, \mathbb{R}^N))$

$$dJ(F(\Omega_0); V) = dJ(F(\Omega_0))(V(0)). \qquad \square$$

Definition 3.2 $J : \mathcal{X}(\Omega_0) \to \mathbb{R}$ is *Hadamard differentiable* at $F(\Omega_0)$, $F \in \mathcal{F}(\Theta)$, if

(i) it is Hadamard semidifferentiable at $F(\Omega_0)$
(ii) and $dJ(F(\Omega_0)) : \mathcal{D}(\mathbb{R}^N, \mathbb{R}^N) \to \mathbb{R}$ is linear and continuous.

□

This is to be compared with the definitions of *Eulerian semiderivative* and *Gradient* given in the thesis of J.-P. Zolésio [26] in 1979.

Definition 3.3 (Definition 3.1) Let $\Theta = C_0^k(\mathbb{R}^N, \mathbb{R}^N)$. The function $J : \mathcal{X}(\Omega_0) = \{F(\Omega_0) : F \in \mathcal{F}(\Theta)\} \to \mathbb{R}$ is *Hadamard semidifferentiable* at $F(\Omega_0)$, $F \in \mathcal{F}(\Theta)$, if

(i) for all $V \in C^0([0, \tau]; \mathcal{D}(\mathbb{R}^N, \mathbb{R}^N))$

$$dJ(F(\Omega_0); V) \overset{\text{def}}{=} \lim_{t \searrow 0} \frac{J(T_t(V)(F(\Omega_0)) - J(F(\Omega_0)}{t} \quad \text{exists,}$$

$$\frac{dT_t}{dt} = V(t) \circ T_t, \; T_0 = F,$$

Definition of the Eulerian semiderivative *in Zolésio thesis [26, p. 12].*

(ii) and there exists a function $dJ(F(\Omega_0)) : \mathcal{D}(\mathbb{R}^N, \mathbb{R}^N) \to \mathbb{R}$ such that for all $V \in C^0([0, \tau]; \mathcal{D}(\mathbb{R}^N, \mathbb{R}^N))$

$$dJ(F(\Omega_0); V) = dJ(F(\Omega_0))(V(0)).$$ □

Definition 3.4 (Definition 3.2) $J : \mathcal{X}(\Omega_0) \to \mathbb{R}$ is *Hadamard differentiable* at $F(\Omega_0)$, $F \in \mathcal{F}(\Theta)$, if

- it is *Hadamard semidifferentiable* at $F(\Omega_0)$
- and the function $dJ(F(\Omega_0)) : \mathcal{D}(\mathbb{R}^N, \mathbb{R}^N) \to \mathbb{R}$ is linear and continuous.

Definition of the Gradient of J *in the 1979 thesis of Zolésio* [26, p. 13] □

4 State Constrained Objective Functions: Averaged Adjoint

In this section we recall the framework of [24, 25] and its generalization to the multivalued case in [8, 9].

4.1 Abstract Framework

4.1.1 Preliminaries

In this paper, a *Lagrangian* is a function of the form

$$(t, x, y) \mapsto G(t, x, y) : [0, \tau] \times X \times Y \to \mathbb{R}, \quad \tau > 0,$$

where Y is a *vector space*, X is a non empty subset of a vector space, and $y \mapsto G(t, x, y)$ is *affine*. Associate with the *parameter t* the *parametrized minimax*

$$t \mapsto g(t) \overset{\text{def}}{=} \inf_{x \in X} \sup_{y \in Y} G(t, x, y) : [0, \tau] \to \mathbb{R}. \tag{4.1}$$

When the limits exist we shall use the following compact notation:

$$dg(0) \stackrel{\text{def}}{=} \lim_{t \searrow 0} \frac{g(t) - g(0)}{t} \qquad \begin{cases} \underline{d}g(0) \stackrel{\text{def}}{=} \liminf\limits_{t \searrow 0} \left(g(t) - g(0)\right)/t \\ \overline{d}g(0) \stackrel{\text{def}}{=} \limsup\limits_{t \searrow 0} \left(g(t) - g(0)\right)/t \end{cases}$$

$$d_t G(0, x, y) \stackrel{\text{def}}{=} \lim_{t \searrow 0} \frac{G(t, x, y) - G(0, x, y)}{t}$$

$$\varphi \in X, \quad d_x G(t, x, y; \varphi) \stackrel{\text{def}}{=} \lim_{\theta \searrow 0} \frac{G(t, x + \theta\varphi, y) - G(t, x, y)}{\theta}$$

$$\psi \in Y, \quad d_y G(t, x, y; \psi) \stackrel{\text{def}}{=} \lim_{\theta \searrow 0} \frac{G(t, x, y + \theta\psi) - G(t, x, y)}{\theta}.$$

The notation $t \searrow 0$ and $\theta \searrow 0$ means that t and θ go to 0 by strictly positive values $t > 0$ and $\theta > 0$.

4.1.2 Minimax and Averaged Adjoint

Since $G(t, x, y)$ is affine in y, for all $(t, x) \in [0, \tau] \times X$,

$$\forall y, \psi \in Y, \quad d_y G(t, x, y; \psi) = G(t, x, \psi) - G(t, x, 0) = d_y G(t, x, 0; \psi).$$

The following variational equation is the *state equation* at $t \geq 0$:

$$\boxed{\text{to find } x^t \in X \text{ such that for all } \psi \in Y,\ d_y G(t, x^t, 0; \psi) = 0.}$$

The set of solutions (*states*) x^t at $t \geq 0$ is denoted

$$E(t) \stackrel{\text{def}}{=} \left\{x^t \in X : \forall \varphi \in Y,\ d_y G(t, x^t, 0; \varphi) = 0\right\}$$

The standard adjoint state equation at $t \geq 0$:

$$\boxed{\text{to find } p^t \in Y \text{ such that } \forall \varphi \in X, \quad d_x G(t, x^t, p^t; \varphi) = 0},$$

$$Y(t, x^t) \stackrel{\text{def}}{=} \text{ set of solutions.}$$

Under appropriate conditions and uniqueness of the pair (x^t, p^t),

$$\boxed{dg(0) = d_t G(0, x^0, p^0),}$$

where (x^0, p^0) is the solution of the coupled state-adjoint state equations at $t = 0$. Finally, consider the following sets

$$\text{states:} \quad E(t) \stackrel{\text{def}}{=} \left\{x^t \in X : \forall \varphi \in Y,\ d_y G(t, x^t, 0; \varphi) = 0\right\}$$

$$\text{minimizers:} \quad X(t) \stackrel{\text{def}}{=} \left\{x^t \in X : g(t) = \inf_{x \in X} \sup_{y \in Y} G(t, x, y) = \sup_{y \in Y} G(t, x^t, y)\right\}.$$

Lemma 4.1 (Constrained Infimum and Minimax)

(i) $\inf_{x \in X} \sup_{y \in Y} G(t, x, y) = \inf_{x \in E(t)} G(t, x, 0)$.
(ii) *The minimax* $g(t) = +\infty$ *if and only if* $E(t) = \varnothing$. *Hence* $X(t) = X$.
(iii) *If* $E(t) \neq \varnothing$, *then*

$$X(t) = \{x^t \in E(t) : G(t, x^t, 0) = \inf_{x \in E(t)} G(t, x, 0)\} \subset E(t) \tag{4.2}$$

and $g(t) < +\infty$.

Hypothesis (H0) *Let* X *be a vector space.*

(i) *For all* $t \in [0, \tau]$, $x^0 \in X(0)$, $x^t \in X(t)$, *and* $y \in Y$, *the function*

$$s \mapsto G(t, x^0 + s(x^t - x^0), y) : [0, 1] \to \mathbb{R} \tag{4.3}$$

is absolutely continuous. This implies that, for almost all s, *the derivative exists and is equal to* $d_x G(t, x^0 + s(x^t - x^0), y; x^t - x^0)$ *and that it is the integral of its derivative. In particular,*

$$\boxed{G(t, x^t, y) = G(t, x^0, y) + \int_0^1 d_x G(t, x^0 + s(x^t - x^0), y; x^t - x^0)\, ds.} \tag{4.4}$$

(ii) *For all* $t \in [0, \tau]$, $x^0 \in X(0)$, $x^t \in X(t)$, $y \in Y$, $\varphi \in X$, *and almost all* $s \in (0, 1)$, $d_x G(t, x^0 + s(x^t - x^0), y; \varphi)$ *exists and the function* $s \mapsto d_x G(t, x^0 + s(x^t - x^0), y; \varphi)$ *belongs to* $L^1(0, 1)$.

Recall the definition of the *standard adjoint* at $t \geq 0$:

$$\boxed{\text{to find } p^t \in Y \text{ such that } \forall \varphi \in X, \quad d_x G(t, x^t, p^t; \varphi) = 0.}$$

Definition 4.1 (K. Sturm) Given $x^0 \in X(0)$ and $x^t \in X(t)$, the *averaged adjoint state equation*:

$$\boxed{\text{to find } y^t \in Y,\ \forall \varphi \in X, \quad \int_0^1 d_x G(t, x^0 + s(x^t - x^0), y^t; \varphi)\, ds = 0.} \tag{4.5}$$

The set of solutions will be denoted $Y(t, x^0, x^t)$. The *standard adjoint* p^t at $t \geq 0$ is a solution of the equation

$$\boxed{\text{to find } p^t \in Y \text{ such that } \forall \varphi \in X, \quad d_x G(t, x^t, p^t; \varphi) = 0.} \qquad \square$$

Clearly, $Y(0, x^0, x^0)$ reduces to the set of *standard adjoint states* p^0 at $t = 0$

$$\boxed{Y(0, x^0) \overset{\text{def}}{=} \left\{ p^0 \in Y : \forall \varphi \in X,\ d_x G(0, x^0, p^0; \varphi) = 0 \right\}.} \tag{4.6}$$

An *important consequence* of the introduction of the averaged adjoint state is the following identity: for all $x^0 \in X(0)$, $x^t \in X(t)$, and $y^t \in Y(t, x^0, x^t)$,

$$\boxed{g(t) = G(t, x^t, 0) = G(t, x^t, y^t) = G(t, x^0, y^t).} \tag{4.7}$$

Since for all y^0, $g(0) = G(0, x^0, 0) = G(0, x^0, y^0)$,

$$g(t) - g(0) = G(t, x^0, y^t) - G(0, x^0, y^0)$$

$$\Rightarrow \boxed{dg(0) = \lim_{t \searrow 0} \frac{g(t) - g(0)}{t} = \lim_{t \searrow 0} \frac{G(t, x^0, y^t) - G(0, x^0, y^0)}{t}.}$$

4.2 Original Condition of Sturm

Theorem 4.1 ([24], [25, Thm. 3.1]) *Consider the Lagrangian functional*

$$(t, x, y) \mapsto G(t, x, y) : [0, \tau] \times X \times Y \to \mathbb{R}, \quad \tau > 0,$$

where X and Y are vector spaces and the function $y \mapsto G(t, x, y)$ is affine. Let (H0) *and the following hypotheses be satisfied:*

(H1) *for all $t \in [0, \tau]$, $g(t)$ is finite, $X(t) = \{x^t\}$ and $Y(t, x^0, x^t) = \{y^t\}$ are singletons;*
(H2) *$d_t G(t, x^0, y)$ exists for all $t \in [0, \tau]$ and all $y \in Y$;*
(H3) *the following limit exists*

$$\boxed{\lim_{s \searrow 0,\, t \searrow 0} d_t G(s, x^0, y^t) = d_t G(0, x^0, y^0)}. \tag{4.8}$$

Then, $dg(0)$ exists and

$$dg(0) = d_t G(0, x^0, y^0).$$

Condition (H3) is typical of what can be found in the literature (e.g. [2]).

Proof From Hypothesis (H2), $d_t G(t, x^0, y)$ exists for all $t \in [0, \tau]$ and $y \in Y$. Hence, there exists $\theta_t \in (0, 1)$ such that

$$\begin{aligned} G(t, x^0, y^t) - G(0, x^0, y^0) &= G(0, x^0, y^t) + t\, d_t G(\theta_t t, x^0, y^t) - G(0, x^0, y^0) \\ &= \underbrace{d_y G(0, x^0, 0; y^t - y^0)}_{=0} + t\, d_t G(\theta_t t, x^0, y^t) \\ &= t\, d_t G(\theta_t t, x^0, y^t) \\ \Rightarrow \quad \frac{G(t, x^0, y^t) - G(0, x^0, y^0)}{t} &= d_t G(\theta_t t, x^0, y^t) \end{aligned}$$

since $d_y G(0, x^0, 0; y^t - y^0) = 0$. From hypothesis (H3)

$$\lim_{s \searrow 0,\, t \searrow 0} d_t G(s, x^0, y^t) = d_t G(0, x^0, y^0). \tag{4.9}$$

$$\Rightarrow \boxed{dg(0) = \lim_{t \searrow 0} \frac{G(t, x^0, y^t) - G(0, x^0, y^0)}{t} = d_t G(0, x^0, y^0).} \tag{4.10}$$

□

4.3 Sturm's Theorem Revisited and Extended: A New Condition

This is an extension of [24] and [25, Thm. 3.1] with only a local differentiability condition at $t = 0$. To our best knowledge, the *extra term* $R(0, x^0, y^0)$ is new (an example of a topological derivative will be given later).

Theorem 4.2 (Singleton Case, [8, 9]) *Consider the Lagrangian functional*

$$(t, x, y) \mapsto G(t, x, y) : [0, \tau] \times X \times Y \to \mathbb{R}, \quad \tau > 0,$$

where X and Y are vector spaces and the function $y \mapsto G(t, x, y)$ is affine. Let (H0) *and the following hypotheses be satisfied:*

(H1) *for all $t \in [0, \tau]$, $g(t)$ is finite, $X(t) = \{x^t\}$ and $Y(t, x^0, x^t) = \{y^t\}$ are singletons;*

(H2') *$d_t G(0, x^0, y^0)$ exists;*

(H3') *the following limit exists*

$$R(0, x^0, y^0) \overset{\text{def}}{=} \lim_{t \searrow 0} d_y G\left(t, x^0, 0; \frac{y^t - y^0}{t}\right). \tag{4.11}$$

Then, $dg(0)$ exists and

$$dg(0) = d_t G(0, x^0, y^0) + R(0, x^0, y^0).$$

Proof Recalling that $g(t) = G(t, x^t, y^t) = G(t, x^0, y^t)$,

$$\begin{aligned} g(t) - g(0) &= G(t, x^0, y^t) - G(0, x^0, y^0) \\ &= G(t, x^0, y^0) + d_y G(t, x^0, 0; y^t - y^0) - G(0, x^0, y^0) \end{aligned}$$

$$\Rightarrow \quad \frac{g(t) - g(0)}{t} = d_y G\left(t, x^0, 0; \frac{y^t - y^0}{t}\right) + \frac{G(t, x^0, y^0) - G(0, x^0, y^0)}{t}$$

$$\Rightarrow dg(0) = \lim_{t \searrow 0} d_y G\left(t, x^0, 0; \frac{y^t - y^0}{t}\right) + d_t G(0, x^0, y^0)$$

from hypotheses (H2') and (H3').

Condition (H3') is optimal since under hypotheses (H1)

$$dg(0) \text{ exists} \iff \lim_{t \searrow 0} d_y G\left(t, x^0, 0; \frac{y^t - y^0}{t}\right) \text{ exists.}$$

Hypotheses (H2') and (H3') are weaker and more general than (H2) and (H3):

(H2') It is only assumed that $d_t G(0, x^0, y^0)$ exists. Hypothesis (H2) assumes that $d_t G(t, x^0, y)$ exists for all $t \in [0, \tau]$ and $y \in Y$.

(H3') Hypothesis (H3) assumes that

$$\lim_{s \searrow 0,\, t \searrow 0} d_t G(s, x^0, y^t) = d_t G(0, x^0, y^0) \tag{4.12}$$

which implies that

$$R(0, x^0, y^0) = \lim_{t \searrow 0} d_y G\left(t, x^0, 0; \frac{y^t - y^0)}{t}\right) = 0. \tag{4.13}$$

Hence, condition (H3') with $R(0, x^0, y^0) = 0$ is weaker and indeed more general than (H3) when the limit is not zero.

4.4 A New Condition with the Standard Adjoint

The use of the averaged adjoint naturally exhibited the possible occurrence of an extra term and led to a relatively simple expression for it. However, this extra term can also be obtained by using the standard adjoint.

Recalling that $g(t) = G(t, x^t, y)$ and $g(0) = G(0, x^0, y)$ for any $y \in Y$, then for the standard adjoint state p^0 at $t = 0$

$$g(t) - g(0) = G(t, x^t, p^0) - G(t, x^0, p^0) + \Big(G(t, x^0, p^0) - G(0, x^0, p^0)\Big).$$

Dividing by $t > 0$

$$\begin{aligned}\frac{g(t) - g(0)}{t} &= \frac{G(t, x^t, p^0) - G(t, x^0, p^0)}{t} + \frac{G(t, x^0, p^0) - G(0, x^0, p^0)}{t} \\ &= \int_0^1 d_x G\left(t, (1-\theta)x^0 + \theta x^t, p^0; \frac{x^t - x^0}{t}\right) d\theta \\ &\quad + \frac{G(t, x^0, p^0) - G(0, x^0, p^0)}{t}.\end{aligned}$$

Therefore, in view of Hypothesis (H2'), the limit $dg(0)$ exists if and only if the limit of the first term exists

$$\Rightarrow dg(0) = \lim_{t \searrow 0} \int_0^1 d_x G\left(t, (1-\theta)x^0 + \theta x^t, p^0; \frac{x^t - x^0}{t}\right) d\theta + d_t G(0, x^0, p^0)$$

and the existence of the limit of the first term can replace hypothesis (H3'). As a result, we have two ways of expressing hypothesis (H3') since

$$\lim_{t \searrow 0} \int_0^1 d_x G\left(t, (1-\theta)x^0 + \theta x^t, p^0; \frac{x^t - x^0}{t}\right) d\theta = \lim_{t \searrow 0} d_y G\left(t, x^0, 0; \frac{y^t - y^0}{t}\right).$$

4.5 Another Form of Hypothesis (H3')

Since $d_x G$ and $d_x d_y G$ both exist, Hypothesis (H3') can be rewritten as follows

$$\begin{aligned}&d_y G\left(t, x^0, 0; \frac{y^t - y^0}{t}\right) \\ &= d_y G\left(t, x^0, 0; \frac{y^t - y^0}{t}\right) - d_y G\left(t, x^t, 0; \frac{y^t - y^0}{t}\right) \\ &= \int_0^1 d_x d_y G\left(t, \theta x^0 + (1-\theta)x^t, 0; \frac{y^t - y^0}{t^\alpha}; \frac{x^0 - x^t}{t^{1-\alpha}}\right) d\theta,\end{aligned}$$

for some $\alpha \in [0, 1]$. For instance with $\alpha = 1/2$, it would be sufficient to find bounds on the differential quotients

$$\frac{y^t - y^0}{t^{1/2}} \text{ and } \frac{x^t - x^0}{t^{1/2}}$$

which is less demanding than finding a bound on $(x^t - x^0)/t$ or $(y^t - y^0)/t$.

When the integral can be taken inside, the expressions simplify

$$d_y G\left(t, x^0, 0; \frac{y^t - y^0}{t}\right) = d_x d_y G\left(t, \frac{x^0 + x^t}{2}, 0; \frac{y^t - y^0}{t^\alpha}; \frac{x^0 - x^t}{t^{1-\alpha}}\right)$$

$$\boxed{\lim_{t \searrow 0} d_y G\left(t, x^0, 0; \frac{y^t - y^0}{t}\right) = \lim_{t \searrow 0} d_x d_y G\left(t, \frac{x^0 + x^t}{2}, 0; \frac{y^t - y^0}{t^\alpha}; \frac{x^0 - x^t}{t^{1-\alpha}}\right).}$$

4.6 Back to the Simple Illustrative Example from PDE Control

Given the direction $b \in L^2(\Omega)$, we want to compute $df(a; b) = \lim_{t \searrow 0}(f(a + tb) - f(a))/t$. The *state* $u^t \in H_0^1(\Omega)$ at $t > 0$ is solution of

$$\int_\Omega \nabla u^t \cdot \nabla \psi - (a + tb)\, \psi \, dx = 0, \quad \forall \psi \in H_0^1(\Omega), \tag{4.14}$$

and the *associated Lagrangian* is

$$L(t, \varphi, \psi) \overset{\text{def}}{=} \int_\Omega \frac{1}{2} |\varphi - g|^2 \, dx + \int_\Omega \nabla \varphi \cdot \nabla \psi - (a + tb)\, \psi \, dx.$$

It is readily seen that

$$g(t) \overset{\text{def}}{=} f(a + tb) = \inf_{\varphi \in H_0^1(\Omega)} \sup_{\psi \in H_0^1(\Omega)} L(t, \varphi, \psi)$$

$$dg(0) \overset{\text{def}}{=} \lim_{t \searrow 0} \frac{g(t) - g(0)}{t} = df(a; b).$$

Recall that

$$L(t, \varphi, \psi) \overset{\text{def}}{=} \int_\Omega \frac{1}{2} |\varphi - g|^2 + \nabla \varphi \cdot \nabla \psi - (a + tb)\, \psi \, dx.$$

It is readily seen that

$$d_y L(t, \varphi, \psi; \psi') = \int_\Omega \nabla\varphi \cdot \nabla\psi' - (a + tb)\, \psi'\, dx$$

$$d_x L(t, \varphi, \psi; \varphi') = \int_\Omega (\varphi - g)\, \varphi' + \nabla\varphi' \cdot \nabla\psi\, dx, \quad d_t L(t, \varphi, \psi) = -\int_\Omega b\, \psi\, dx.$$

Observe that the *derivative of the state* $\dot{u} \in H_0^1(\Omega)$ exists:

$$\int_\Omega \nabla \left(\frac{u^t - u^0}{t}\right) \cdot \nabla\psi - b\, \psi\, dx = 0, \quad \forall \psi \in H_0^1(\Omega), \tag{4.15}$$

implies that $(u^t - u^0)/t = \dot{u} \in H_0^1(\Omega)$ solution of

$$\int_\Omega \nabla\dot{u} \cdot \nabla\psi - b\, \psi\, dx = 0, \quad \psi \in H_0^1(\Omega). \tag{4.16}$$

The *averaged adjoint* $y^t \in H_0^1(\Omega)$ is solution of

$$\int_\Omega \left(\frac{u^t + u^0}{2}\right) \varphi + \nabla y^t \cdot \nabla\varphi\, dx = 0, \quad \forall \varphi \in H_0^1(\Omega).$$

$$\int_\Omega \left(\frac{u^t + u^0}{2}\right) \varphi + \nabla y^t \cdot \nabla\varphi\, dx = 0, \quad \forall \varphi \in H_0^1(\Omega).$$

$$\text{adjoint at } t = 0 : \int_\Omega u^0\, \varphi + \nabla y^0 \cdot \nabla\varphi\, dx = 0, \quad \forall \varphi \in H_0^1(\Omega),$$

$$\Rightarrow \int_\Omega \frac{1}{2}\left(\frac{u^t - u^0}{t}\right) \varphi + \nabla \left(\frac{y^t - y^0}{t}\right) \cdot \nabla\varphi\, dx = 0, \quad \forall \varphi \in H_0^1(\Omega). \tag{4.17}$$

It remains to check that the limit in (4.11) exists: $d_y G(t, x^0, 0; (y^t - y^0)/t) \to 0$

$$\int_\Omega \nabla u^0 \cdot \nabla \left(\frac{y^t - y^0}{t}\right) - (a + tb) \left(\frac{y^t - y^0}{t}\right) dx = -t \int_\Omega b \left(\frac{y^t - y^0}{t}\right) dx$$

$$= -t \int_\Omega \nabla \left(\frac{u^t - u^0}{t}\right) \cdot \nabla \left(\frac{y^t - y^0}{t}\right) dx$$

$$= \frac{t}{2} \int_\Omega \left|\frac{u^t - u^0}{t}\right|^2 dx = \frac{t}{2} \int_\Omega |\dot{u}|^2\, dx \to 0$$

as $t \to 0$ using (4.16) and (4.17). Therefore, by Theorem 4.2,

$$df(a; b) = -\int_\Omega b\, y^0\, dx, \quad y^0 \in H_0^1(\Omega), \tag{4.18}$$

$$\int_\Omega (u-g)\varphi + \nabla y^0 \cdot \nabla\varphi\, dx = 0, \quad \forall \varphi \in H_0^1(\Omega). \tag{4.19}$$

5 Example of a Topological Derivative: Non-zero Extra Term

5.1 Topological Derivative

The *topological derivative* rigorously introduced by J. Sokołowski and A. Żochowski [23] induces topological changes. For instance, let f be an *objective function* defined on a family of open subsets of $\mathbb{R}^N$. Given a point a in the open set Ω, let $\overline{B}_r(a)$ be a closed ball of radius r and center a such that $\overline{B}_r(a) \subset \Omega$. Consider the perturbed domain $\Omega_r \overset{\text{def}}{=} \Omega\backslash\overline{B}_r(a)$: Ω minus the hole $\overline{B}_r(a)$. In this simple case the *topological derivative* is defined as

$$df(0) \overset{\text{def}}{=} \lim_{r\searrow 0} \frac{f(\Omega_r) - f(\Omega)}{|\overline{B}_r(a)|}, \tag{5.1}$$

where $|\overline{B}_r(a)|$ is the volume of $\overline{B}_r(a)$ in $\mathbb{R}^N$.

When f is of the form $f(\Omega) = \int_\Omega \varphi\, dx$, the application of the Lebesgue differentiation theorem gives $df(0) = -\varphi(a)$. Of course, many other types of topological perturbations can be considered (see the recent paper [6]).

5.2 One-Dimensional Example: Objective Function and State Equation

Given ε, $0 < \varepsilon < 1$, $a > 0$, and the domain $\Omega = (-a, a)$, consider the following state equation: to find $u \in W^{1,2-\varepsilon}(-a, a)$ such that

$$\forall \varphi \in W^{1,\frac{2-\varepsilon}{1-\varepsilon}}(-a,a) \quad \boxed{\int_{-a}^{a} \frac{du}{dx}\frac{d\varphi}{dx} + u\,\varphi\, dx = \int_{-a}^{a} \frac{d}{dx}\sqrt{|x|}\,\frac{d\varphi}{dx} + \sqrt{|x|}\,\varphi\, dx.}$$

Here, $X = W^{1,2-\varepsilon}(-a,a)$ and $Y = W^{1,\frac{2-\varepsilon}{1-\varepsilon}}(-a,a)$ are reflexive Banach spaces since $2-\varepsilon > 1$ and $\frac{2-\varepsilon}{1-\varepsilon} > 1$. The elements of X will be denoted u and $x \in (-a, a)$

will be the space variable. There exists a unique[1] solution $u(x) = \sqrt{|x|}$, $-a \le x \le a$, and the injections $W^{1,2-\varepsilon}(-a,a) \to C^0[-a,a]$ and $W^{1,\frac{2-\varepsilon}{1-\varepsilon}}(-a,a) \to C^0[-a,a]$ are continuous. As a result, the following *objective function* is well-defined:

$$\boxed{f(\Omega) \stackrel{\text{def}}{=} |u(a)|^2 + |u(-a)|^2 - 2\,|u(0)|^2.}$$

5.3 Perturbed Problems via Dilation of the Point a

Let $\overline{B}_r(0) \subset \mathbb{R}$ be the closed ball of radius r in 0. Denote by $t = |\overline{B}_r(0)| = 2r$ its *volume*. The *perturbed domain* $\Omega_r = \Omega\backslash\overline{B}_r(0) = (-a,-r)\cup(r,a)$ has 2 connected components and it is not possible to construct a bijection between Ω and Ω_r. The *perturbed problems* are parametrized by r, $0 < r < a/2$: to find $u_r \in W^{1,2-\varepsilon}(\Omega_r)$ such that

$$\forall \varphi \in W^{1,\frac{2-\varepsilon}{1-\varepsilon}}(\Omega_r), \quad \int_{\Omega_r} \frac{du_t}{dx}\frac{d\varphi}{dx} + u_t\,\varphi\,dx = \int_{\Omega_r} \frac{d\sqrt{|x|}}{dx}\frac{d\varphi}{dx} + \sqrt{|x|}\,\varphi\,dx$$

with the objective function

$$j(r) \stackrel{\text{def}}{=} |u_r(a)|^2 - |u_r(r)|^2 + |u_r(-a)^2 - |u_r(-r)|^2.$$

The function $u_r(x) = \sqrt{|x|}$ is the unique solution and

$$j(r) = 2a - 2r \quad \Rightarrow dj(0) \stackrel{\text{def}}{=} \lim_{r\searrow 0} \frac{1}{2r}(j(r) - j(0)) = -1.$$

By construction, $\Omega_r = T_r(\Omega\backslash\{0\})$, where the bijection T_r is defined as

$$x \mapsto T_r(x) \stackrel{\text{def}}{=} \left\{ \begin{array}{ll} x - r\left(1 + \dfrac{x}{a}\right), & x \in (-a,0) \\[2ex] x + r\left(1 - \dfrac{x}{a}\right), & x \in (0,a) \end{array} \right\} : \Omega\backslash\{0\} \to \Omega_r = \Omega\backslash\overline{B}_r(0)$$

[1] Given measurable functions $k_1, k_2 : [-a,a] \to \mathbb{R}$ such that $\alpha \le k_i(x) \le \beta$ for some constants $\alpha > 0$ and $\beta > 0$, and real numbers $1 < p < \infty$, $p^{-1} + q^{-1} = 1$, associate with the continuous bilinear mapping

$$\varphi, \psi \mapsto b(\varphi,\psi) \stackrel{\text{def}}{=} \int_{-a}^{a} k_1(x)\frac{d\varphi}{dx}\frac{d\psi}{dx} + k_2(x)\,\varphi\,\psi\,dx : W^{1,p}(-a,a) \times W^{1,q}(-a,a) \to \mathbb{R},$$

the continuous linear operator $A : W^{1,p}(-a,a) \to W^{1,q}(-a,a)'$ which is a topological isomorphism for all $p \in (1,\infty)$ [1]. Here, $p = 2 - \varepsilon$ and $q = (2-\varepsilon)/(1-\varepsilon)$.

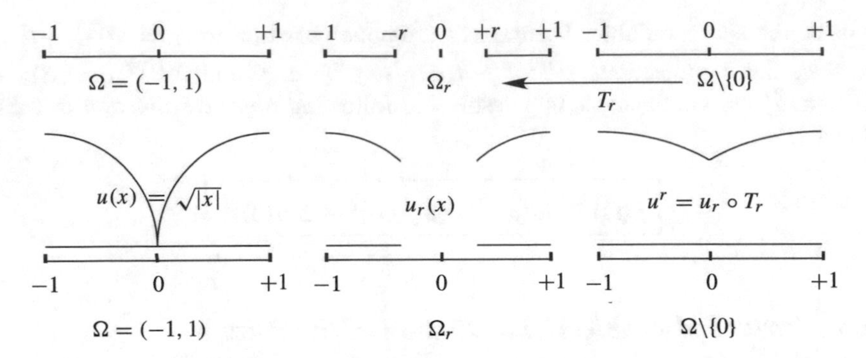

Fig. 1 For $a = 1$, the function u has a cusp at the point 0

and notice that $T_r(0^-) = -r$, $T_r(0^+) = r$, $T_r(a^-) = a$, and $T_r(-a^+) = -a$ (see Fig. 1). Prior to proceeding further, it is advantageous to simplify the computations by observing that the function $u^r(x) = \sqrt{|T_r(x)|}$ is symmetrical with respect to $x = 0$, that is, $u^r(-x) = u^r(x)$ and

$$j(r) = 2\left[u^r(a)^2 - u^r(0^+)^2\right]. \tag{5.2}$$

As a result

$$dj(0) = \lim_{r\searrow 0} \frac{j(r) - j(0)}{2r} = \lim_{r\searrow 0} \frac{u^r(a)^2 - u^r(0^+)^2}{r}$$

By changing the variable r to t, it is sufficient to apply Theorem 4.2 to the following problem on $(0, a)$: to find $u^t \in W^{1,2-\varepsilon}(0, a)$ such that for all $\varphi \in W^{1,\frac{2-\varepsilon}{1-\varepsilon}}(0, a)$

$$\boxed{\int_0^a \frac{a}{a-t}\frac{du^t}{dx}\frac{d\varphi}{dx} + \frac{a-t}{a}u^t\varphi\, dx = \int_0^a \frac{1}{2\sqrt{T_t(x)}}\frac{d\varphi}{dx} + \frac{a-t}{a}\sqrt{T_t(x)}\,\varphi\, dx}$$

with the objective function

$$\boxed{j^+(t) \overset{\text{def}}{=} u^t(a)^2 - u^t(0^+)^2,} \quad dj^+(0) = \lim_{t\searrow 0}(j^+(t) - j^+(0))/t.$$

5.4 Non-convex Lagrangian and Standard Adjoint Equation

The Lagrangian associated with the perturbed problems is

$$
\begin{aligned}
G(t,\varphi,\psi) \overset{\text{def}}{=} & |\varphi(a)|^2 - |\varphi(0)|^2 \\
& + \int_0^a \left(\frac{a}{a-t}\right) \frac{d\varphi}{dx}\frac{d\psi}{dx} + \left(\frac{a-t}{a}\right)\varphi\,\psi\,dx \\
& - \int_0^a \frac{1}{2\sqrt{T_t(x)}}\frac{d\psi}{dx} + \left(\frac{a-t}{a}\right)\sqrt{T_t(x)}\,\psi\,dx.
\end{aligned}
\tag{5.3}
$$

It is non-convex in the φ variable in view of the presence of the term $-|\varphi(0)|^2$. The standard adjoint p^t is solution of the adjoint equation

$$
\forall \varphi \in W^{1,2-\varepsilon}(0,a), \quad
\begin{cases}
2u^t(a)\,\varphi(a) - 2u^t(0)\,\varphi(0) \\
\displaystyle + \int_0^a \left(\frac{a}{a-t}\right)\frac{d\varphi}{dx}\frac{dp^t}{dx} + \left(\frac{a-t}{a}\right)\varphi\,p^t\,dx = 0.
\end{cases}
$$

In particular this is true for all $\varphi \in H^1(0,a) = W^{1,2}(0,a) \subset W^{1,2-\varepsilon}(0,a)$. Since the differential operator is uniformly coercive for $0 \le t \le a/2$, there exist a unique $p^t \in H^1(0,a)$.

But, in view of the fact that for $0 \le t \le a/2$, u^t is finite for all x, we get more regularity: $p_t \in H^2(0,a) \cap C^\infty(0,a)$ is solution of

$$
\boxed{
\begin{aligned}
& -\frac{a}{a-t}\frac{d^2p^t}{dx^2} + \frac{a-t}{a}p^t = 0 \text{ in } (0,a) \\
& \frac{a}{a-t}\frac{dp^t}{dx}(a) = -2\,u^t(a), \qquad \frac{a}{a-t}\frac{dp^t}{dx}(0) = -2\,u^t(0).
\end{aligned}
}
$$

Since $u^t(0) = \sqrt{t}$ and $u^t(a) = \sqrt{a}$, the explicit solution is

$$
\boxed{
\begin{aligned}
p^t(x) = & \frac{a}{a-t}\frac{2}{e^{a-t} - e^{-(a-t)}} \\
& \left[\sqrt{t}\left(e^{\frac{a-t}{a}(a-x)} + e^{-\frac{a-t}{a}(a-x)}\right) - \sqrt{a}\left(e^{\frac{a-t}{a}x} + e^{-\frac{a-t}{a}x}\right)\right].
\end{aligned}
}
$$

At $t = 0$,

$$
p^0(x) = -2\sqrt{a}\,\frac{e^x + e^{-x}}{e^a - e^{-a}}.
$$

5.5 Computation of the Derivative $d_t G(t, \varphi, \psi)$

The right-hand side t-derivative is

$$\begin{aligned} d_t G(t, \varphi, \psi) = & \int_0^a \frac{a}{(a-t)^2} \frac{d\varphi}{dx} \frac{d\psi}{dx} - \frac{1}{a} \varphi \, \psi \, dx + \frac{1}{a} \int_0^a \sqrt{T_t} \, \psi \, dx \\ & - \frac{1}{2} \int_0^a \left[\frac{-1}{2(T_t)^{3/2}} \frac{d\psi}{dx} + \left(\frac{a-t}{a} \right) \frac{1}{\sqrt{T_t}} \psi \right] dx \\ & + \frac{1}{2a} \int_0^a \left[\frac{-x}{2(T_t)^{3/2}} \frac{d\psi}{dx} + \left(\frac{a-t}{a} \right) \frac{x}{\sqrt{T_t}} \psi \right] dx. \end{aligned}$$

At $t = 0$, substitute $u^0(x) = \sqrt{x}$ and p^0 and Integrate by parts

$$\begin{aligned} & d_t G(0, u^0, p^0) \\ & = \int_0^a \frac{1}{a} \frac{d\sqrt{x}}{dx} \frac{dp^0}{dx} - \frac{1}{a} \sqrt{x} \, p^0 \, dx + \frac{1}{a} \int_0^a \sqrt{x} \, p^0 \, dx \\ & \quad - \frac{1}{2} \int_0^a \left[\frac{-1}{2x^{3/2}} \frac{dp^0}{dx} + \frac{1}{\sqrt{x}} p^0 \right] dx + \frac{1}{2a} \int_0^a \left[\frac{-1}{2\sqrt{x}} \frac{dp^0}{dx} + \sqrt{x} p^0 \right] dx \\ & = \frac{1}{2a} \int_0^a \frac{d\sqrt{x}}{dx} \frac{dp^0}{dx} + \sqrt{x} p^0 \, dx - \frac{1}{2} \int_0^a \left[\frac{d}{dx} \frac{1}{\sqrt{x}} \frac{dp^0}{dx} + \frac{1}{\sqrt{x}} p^0 \right] dx = 0. \end{aligned}$$

5.6 Averaged Adjoint State Equation

Go back to the Lagrangian (5.3) of the perturbed problem and compute

$$\begin{aligned} d_x G(t, \bar{\varphi}, \psi; \varphi) = & 2 \, \bar{\varphi}(a) \, \varphi(a) - 2 \, \bar{\varphi}(0) \, \varphi(0) \\ & + \int_0^a \left(\frac{a}{a-t} \right) \frac{d\varphi}{dx} \frac{d\psi}{dx} + \left(\frac{a-t}{a} \right) \varphi \, \psi \, dx. \end{aligned}$$

The averaged adjoint state equation for y^t must satisfy the equation:

$$\begin{aligned} \forall \varphi \in W^{1,2-\varepsilon}(0, a), \quad 0 = & \int_0^1 d_x G(t, u^0 + s(u^t - u^0), y^t; \varphi) \\ = & (u^t(a) + u^0(a)) \, \varphi(a) - (u^t(0) + u^0(0)) \, \varphi(0) \\ & + \int_0^a \left(\frac{a}{a-t} \right) \frac{d\varphi}{dx} \frac{dy^t}{dx} + \left(\frac{a-t}{a} \right) \varphi \, y^t \, dx. \end{aligned}$$

Its solution $y^t \in H^2(0,a) \cap C^\infty(0,a)$ satisfies the following equations

$$\text{averaged adjoint state}\quad \boxed{\begin{aligned} -\left(\frac{a}{a-t}\right)\frac{d^2y^t}{dx^2} + \left(\frac{a-t}{a}\right)y^t = 0, \quad \text{in } (0,a) \\ \left(\frac{a}{a-t}\right)\frac{dy^t}{dx}(0) = -(u^t(0)+u^0(0)) \text{ at } x=0 \\ \left(\frac{a}{a-t}\right)\frac{dy^t}{dx}(a) = -(u^t(a)+u^0(a)) \text{ at } x=a. \end{aligned}}$$

Its explicit expression with $u^t(0) = \sqrt{t}$ and $u^t(a) = \sqrt{a}$ is

$$y^t(x) = \frac{a}{a-t}\frac{1}{e^{a-t}-e^{-(a-t)}} \left[\sqrt{t}\left(e^{\frac{a-t}{a}(a-x)} + e^{-\frac{a-t}{a}(a-x)}\right) - 2\sqrt{a}\left(e^{\frac{a-t}{a}x} + e^{-\frac{a-t}{a}x}\right)\right].$$

The condition (H3') to be checked is the *existence of the limit*

$$\lim_{t\searrow 0} d_yG\left(t,u^0,0;\frac{y^t-y^0}{t}\right).$$

So, for $\psi = (y^t - y^0)/t \in H^2(0,a)$,

$$\begin{aligned} &d_yG(t,u^0,0;\psi) \\ &= \int_0^a \left(\frac{a}{a-t}\right)\frac{du^0}{dx}\frac{d\psi}{dx} + \left(\frac{a-t}{a}\right)u^0\,\psi\,dx \\ &\quad - \int_0^a \frac{1}{2\sqrt{T_t(x)}}\frac{d\psi}{dx} + \left(\frac{a-t}{a}\right)\sqrt{T_t(x)}\,\psi\,dx \\ &= \int_0^a \left(\frac{a}{a-t}\right)\frac{du^0}{dx}\frac{d\psi}{dx} + \left(\frac{a-t}{a}\right)u^0\,\psi\,dx \\ &\quad - \int_0^a \frac{a}{a-t}\frac{d\sqrt{T_t(x)}}{dx}\frac{d\psi}{dx} + \frac{a-t}{a}\sqrt{T_t(x)}\,\psi\,dx \\ &= \int_0^a \left(\frac{a}{a-t}\right)\frac{d(u^0-u^t)}{dx}\frac{d\psi}{dx} + \left(\frac{a-t}{a}\right)(u^0-u^t)\,\psi\,dx \\ &= (u^0-u^t)\left(\frac{a}{a-t}\right)\frac{d}{dx}\left(\frac{y^t-y^0}{t}\right)\Big|_{x=0}^{a} \to -1. \end{aligned}$$

5.7 Material Derivative

Theorem 5.1

(i) *For* $0 < \varepsilon < 1$ *and* $t \in [0, a/2]$,

$$\|u^t\|_{W^{1,2-\varepsilon}(0,a)} \leq c(\varepsilon, a), \quad \|u^t - u^0\|_{C^0[0,a]} \leq \sqrt{t},$$
$$u^t \to u^0 \text{ in } W^{1,2-\varepsilon}(0,a)\text{-weak}$$

for some bound $c(\varepsilon, a)$ *and the rate of convergence is sharp.*

(ii) *For* $x \in (0, a)$, *the* material derivative *is given by*

$$\dot{u}(x) \stackrel{\text{def}}{=} \lim_{t \searrow 0} \frac{u^t(x) - u^0(x)}{t} = \frac{1}{2}\left(\frac{1}{\sqrt{|x|}} - \frac{\sqrt{|x|}}{a}\right) \geq 0,$$

$\dot{u} \in L^{2-\varepsilon}(0,a)$ *for* $0 < \varepsilon \leq 1$, *but* $\dot{u} \notin L^2(0,a)$. *Moreover, as* $t \to 0$,

$$\left\|(u^t - u^0)/t - \dot{u}\right\|_{L^{2-\varepsilon}(0,a)} \to 0. \tag{5.4}$$

(iii) *As for the derivative of* $\dot{u}$,

$$\frac{d\dot{u}}{dx}(x) = -\frac{1}{4\sqrt{|x|}} \begin{cases} 1/x + 1/a, & x \in (0,a) \\ 1/x - 1/a, & x \in (-a,0) \end{cases} \quad \frac{d\dot{u}}{dx}(0^+) = -\infty,$$

$\frac{d\dot{u}}{dx} \notin L^1(0,a)$, *and, a fortiori,* $\frac{d\dot{u}}{dx} \notin L^{2-\varepsilon}(0,a)$.

Remark 5.1 From part (iii)

$$\frac{d\dot{u}}{dx} \notin L^1(0,a), \text{ and, a fortiori, } \frac{d\dot{u}}{dx} \notin L^{2-\varepsilon}(0,a).$$

and we cannot apply the chain rule to get $dj(0^+)$ since the expression is undetermined: $2\,u^0(a)\,\dot{u}(a) - 2\,u^0(0)\,\dot{u}(0) = 2\,u^0(a)\,\dot{u}(a) - 2\,[0\,(-\infty)]$!

6 Multivalued Case: Two Theorems without and with the Extra Term

We give two theorems for the existence and expressions of $dg(0)$ in the multivalued case where only a right-hand side derivative of g is expected. New conditions and quadratic examples were given in [9] without the extra term. Complete conditions including the extra term were published in [8] at an IFAC meeting in 2016 prior to the publication of [9] due to longer publication delays in the Journal of Convex Analysis.

In this paper, we give the latest version from [8]. The first theorem is a mild generalization of the singleton case. Yet, it can be applied to PDE problems with non-homogeneous Dirichlet boundary conditions where non-unique extensions are used (cf. [12]). A new non-convex multivalued example will be given for the second more general theorem.

6.1 First Theorem: Mild Generalization

Theorem 6.1 (A First Extension) *Given X, Y, and G, let* (H0) *and the following hypotheses be satisfied:*

(H1) *for all t in $[0, \tau]$, $X(t) \neq \varnothing$ and $g(t)$ is finite, and for all $x^t \in X(t)$ and $x^0 \in X(0)$, $Y(t, x^0, x^t) \neq \varnothing$;*
(H2) *for all $x \in X(0)$ and $y \in Y(0, x)$, $d_t G(0, x, y)$ exists;*
(H3) *there exist $\hat{x}^0 \in X(0)$, $\hat{y}^0 \in Y(0, \hat{x}^0)$, and $R(0, \hat{x}^0, \hat{y}^0)$ such that for each sequence $t_n \to 0$, $0 < t_n \leq \tau$, there exist a subsequence $\{t_{n_k}\}$ of $\{t_n\}$, $x^{t_{n_k}} \in X(t_{n_k})$, and $y^{t_{n_k}} \in Y(t_{n_k}, \hat{x}^0, x^{t_{n_k}})$ such that*

$$\lim_{k \to \infty} d_y G\left(t_{n_k}, \hat{x}^0, 0; (y^{t_{n_k}} - \hat{y}^0)/t_{n_k}\right) = R(0, \hat{x}^0, \hat{y}^0).$$

Then, $dg(0)$ exists and there exist $\hat{x}^0 \in X(0)$ and $\hat{y}^0 \in Y(0, \hat{x}^0)$ such that

$$dg(0) = d_t G(0, \hat{x}^0, \hat{y}^0) + R(0, \hat{x}^0, \hat{y}^0).$$

When $X(0) = \{x^0\}$ and $Y(0, x^0) = \{y^0\}$ are singletons, the above hypotheses are equivalent to the ones of Theorem 4.2.

6.2 Second Theorem: General Case

Theorem 6.2 (General Case) *Given X, Y, and G, let* (H0) *and the following hypotheses be satisfied:*

(H1) *$\forall t \in [0, \tau]$, $X(t) \neq \varnothing$, $g(t)$ is finite, and $\forall x^t \in X(t)$ and $x^0 \in X(0)$, $Y(t, x^0, x^t) \neq \varnothing$;*
(H2) *for all $x \in X(0)$ and $y \in Y(0, x)$, $d_t G(0, x, y)$ exists and, for each $x \in X(0)$, there exists a function $y \mapsto R(0, x, y) : Y(0, x) \to \mathbb{R}$ satisfying (H3) and (H4) below;*

(H3) *for each sequence* $t_n \to 0$, $0 < t_n \le \tau$, $\exists x^0 \in X(0)$ *such that for all* $y^0 \in Y(0, x^0)$, $\exists$*a subsequence* $\{t_{n_k}\}$ *of* $\{t_n\}$, $x^{t_{n_k}} \in X(t_{n_k})$, *and* $y^{t_{n_k}} \in Y(t_{n_k}, x^0, x^{t_{n_k}})$ *such that*

$$\liminf_{k\to\infty} d_y G\left(t_{n_k}, x^0, 0; (y^{t_{n_k}} - y^0)/t_{n_k}\right) \ge R(0, x^0, y^0);$$

(H4) *for each sequence* $t_n \to 0$, $0 < t_n \le \tau$ *and all* $x^0 \in X(0)$, *there exist* $y^0 \in Y(0, x^0)$, *a subsequence* $\{t_{n_k}\}$ *of* $\{t_n\}$, $x^{t_{n_k}} \in X(t_{n_k})$, *and* $y^{t_{n_k}} \in Y(t_{n_k}, x^0, x^{t_{n_k}})$ *such that*

$$\limsup_{k\to\infty} d_y G\left(t, x^0, 0; (y^{t_{n_k}} - y^0)/t_{n_k}\right) \le R(0, x^0, y^0).$$

Then, $dg(0)$ *exists and there exists* $\hat{x}^0 \in X(0)$ *and* $\hat{y}^0 \in Y(0, \hat{x}^0)$ *such that*

$$\begin{aligned} dg(0) &= d_t G(0, \hat{x}^0, \hat{y}^0) + R(0, \hat{x}^0, \hat{y}^0) \\ &= \sup_{y \in Y(0,\hat{x}^0)} d_t G(0, \hat{x}^0, y) + R(0, \hat{x}^0, y) \\ &= \inf_{x\in X(0)} \sup_{y\in Y(0,x)} d_t G(0, x, y) + R(0, x, y). \end{aligned}$$

6.3 Second Theorem: A Non-convex Example Where $X(0)$ is Not a Singleton

Consider the objective function and the constraint set

$$f(x) \stackrel{\text{def}}{=} Qx \cdot x, \quad U \stackrel{\text{def}}{=} \{x \in \mathbb{R}^n : Ax \cdot x = 1\}, \quad \inf f(U), \tag{6.1}$$

where Q is an arbitrary symmetrical $n \times n$ matrix and $A > 0$ is a symmetrical $n \times n$ positive definite matrix. $U \neq \varnothing$ is compact and the function f is not necessarily convex. This minimization problem is equivalent to the generalized eigenvalue problem

$$\boxed{\lambda(Q, A) \stackrel{\text{def}}{=} \inf_{x\neq 0} \frac{Qx \cdot x}{Ax \cdot x},} \tag{6.2}$$

where the minimizer $\hat{x}$ is solution of the problem

$$[Q - \lambda(Q, A)A]\hat{x} = 0, \quad A\hat{x} \cdot \hat{x} = 1. \tag{6.3}$$

The semidifferential of $\lambda(Q, A)$ with respect to Q in a direction Q' and A in the direction A' can be found in [4, pp. 166–168] for symmetrical matrices:

$$\begin{aligned} d\lambda(Q, A; Q', A') &= \inf_{x \in X(0)} Q'x \cdot x\,(Ax \cdot x) - (Qx \cdot x)\, A'x \cdot x \\ &= \inf_{x \in X(0)} Q'x \cdot x - \lambda(Q, A)\, A'x \cdot x, \end{aligned} \tag{6.4}$$

$$\text{minimizers } X(0) \overset{\text{def}}{=} \left\{x \in \mathbb{R}^n : [Q - \lambda(Q, A)A]x = 0 \text{ and } Ax \cdot x = 1\right\} \tag{6.5}$$

$$\text{states } E(0) \overset{\text{def}}{=} \left\{x \in \mathbb{R}^n : Ax \cdot x = 1\right\}. \tag{6.6}$$

For $t \geq 0$, $x \in \mathbb{R}^n$, and $y \in \mathbb{R}$, introduce the Lagrangian

$$G(t, x.y) \overset{\text{def}}{=} (Q + tQ')x \cdot x + y\,[(A + tA')x \cdot x - 1] \tag{6.7}$$

$$g(t) \overset{\text{def}}{=} \inf_{x \in \mathbb{R}^n} \sup_{y \in \mathbb{R}} G(t, x, y), \quad dg(0) \overset{\text{def}}{=} \frac{g(t) - g(0)}{t}, \tag{6.8}$$

where A' and Q' are symmetrical matrices. Set $Q(t) = Q + tQ'$ and $A(t) = A + tA'$. It is easy to check that

$$d_t G(t, x, y) = Q'x \cdot x + y\, A'x \cdot x \tag{6.9}$$

$$d_x G(t, x, y; x') = 2\,[Q(t) + y\, A(t)]\,x \cdot x' \tag{6.10}$$

$$d_y G(t, x, y; y') = y'\,[A(t)x \cdot x - 1]. \tag{6.11}$$

Since A is positive definite, there exists $\alpha > 0$ such that for all $x \in \mathbb{R}^n$, $Ax \cdot x \geq \alpha \|x\|^2$. Hence, there exists $\tau > 0$ such that for all $0 \leq t \leq \tau$

$$\forall t,\ 0 \leq t \leq \tau,\ \forall x \in \mathbb{R}^n, \quad A(t)x \cdot x \geq \frac{\alpha}{2} \|x\|^2$$

and for such t, the set of constraints $E(t) \overset{\text{def}}{=} \{x : A(t)x \cdot x = 1\} \neq \varnothing$ is compact. So there exist minimizers $x^t \in \mathbb{R}^n$ and $X(t)$ is not empty for $0 \leq t \leq \tau$

$$\lambda^t \overset{\text{def}}{=} \inf_{A(t)x \cdot x = 1} Q(t)x \cdot x = Q(t)x^t \cdot x^t \tag{6.12}$$

To summarize,

$$d_t G(t, x, y) = Q'x \cdot x + y\, A'x \cdot x \tag{6.13}$$

$$\left[Q(t) + y^t\, A(t)\right] \frac{x^t + x^0}{2} = 0 \text{ (average adjoint equation)} \tag{6.14}$$

$$\forall y', \quad d_y G(t, x^t, 0; y') = y'\,[A(t)x^t \cdot x^t - 1] = 0 \text{ (state equation)} \tag{6.15}$$

$$d_y G\left(t, x^0, 0; \frac{y^t - y^0}{t}\right) = \frac{y^t - y^0}{t}\,[A(t)x^0 \cdot x^0 - 1]. \tag{6.16}$$

From the Lagrange Multiplier rule, the standard adjoint is solution of

$$\boxed{\left[Q(t)+p^t\,A(t)\right]x^t=0} \quad \Rightarrow p^t=-Q(t)x^t\cdot x^t=-\lambda^t. \tag{6.17}$$

The set of minimizers is

$$X(t)=\left\{x\in\mathbb{R}^n : [Q(t)+p^tA(t)x=0 \text{ and } A(t)x\cdot x=1\right\}. \tag{6.18}$$

For all $x^t\in X(t)$, $x^t\neq 0$ and $-x^t\in X(t)$. So $X(t)$ is not a singleton. However,

$$\forall x^t\in X(t),\quad Y(t,x^t)=\{-\lambda^t\}$$

and $Y(t,x^t)$ is a singleton independent of the choice of $x^t\in X(t)$.

Given $x^0\in X(0)$ and $x^t\in X(t)$, the *averaged adjoint* is solution of the equation:

$$\begin{aligned}
\forall x',\quad 0 &= \int_0^1 d_xG(t,x^0+s(x^t-x^0),y^t;x')ds\\
&=2\int_0^1\left[Q(t)+y^t\,A(t)\right](x^0+s(x^t-x^0))\cdot x'\,ds\\
&=2\left[Q(t)+y^t\,A(t)\right]\frac{x^t+x^0}{2}\cdot x'\\
&\Rightarrow \boxed{\left[Q(t)+y^t\,A(t)\right]\frac{x^t+x^0}{2}=0.}
\end{aligned} \tag{6.19}$$

$$\Rightarrow Y(t,x^0,x^t)=\begin{cases}\left\{-\dfrac{Q(t)\frac{x^t+x^0}{2}\cdot\frac{x^t+x^0}{2}}{A(t)\frac{x^t+x^0}{2}\cdot\frac{x^t+x^0}{2}}\right\}, & \text{if } x^t+x^0\neq 0\\[2ex] \mathbb{R}, & \text{if } x^t+x^0=0\end{cases} \tag{6.20}$$

Therefore, $Y(t,x^0,x^t)\neq\varnothing$.

Lemma 6.1 (i) *For all t, $0\le t\le\tau$,*

$$\forall x^t\in X(t),\quad Y(t,x^t,x^t)=\{-\lambda^t\} \tag{6.21}$$

where λ^t is the minimum of the objective function $Q(t)x\cdot x$ with respect to $E(t)=\{x\in\mathbb{R}^n : A(t)x\cdot x=1\}$ as seen in (6.12).

(ii) *For each sequence $\{t_n : 0<t_n\le\tau\}$, there exist $\bar{x}\in X(0)$, $x^{t_n}\in X(t_n)$, and $y^{t_n}\in Y(t_n,\bar{x},x^{t_n})$ such that*

$$x^{t_n}\to\bar{x},\quad \lambda^{t_n}\to\lambda^0,\quad \textit{and}\quad y^{t_n}\to y^0=-\lambda^0, \tag{6.22}$$

and the set $Y(t_n,\bar{x},x^{t_n})=\{y^{t_n}\}$ is a singleton.

(iii) *As $t \searrow 0$, the quotient*

$$\frac{\lambda^t - \lambda^0}{t} \tag{6.23}$$

is bounded.

(iv) *For the sequences of part (ii), the quotients*

$$\frac{\lambda^{t_n} - \lambda^0}{t_n} \quad \text{and} \quad \frac{y^{t_n} - y^0}{t_n} \tag{6.24}$$

are bounded.

Theorem 6.3 *Given symmetrical $n \times n$ matrices A, A', Q, and Q' such that A is positive definite, there exists at least one x^0 such that $Ax^0 \cdot x^0 = 1$ and*

$$\lambda(Q, A) = \inf_{Ax \cdot x = 1} Qx \cdot x = Qx^0 \cdot x^0. \tag{6.25}$$

Moreover

$$\boxed{\begin{aligned} d\lambda(Q, A; Q', A') &\overset{\text{def}}{=} \lim_{t \searrow 0} \frac{\lambda(Q + tQ', A + tA') - \lambda(Q, A)}{t} \\ &= \inf_{x^0 \in X(0)} \left[Q' - \lambda(Q, A)\, A'\right] x^0 \cdot x^0, \end{aligned}} \tag{6.26}$$

$$X(0) \overset{\text{def}}{=} \left\{x \in \mathbb{R}^n : Ax \cdot x = 1 \text{ and } [Q - \lambda(Q, A)A]x = 0\right\}. \tag{6.27}$$

If $\lambda(Q, A)$ is not simple the dimension of the space $X(0)$ is greater or equal to 2 and we only have a semi-différential.

Proof (i) *Hypothesis (H1)*. We have seen that for all $0 \leq t \leq \tau$, $X(t) \neq \varnothing$ and that, for all $x^t \in X(t)$, $Y(t, x^t) = \{-\lambda^t\}$. For the averaged adjoint y^t

$$\Rightarrow Y(t, x^0, x^t) = \begin{cases} \left\{ - \dfrac{Q(t)\frac{x^t + x^0}{2} \cdot \frac{x^t + x^0}{2}}{A(t)\frac{x^t + x^0}{2} \cdot \frac{x^t + x^0}{2}} \right\}, & \text{if } x^t + x^0 \neq 0 \\ \mathbb{R}, & \text{if } x^t + x^0 = 0 \end{cases}$$

(ii) *Hypothesis (H2)*. We have seen that $d_t G(t, x, y) = Q'x \cdot x + y\, A'x \cdot x$. So for all $x^0 \in X(0)$ and the singleton $Y(0, x^0) = \{-\lambda^0\}$

$$d_t G(t, x^0, y^0) = Q'x^0 \cdot x^0 - \lambda^0 A'x^0 \cdot x^0.$$

(iii) *Hypothesis (H3)*. For each sequence $t_n \to 0$, $0 < t_n \leq \tau$, choose the sequence $\{x^{t_n}\}$ and its limit $\bar{x} \in X(0)$ from the Lemma (ii) and use the fact that the corresponding sequence $\frac{y^{t_n}-y^0}{t_n}$ is bounded by some constant c from the Lemma (iv):

$$\left| d_y G\left(t_n, \bar{x}, 0; \frac{y^{t_n}-y^0}{t_n}\right) \right| = \left| \frac{y^{t_n}-y^0}{t} \;[A(t_n)\bar{x}\cdot\bar{x}-1] \right|$$

$$\leq \left| \frac{y^{t_n}-y^0}{t_n} \right| \;|A(t_n)\bar{x}\cdot\bar{x}-1| \leq c\;|A(t_n)\bar{x}\cdot\bar{x}-1| \to c\;|A(0)\bar{x}\cdot\bar{x}-1| = 0$$

(iv) *Hypothesis (H4).* For all $x^0 \in X(0)$ $Y(0,x^0) = \{-\lambda^0\}$ is a singleton independent of $x^0 \in X(0)$. As in (iii), for each sequence $t_n \to 0$, $0 < t_n \leq \tau$, choose the sequence $\{x^{t_n}\}$ and its limit $\bar{x} \in X(0)$ from the Lemma (ii) and use the fact that the corresponding sequence $\frac{y^{t_n}-y^0}{t_n}$ is bounded by some constant c from the Lemma (iv):

$$\begin{aligned} \left| d_y G\left(t_n, x^0, 0; \frac{y^{t_n}-y^0}{t_n}\right) \right| &= \left| \frac{y^{t_n}-y^0}{t} \left[A(t_n)x^0\cdot x^0-1\right] \right| \\ &\leq \left| \frac{y^{t_n}-y^0}{t_n} \right| \left|A(t_n)x^0\cdot x^0-1\right| \\ &\leq c\left|A(t_n)x^0\cdot x^0-1\right| \to c\left|A(0)x^0\cdot x^0-1\right| = 0. \end{aligned}$$

(v) The conclusion follows from Theorem 6.2 where the sup disappears since $Y(0,x^0) = \{-\lambda^0\} = \{-\lambda(Q,A)\}$ is a singleton independent of $x^0 \in X(0)$.

□

Acknowledgements This research has been supported by Discovery Grants from the Natural Sciences and Engineering Research Council of Canada.

References

1. P. Auscher and Ph. Tchamitchian, *Square root problem for divergence operators and related topics*, Astérisque 249. 1998.
2. R. Correa and A. Seeger, *Directional derivatives of a minimax function*, Nonlinear Anal. Theory Methods and Appl. **9** (1985), 13–22.
3. J. M. Danskin, *The theory of max-min, with applications*, SIAM J. on Appl. Math. **14**, no. 4 (1966), 641–644.
4. M. C. Delfour, *Introduction to optimization and semidifferential calculus*, MOS-SIAM Series on Optimization, Society for Industrial and Applied Mathematics, Philadelphia, USA, 2012.
5. M. C. Delfour, *Metrics spaces of shapes and geometries from set parametrized functions*, in "New Trends in Shape Optimization", A. Pratelli and G. Leugering, eds., pp. 57–101, International Series of Numerical Mathematics vol. 166, Birkhäuser Basel, 2015.

6. M. C. Delfour, *Differentials and Semidifferentials for Metric Spaces of Shapes and Geometries*, in "System Modeling and Optimization, (CSMO 2015)," L. Bociu, J. A. Desideri and A. Habbal, eds., pp. 230–239, AICT Series, Springer, 2017.
7. M. C. Delfour, *Topological Derivative: a Semidifferential via the Minkowski Content*, Journal of Convex Analysis 25 (2018), No. 3.
8. M. C. Delfour and K. Sturm, *Minimax differentiability via the averaged adjoint for control/shape sensitivity*, Proc. of the 2nd IFAC Workshop on Control of Systems Governed by Partial Differential quations, IFAC-PaperOnLine **49-8** (2016), pp. 142–149.
9. M. C. Delfour and K. Sturm, *Parametric Semidifferentiability of Minimax of Lagrangians: Averaged Adjoint Approach*, Journal of Convex Analysis **24** (2017), No. 4, 1117–1142.
10. M. C. Delfour and J.-P. Zolésio, *Shape sensitivity analysis via min max differentiability*, SIAM J. on Control and Optimization **26** (1988), 834–862.
11. M. C. Delfour and J.-P. Zolésio, *Velocity method and Lagrangian formulation for the computation of the shape Hessian*, SIAM J. Control Optim. (6) **29** (1991), 1414–1442.
12. M. C. Delfour and J.-P. Zolésio, *Shapes and geometries: metrics, analysis, differential calculus, and optimization*, 2nd edition, SIAM series on Advances in Design and Control, Society for Industrial and Applied Mathematics, Philadelphia, USA 2011.
13. V. F. Dem'janov, *Differentiation of the maximin function. I, II*, (Russian) Ž. Vyčisl. Mat. i Mat. Fiz. **8** (1968), 1186–1195; ibid. **9**, 55–67.
14. V. F. Dem'janov, *Minimax: Directional Differentiation*, Izdat. Leningrad Gos. Univ., Leningrad 1974.
15. V. F. Dem'janov and V. N. Malozemov, *Introduction to the Minimax*, Trans from Russian by D. Louvish, Halsted Press [John Wiley & Sons], New York-Toronto 1974.
16. H. Federer, *Curvature measures*, Trans. Amer. Math. Soc. **93** (1959), 418–419.
17. H. Federer, *Geometric measure theory*, Springer-Verlag, Berlin, Heidelberg, New York, 1969.
18. J. Hadamard, *La notion de différentielle dans l'enseignement*, Scripta Univ. Ab. Bib., Hierosolymitanarum, Jerusalem, 1923. Reprinted in the Mathematical Gazette **19**, no. 236 (1935), 341–342.
19. P. I. Kogut, O. Kupenko, and G. Leugering, *Optimal control in matrix-valued coefficients for nonlinear monotone problems: optimality conditions II*, Z. Anal. Anwend. (Journal for Analysis and its Applications) 34 (2015), no. 2, 199–219.
20. A. M. Micheletti, *Metrica per famiglie di domini limitati e proprietà generiche degli autovalori*, Ann. Scuola Norm. Sup. Pisa (3) **26** (1972), 683–694.
21. A. A. Novotny and J. Sokołowski, *Topological Derivatives in Shape Optimization*, Interaction of Mechanics and Mathematics, Springer, Heidelberg, N. Y., 2013.
22. S. Ya Serovaiskii, *Optimization in a nonlinear elliptic system with a control in the coefficients (Russian)*, Mat. Zametki 54 (1993), no. 2, 85–95, 159; translation in Math. Notes 54 (1993), no. 1–2, 825–832 (1994).
23. J. Sokołowski and A. Żochowski, *On the topological derivative in shape optimization*, SIAM J. Control Optim. **37**, no. 4 (1999), 1251–1272.
24. K. Sturm, *On shape optimization with non-linear partial differential equations*, Doctoral thesis, Technische Universiltät of Berlin, Germany 2014.
25. K. Sturm, *Minimax Lagrangian approach to the differentiability of non-linear PDE constrained shape functions without saddle point assumption*, SIAM J. on Control and Optim., **53** (2015), No. 4, 2017–2039.
26. J. P. Zolésio, *Identification de domaines par déformation*, Thèse de doctorat d'état, Université de Nice, France, 1979.

On Optimization Transfer Operators in Shape Spaces

Volker H. Schulz and Kathrin Welker

Abstract We discuss several approaches towards shape optimization problems in the context of partial differential equations. The general framework used is that of optimization on shape manifolds. Here, transfer operators between tangent spaces and the manifold are crucial and in general non-trivial.

Keywords Shape optimization · Shape spaces · Transfer operators

1 Introduction

The optimization of shapes is a challenging task, in particular since shapes cannot be considered as objects in a linear space. Theoretical and practical approaches have a long history—several approaches are referred below. A particular challenge arises in the context of applications involving partial differential equations (PDE). This field is one of the main practical motivations for the discipline of shape optimization. In contrast to typical differential geometric descriptions of shapes, which are more or less dealing with smooth surfaces, a PDE context with weak formulations of equations in Sobolev spaces introduces a certain perturbation concerning the smoothness of shapes. This is caused, because weak solutions of equations ultimately result in less smooth deformations of smooth shapes. As discussed below, shape deformations can be considered as tangent vectors which are transferred to a shape manifold. These transfer operators together with transfer operators between tangent spaces play a crucial role in the practical implementation of efficient solutions techniques for PDE constrained shape optimization problems. This paper tries to give an overview over various recent approaches towards PDE shape optimization within the framework of optimization on shape manifolds with a particular focus on the transfer operators involved.

V. H. Schulz (✉) · K. Welker
Trier University, Trier, Germany
e-mail: volker.schulz@uni-trier.de; welker@uni-trier.de

V. Schulz, D. Seck (eds.), *Shape Optimization, Homogenization and Optimal Control*, International Series of Numerical Mathematics 169,
https://doi.org/10.1007/978-3-319-90469-6_13

This paper is organized as follows. Different shape space concepts are summarized in Sect. 2. In Sect. 3, we consider a special shape manifold, the manifold of smooth shapes, and enable its connection to shape calculus. In particular, this results in quasi-Newton methods on the infinitely dimensional manifold of smooth shapes. This section also deals with computationally efficient vector transports and associated retractions. Finally, Sect. 4 defines the space of so-called $H^{1/2}$-shapes and gives its connection to shape calculus in order to formulate efficient shape optimization methods. A novel relation of the geodesics in $H^{1/2}$-shape spaces with recently investigated elastic deformations of images is discussed.

2 Overview of Shape Spaces

Shapes and their similarities has been extensively studied in recent decades. David G. Kendall [19] has already introduced the notion of a shape space in 1984. In [19], shapes are characterized by labeled points in the Euclidean space, so-called landmarks, and the author investigates Riemannian structures on this space. However, there is a large number of different shape concepts, e.g., landmark vectors [9, 16, 19, 31, 38], plane curves [26–29], surfaces [3, 4, 20, 22, 25], boundary contours of objects [14, 24, 46], multiphase objects [45], characteristic functions of measurable sets [48] and morphologies of images [11]. In order to answer natural questions like *"How different are shapes?", "Can we determine the measure of their difference?"* or *"Can we infer any information?"* mathematically, we put a metric on the space of shapes. There are various different types of metrics on shape spaces, e.g., inner metrics [3, 4, 27], outer metrics [5, 8, 15, 19, 27], metamorphosis metrics [17, 40], the Wasserstein or Monge-Kantorovic metric on the shape space of probability measures [2, 6, 7], the Weil-Peterson metric [23, 36], current metrics [12, 13, 41] and metrics based on elastic deformations [14, 47]. However, in general, the modelling of both, the shape space and the associated metric, is a challenging task and different approaches lead to diverse models. The suitability of an approach depends on the demands in a given situation. There exists no common shape space or shape metric suitable for all applications. In the setting of PDE constrained shape optimization, one has to deal with polygonal shape representations from a computational point of view. This is because finite element (FE) methods are usually used to discretize the models. In [35], an inner product, which is called Steklov-Poincaré metric, for the application of FE methods is proposed. The Steklov-Poincaré metric correlates shape gradients with H^1-deformations. Under special assumptions, these deformations give shapes of class $H^{1/2}$. In [42], the space of $H^{1/2}$-shapes is defined and its diffeological structure is clarified. The combination of this particular shape space and its associated inner product is an essential step towards applying efficient FE solvers as outlined in [37]. In the following sections, we focus on the space of smooth shapes and the space of $H^{1/2}$-shapes.

3 Optimization Methods on the Space of Smooth Shapes

First, we introduce the space of smooth shapes. In [26], the *set of all two-dimensional smooth shapes* is characterized by

$$B_e(S^1, \mathbb{R}^2) := \mathrm{Emb}(S^1, \mathbb{R}^2)/\mathrm{Diff}(S^1), \tag{3.1}$$

i.e., the obit space of $\mathrm{Emb}(S^1, \mathbb{R}^2)$ under the action by composition from the right by the Lie group $\mathrm{Diff}(S^1)$. Here $\mathrm{Emb}(S^1, \mathbb{R}^2)$ denotes set of all embeddings from the unit circle S^1 into the plane $\mathbb{R}^2$ and $\mathrm{Diff}(S^1)$ is the set of all diffeomorphisms from S^1 into itself. In [21], it is proven that the shape space $B_e(S^1, \mathbb{R}^2)$ is a smooth manifold. For the sake of completeness it should be mentioned that the shape space $B_e(S^1, \mathbb{R}^2)$ together with appropriate inner products is even a Riemannian manifold. In [27], a survey of various suitable inner products is given. Note that the shape space $B_e(S^1, \mathbb{R}^2)$ and its theoretical results can be generalized to higher dimensions (cf. [25]).

The tangent space $T_\Gamma B_e(S^1, \mathbb{R}^2)$ is isomorphic to the set of all smooth normal vector fields along $\Gamma \in B_e(S^1, \mathbb{R}^2)$, i.e.,

$$T_\Gamma B_e(S^1, \mathbb{R}^2) \cong \left\{h \colon h = \alpha n,\ \alpha \in \mathcal{C}^\infty(S^1)\right\}, \tag{3.2}$$

where n denotes the exterior unit normal field to the shape boundary Γ such that $n(\theta) \perp \Gamma_\theta(\theta)$ for all $\theta \in S^1$, where $\Gamma_\theta = \frac{\partial \Gamma}{\partial \theta}$ denotes the circumferential derivative as in [26].

Now, we enable the connection of shape calculus to the geometric concepts of the space B_e of smooth shapes. To be more precise, we want to solve shape optimization problems on B_e. A *shape optimization problem* is given by

$$\min_{\Omega} J(\Omega), \tag{3.3}$$

where $J\colon \mathcal{A} \to \mathbb{R}$, $\Omega \mapsto J(\Omega)$ with $\mathcal{A} \subset \{\Omega \colon \Omega \subset D \subset \mathbb{R}^d,\ D \neq \emptyset\}$. The function J is called a *shape functional*. Often, shape optimization problems are constrained by equations involving an unknown function of two or more variables and at least one partial derivative of this function. When J in (3.3) depends on a solution of a PDE, we call the shape optimization problem *PDE constrained*. To solve shape optimization problems, we need their shape derivatives. For the definition of shape derivatives or a detailed introduction into shape calculus, we refer to the monographs [10, 39]. In the following, the shape derivative of a shape functional J at Ω in direction of a sufficiently smooth vector field V is denoted by $DJ(\Omega)[V]$. Shape derivatives can always be expressed as boundary integrals due to the Hadamard

Structure Theorem [39, Theorem 2.27]. In many cases, the shape derivative arises in two equivalent notational forms:

$$DJ_\Omega[V] := \int_\Omega RV(x)\,dx \qquad \text{(volume formulation)} \tag{3.4}$$

$$DJ_\Gamma[V] := \int_\Gamma r(s)\,\langle V(s), n(s)\rangle\,ds \qquad \text{(surface formulation)} \tag{3.5}$$

Here $r \in L^1(\Gamma)$ and R is a differential operator acting linearly on the vector field V with $DJ_\Omega[V] = DJ(\Omega)[V] = DJ_\Gamma[V]$. In particular, there exists a scalar distribution r on the boundary Γ of the domain Ω under consideration due to the Hadamard Structure Theorem. If we assume $r \in L^1(\Gamma)$, the shape derivative can be expressed on the boundary Γ of Ω (cf. (3.5)). The distribution r is often called the *shape gradient*. However, note that gradients depend always on chosen scalar products defined on the space under consideration. Thus, it rather means that r is the usual L^2-shape gradient. If we want to optimize on the shape manifold B_e, we have to find a representation of the shape gradient with respect to an appropriate inner product. This representation is called the *Riemannian shape gradient* and required to formulate optimization methods on B_e.

In order to deal with surface formulations of shape derivatives in optimization techniques, the Sobolev metric is an appropriate inner product. In the following, we consider the first Sobolev metric g^1 on the shape space B_e. It is given by

$$\begin{aligned} g^1 \colon T_\Gamma B_e(S^1, \mathbb{R}^2) \times T_\Gamma B_e(S^1, \mathbb{R}^2) &\to \mathbb{R}, \\ (h, k) &\mapsto \int_{S^1} \langle (I - A\Delta_\Gamma)\,\alpha, \beta \rangle\, ds, \end{aligned} \tag{3.6}$$

where $h = \alpha n$, $k = \beta n$ denote elements of the tangent space $T_\Gamma B_e(S^1, \mathbb{R}^2)$, $A > 0$ and Δ_Γ denotes the Laplace-Beltrami operator on the surface Γ. For the definition of the Sobolev metric g^1 in higher dimensions we refer to [3].

Now, we have to detail the Riemannian shape gradient with respect to g^1. The shape derivative can be expressed as

$$DJ_\Gamma[V] = \int_\Gamma \alpha r\, ds \tag{3.7}$$

if $V\big|_{\partial\Omega} = \alpha n$. In order to get an expression of the Riemannian shape gradient with respect to the Sobolev metric g^1, we look at the isomorphism (3.2). Due to this isomorphism, a tangent vector $h \in T_\Gamma B_e$ is given by $h = \alpha n$ with $\alpha \in \mathcal{C}^\infty(\Gamma)$. This leads to the following definition.

Definition 3.1 The Riemannian shape gradient of a shape differentiable objective function J in terms of the Sobolev metric g^1 is given by

$$\text{grad}(J) = qn \;\text{ with }\; (I - A\Delta_\Gamma)q = r, \tag{3.8}$$

where $\Gamma \in B_e$, $A > 0$, $q \in \mathcal{C}^\infty(\Gamma)$ and r is the L^2-shape gradient given in (3.5).

The Riemannian shape gradient is required to formulate optimization methods in the shape space B_e. In order to formulate a Newton method in B_e, we have to detail the Riemannian shape Hessian. It is based on the Riemannian connection ∇ associated to the Sobolev metric g^1. In differential geometry, the Riemannian connection is often written in terms of the Christoffel symbols. In [3], Christoffel symbols associated with the Sobolev metrics are provided. However, in order to provide a relation with shape calculus, another representation of the covariant derivative in terms of the Sobolev metric g^1 is needed. This special representation is given in [42]. We replicate the relevant theorem.

Theorem 3.2 *Let $A > 0$, let $D_s := \frac{\partial_\theta}{|c_\theta|}$ be the arc length derivative of a curve $c \in \text{Emb}(S^1, \mathbb{R}^2)$ and let $h, m \in T_c\text{Emb}(S^1, \mathbb{R}^2)$ denote vector fields along $c \in \text{Emb}(S^1, \mathbb{R}^2)$. Moreover, $L_1 := I - AD_s^2$ is a differential operator on $\mathcal{C}^\infty(S^1, \mathbb{R}^2)$ and L_1^{-1} denotes its inverse operator. The covariant derivative associated with the Sobolev metric g^1 can be expressed as*

$$\nabla_m h = L_1^{-1}(K_1(h)) \text{ with } K_1 := \frac{1}{2}\langle D_s m, v\rangle \left(I + AD_s^2\right), \tag{3.9}$$

where $v = \frac{c_\theta}{|c_\theta|}$ denotes the unit tangent vector.

The Riemannian connection with respect to the Sobolev metric, which is given in Theorem 3.2, makes it possible to specify the Riemannian shape Hessian of an optimization problem. In analogy to [1], we define the Riemannian shape Hessian as follows:

Definition 3.3 In the setting above, the Riemannian shape Hessian of a two times shape differentiable objective function J is defined as the linear mapping

$$T_\Gamma B_e \to T_\Gamma B_e, h \mapsto \text{Hess}(J)[h] := \nabla_h \text{grad}(J). \tag{3.10}$$

In order to formulate optimization methods on the shape manifold B_e we need the concept of *retractions* and *vector transports*. In general, the calculations of optimization methods have to be performed in tangent spaces because manifolds are not necessarily linear spaces. This means, points from a tangent space have to be mapped to the manifold in order to get a new iterate. The computation of the exponential map, which is the theoretically superior choice of such a mapping, is prohibitively expensive in the most applications. However, in [1], it is shown that the retraction specified in the following definition is a first-order approximation and sufficient.

Definition 3.4 A retraction on a manifold M is a smooth mapping $\mathcal{R}\colon TM \to M$ with the following properties:

(i) $\mathcal{R}_p(0_p) = p$, where $\mathcal{R}_p$ denotes the restriction of $\mathcal{R}$ to T_pM and 0_p denotes the zero element of T_pM.
(ii) $d\mathcal{R}_p(0_p) = \mathrm{id}_{T_pM}$, where id_{T_pM} denotes the identity mapping on T_pM and $d\mathcal{R}_p(0_p)$ denotes the pushforward of $0_p \in T_pM$ by $\mathcal{R}$.

Condition (ii) is called the *local rigidity condition*. Equivalently, for every $v \in T_pM$, the curve $\gamma\colon \mathbb{R} \to M, t \mapsto \mathcal{R}_p(tv)$ satisfies $\dot{\gamma}(0) = v$. Moving along this curve γ is thought of as moving in direction v while being constrained to stay on M. Note that every manifold which admits a Riemannian metric also admits a retraction defined by the exponential mapping (cf. [1, Section 5.4]). In [1], the local rigidity condition ensures the quadratic convergence of the Newton method (Algorithm 1 below) by a perturbation argument. It is shown that the retraction step differs from the geodesic step by an error of only higher order.

Now we get the concept of the so-called *vector transport*, which specifies how to transport a tangent vector η form a point $p \in M$ to a point $\mathcal{R}_p(\eta)$, where $\mathcal{R}$ denotes a retraction on M.

Definition 3.5 Let (M, g) be a Riemannian manifold and let $\oplus$ denote the Whitney sum, i.e.,

$$TM \oplus TM := \{(\xi_p, \eta_p)\colon \xi_p, \eta_p \in T_pM,\ p \in M\}. \tag{3.11}$$

A vector transport on M is a differentiable mapping

$$\mathcal{T}\colon TM \oplus TM \to TM,\ (\xi_p, \eta_p) \mapsto \mathcal{T}_{\xi_p}(\eta_p)$$

satisfying the following properties for all $p \in M$:

(i) *Underlying retraction:* There exists a retraction $\mathcal{R}\colon TM \to M$, called the retraction associated with $\mathcal{T}$, such that $\mathcal{T}_{\xi_p}(\eta_p) \in T_{\mathcal{R}_p(\xi_p)}M$.
(ii) *Consistency:* $\mathcal{T}_{0_p}(\eta_p) = \eta_p$ for all $\eta_p \in T_pM$, where 0_p denotes the zero element of T_pM.
(iii) *Linearity:* $\mathcal{T}_{\xi_p}(\lambda\eta_p + \mu\zeta_p) = \lambda\mathcal{T}_{\xi_p}(\eta_p) + \mu\mathcal{T}_{\xi_p}(\zeta_p)$ for all $\eta_p, \zeta_p \in T_pM$ and all $\lambda, \mu \in \mathbb{R}$.

The choice of a computationally efficient vector transport and associated retraction is an important decision in the design of numerical algorithms on non-linear manifolds. If we consider the space of smooth shapes, i.e., $B_e(S^1, \mathbb{R}^2)$, we can consider the following vector transport and retraction:

Example (Retraction and Vector Transport on B_e) Let $\eta_c := \alpha n_c$, $\xi_c := \beta n_c$ be two elements of the tangent space

$$T_cB_e(S^1, \mathbb{R}^2) \cong \{h\colon h = \gamma n_c, \gamma \in \mathcal{C}^\infty(S^1)\}$$

and

$$\begin{aligned} c+\eta_c\colon S^1 &\to \mathbb{R}^2, \\ \theta &\mapsto c(\theta)+\alpha(\theta)n_{c(\theta)}. \end{aligned} \tag{3.12}$$

Then,

$$\begin{aligned} \mathcal{T}\colon TB_e(S^1,\mathbb{R}^2)\oplus TB_e(S^1,\mathbb{R}^2) &\to TB_e(S^1,\mathbb{R}^2), \\ (\alpha n_c, \beta n_c) &\mapsto \mathcal{T}_{\alpha n_c}(\beta n_c) := \beta n_{c+\alpha n_c} \end{aligned} \tag{3.13}$$

is a vector transport in B_e. In particular, α, β in (3.13) are two smooth functions from S^1 into $\mathbb{R}$. There exists a retraction associated with $\mathcal{T}$ defined in (3.13):

$$\begin{aligned} \mathcal{R}\colon TB_e(S^1,\mathbb{R}^2) &\to B_e(S^1,\mathbb{R}^2), \\ \eta_c &\mapsto \mathcal{R}_c(\eta_c) := c+\eta_c, \end{aligned} \tag{3.14}$$

where $\eta_c \in T_cB_e(S^1,\mathbb{R}^2)$. It is easy to verify that (i)–(ii) of Definition 3.4 hold for $\mathcal{R}$ defined in (3.14).

Both, the Riemannian shape gradient and the Riemannian shape Hessian, are required to formulate optimization methods in the shape space B_e. The Newton method on the space of smooth shapes is formulated in Algorithm 1.

Algorithm 1 requires the Hessian in each iteration, but it is a time-consuming process to derive the Hessian. In contrast to Algorithm 1, quasi-Newton methods only need an approximation of the Hessian. Such an approximation is realized, e.g., by a limited-memory Broyden-Fletcher-Goldfarb-Shanno (BFGS) update. An inverse limited-memory BFGS update in (B_e, g^1) is formulated in Algorithm 2, which is conceptually similar to the double loop algorithm in finite dimensional

Algorithm 1 Newton method in B_e

Require: Objective function J on B_e; retraction $\mathcal{R}$ on B_e; affine connection ∇ on B_e.
Goal: Find the solution of $\min_{x\in B_e} J(x)$.
Input: $x_0 \in B_e$

for $k = 0, 1, \dots$ **do**

[1] Compute the increment $\Delta x_k \in T_{x_k}B_e$ as solution of

$$\operatorname{Hess} J(x^k)\Delta x_k = -\operatorname{grad} J(x^k). \tag{3.15}$$

[2] Set

$$x_{k+1} := \mathcal{R}_{x_k}(\Delta x_k). \tag{3.16}$$

end for

Algorithm 2 Inverse limited-memory BFGS update in (B_e, g^1)

$\rho_j \leftarrow g^1(y_j, s_j)^{-1}$
$q \leftarrow \text{grad}J(c_j)$
for $i = j-1, \ldots, j-m$ **do**
  $s_i \leftarrow \mathcal{T}_{s_{j-1}} s_i$
  $y_i \leftarrow \mathcal{T}_{s_{j-1}} y_i$
  $\alpha_i \leftarrow \rho_i g^1(s_i, q)$
  $q \leftarrow q - \alpha_i y_i$
end for
$q \leftarrow \frac{g^1(y_{j-1}, s_{j-1})}{g^1(y_{j-1}, y_{j-1})} q$
for $i = j-m, \ldots, j-1$ **do**
  $\beta_i \leftarrow \rho_i g^1(y_i, q)$
  $q \leftarrow q + (\alpha_i - \beta_i) s_i$
end for
return $s_j = q$

Euclidean spaces. Yet, the inner products are now given by the Sobolev metric g^1 and vector transports $\mathcal{T}$ like (3.13) are considered. Moreover, in the j-th iteration, $\Delta s_j = \mathcal{T}_{s_j} s_j \in T_{c_{j+1}} B_e$ denotes the distance between two iterated shapes and $y_j = \text{grad}J(c_{j+1}) - \mathcal{T}_{s_j} \text{grad}J(c_j) \in T_{c_{j+1}} B_e$ denotes the difference of iterated Riemannian shape gradients. In standard formulations, update formulas require the storage of the whole convergence history up to the current iteration. Limited-memory update techniques have been developed in order to reduce the amount of storage (cf. [30]). Here, in Algorithm 2, m Riemannian shape gradients are stored. In [32], superlinear convergence properties for BFGS quasi-Newton methods on manifolds are analyzed for the case that $\mathcal{T}_{s_j}$ is an isometry.

However, the methods in (B_e, g^1) are based on surface expressions of shape derivatives. E.g., in a limited-memory BFGS method, a representation of the shape gradient with respect to the Sobolev metric g^1 has to be computed and applied as a Dirichlet boundary condition in the linear elasticity mesh deformation. This involves two operations, which are non-standard in FE tools and, thus, lead to additional coding effort. To explain all this goes beyond the scope of this paper. Thus, we refer to [34] for the limited-memory BFGS method in B_e and more details about the two non-standard operations. In order to reduce the effort, the next section deals with volume expressions of shape derivatives.

4 Optimization Methods on the Space of $H^{1/2}$-Shapes

If we consider Sobolev metrics, we have to deal with surface formulations of shape derivatives. An intermediate and equivalent result in the process of deriving these expressions is the volume expression (3.4). In the case of the volume formulation, the shape manifold B_e and the corresponding inner products g^1 are not appropriate.

One possible approach to use volume forms is to consider Steklov-Poincaré metrics defined in the sequel.

Let $\Omega \subset X \subset \mathbb{R}^d$ be a compact domain with C^∞-boundary $\Gamma := \partial\Omega$, where X denotes a bounded domain with Lipschitz-boundary $\Gamma_{\text{out}} := \partial X$. In particular, this means $\Gamma \in B_e(S^{d-1}, \mathbb{R}^d)$. We consider the following scalar products, the so-called *Steklov-Poincaré metrics* (cf. [35]):

$$\begin{aligned} g^S \colon H^{1/2}(\Gamma) \times H^{1/2}(\Gamma) &\to \mathbb{R}, \\ (\alpha, \beta) &\mapsto \int_\Gamma \alpha(s) \cdot [(S^{pr})^{-1}\beta](s)\, ds \end{aligned} \tag{4.1}$$

Here S^{pr} denotes the projected Poincaré-Steklov operator which is given by

$$S^{pr} \colon H^{-1/2}(\Gamma_{\text{out}}) \to H^{1/2}(\Gamma_{\text{out}}),\ \alpha \mapsto (\gamma_0 U)^T n, \tag{4.2}$$

where $\gamma_0 \colon H_0^1(X, \mathbb{R}^d) \to H^{1/2}(\Gamma_{\text{out}}, \mathbb{R}^d)$, $U \mapsto U|_{\Gamma_{\text{out}}}$ and $U \in H_0^1(X, \mathbb{R}^d)$ solves the Neumann problem

$$a(U, V) = \int_{\Gamma_{\text{out}}} \alpha \cdot (\gamma_0 V)^T n\, ds \quad \forall V \in H_0^1(X, \mathbb{R}^d) \tag{4.3}$$

with $a(\cdot, \cdot)$ being a symmetric and coercive bilinear form.

In the following, we state the connection of B_e with respect to the Steklov-Poincaré metric g^S to shape calculus. As already mentioned, the shape derivative can be expressed as the surface integral (3.5) due to the Hadamard Structure Theorem. Recall that the shape derivative can be written more concisely (cf. (3.7)). Due to isomorphism (3.2) and expression (3.7), we can state the connection of the shape space B_e with respect to the Steklov-Poincaré metric g^S to shape calculus:

Definition 4.1 Let r denote the (standard) L^2-shape gradient given in (3.5). Moreover, let S^{pr} be the projected Poincaré-Steklov operator and let γ_0 be as above. The shape gradient of a shape differentiable function J in terms of the Steklov-Poincaré metric g^S is given by

$$\operatorname{grad}(J) = S^{pr} r = (\gamma_0 U)^T n, \tag{4.4}$$

where $U \in H_0^1(X, \mathbb{R}^d)$ solves

$$a(U, V) = \int_\Gamma r \cdot (\gamma_0 V)^T n\, ds \quad \forall V \in H_0^1(X, \mathbb{R}^d). \tag{4.5}$$

Now, the shape gradient with respect to Steklov-Poincaré metric is defined. This enables the formulation of optimization methods in B_e which involve volume formulations of shape derivatives. Note that $U \in H_0^1(X, \mathbb{R}^d)$ in Definition 4.1 solves

$$a(U, V) = \int_\Gamma r \cdot (\gamma_0 V)^T n \, ds = DJ_\Gamma[V] = DJ_\Omega[V] \quad \forall V \in H_0^1(X, \mathbb{R}^d), \tag{4.6}$$

which gives the mesh deformation U and the gradient representation $\mathrm{grad}(J) = (\gamma_0 U)^T n$ all at once. This is very attractive from a computational point of view. The computation of a representation of the shape gradient with respect to the chosen inner product of the tangent space is moved into the mesh deformation itself. The elliptic operator $a(\cdot, \cdot)$ is used as an inner product and a mesh deformation. This leads to only one linear system, which has to be solved.

In Algorithm 3, an inverse limited-memory BFGS quasi-Newton method for shape optimization in terms of g^S is formulated. BFGS update formulas need the evaluation of scalar products, where at least one argument is a *gradient-type* vector. This is a vector which indeed arises as a gradient on the variable boundary. In contrast, a *deformation-type* vector describes an arbitrary boundary deformation in normal direction. According to the Steklov-Poincaré metric, we can assume that a gradient-type vector $u \in T_c B_e$ can be written as

$$u = (\gamma_0 U)^T n \tag{4.7}$$

for some vector field $U \in H_0^1(X, \mathbb{R}^d)$ and γ_0 as above. The other argument v is either of gradient-type or of deformation-type, which can also be assumed to be of the form (4.7), i.e., $v = (\gamma_0 V)^T n$ for some $V \in H_0^1(X, \mathbb{R}^d)$ and γ_0 as above. If u is a gradient of a shape objective function J, we observe

$$g^S(u, v) = DJ_\Gamma[V] = DJ_X[V] = a(U, V). \tag{4.8}$$

This observation can be used to reformulate the scalar product $g^S(\cdot, \cdot)$ on the boundary equivalently as $a(\cdot, \cdot)$ for domain representations. In the sequel, we only consider domain representations $U_j \in H_0^1(X, \mathbb{R}^d)$ of $\mathrm{grad} J(c_j) \in H^{1/2}(\Gamma)$, mesh deformations $S_j \in H_0^1(X, \mathbb{R}^d)$ and differences $Y_j := U_{j+1} - \mathcal{T}_{S_j} U_j \in H_0^1(X, \mathbb{R}^d)$, where $\mathcal{T}$ denotes an appropriate vector transport. With this notation we formulate the double-loop of an inverse limited-memory BFGS quasi-Newton method in Algorithm 3. The resulting vector $q \in H_0^1(X, \mathbb{R}^d)$ is simultaneously a shape deformation as well as a deformation of the domain mesh.

Numerical investigations have shown that the optimization techniques based on Steklov-Poincaré metrics (cf. Algorithm 3) also work on shapes with kinks in the boundary (cf. [33, 35, 37]). In particular, in [33], the difference of the mesh quality by using the limited-memory BFGS shape optimization Algorithm 2 in B_e and the analogous Algorithm 3 in $\mathcal{B}^{1/2}$ is illustrated. It can be predicated that the shape space B_e containing smooth shapes unnecessarily limits the application of these

Algorithm 3 Inverse limited-memory BFGS update in terms of g^S

$\rho_j \leftarrow g^S\left((\gamma_0 Y_j)^T n, (\gamma_0 S_j)^T n\right)^{-1} = a(Y_j, S_j)^{-1}$
$q \leftarrow U_j$
for $i = j-1, \ldots, j-m$ **do**
  $S_i \leftarrow \mathcal{T}_{S_{j-1}} S_i$
  $Y_i \leftarrow \mathcal{T}_{S_{j-1}} Y_i$
  $\alpha_i \leftarrow \rho_i\, g^S\left((\gamma_0 S_i)^T n, (\gamma_0 q)^T n\right) = \rho_i\, a(S_i, q)$
  $q \leftarrow q - \alpha_i Y_i$
end for
$q \leftarrow \frac{g^S\left((\gamma_0 Y_{j-1})^T n, (\gamma_0 S_{j-1})^T n\right)}{g^S\left((\gamma_0 Y_{j-1})^T n, (\gamma_0 Y_{j-1})^T n\right)} U_j = \frac{a(Y_{j-1}, S_{j-1})}{a(Y_{j-1}, Y_{j-1})} U_j$
for $i = j-m, \ldots, j-1$ **do**
  $\beta_i \leftarrow \rho_i\, g^S\left((\gamma_0 Y_i)^T n, (\gamma_0 q)^T n\right) = \rho_i\, a(Y_i, q)$
  $q \leftarrow q + (\alpha_i - \beta_i) Y_i$
end for
return $S_j = q$

methods based on Steklov-Poincaré metrics. In [35], the definition of smooth shapes is firstly extended to so-called $H^{1/2}$-*shapes*. In the sequel, it is clarified what we mean by $H^{1/2}$-shapes. However, only a first try of a definition is given in [35]. From a theoretical point of view there are several open questions about this shape space. The most important question is how the structure of this shape space is. If we do not know the structure, there is no chance to get control over the space. In order to clarify the diffeological structure of this shape space, its definition is adapted and refined in [42]. In the following, we repeat the definition of this shape space and give the relevant results about it.

We would like to recall that a shape in the sense of the shape space B_e is given by the image of an embedding from the unit sphere S^{d-1} into the Euclidean space $\mathbb{R}^d$. In view of our generalization, it has technical advantages to consider so-called *Lipschitz shapes* which are defined as follows:

Definition 4.2 A d-dimensional Lipschitz shape Γ_0 is defined as the boundary $\Gamma_0 = \partial\mathcal{X}_0$ of a compact Lipschitz domain $\mathcal{X}_0 \subset \mathbb{R}^d$ with $\mathcal{X}_0 \neq \emptyset$. The set $\mathcal{X}_0$ is called a Lipschitz set.

General shapes—in this terminology—arise from H^1-deformations of a Lipschitz set $\mathcal{X}_0$. These H^1-deformations, evaluated at a Lipschitz shape Γ_0, give deformed shapes Γ if the deformations are injective and continuous. These shapes are called of class $H^{1/2}$. The set containing all these shapes is defined as follows:

Definition 4.3 Let $\Gamma_0 \subset \mathbb{R}^d$ be a d-dimensional Lipschitz shape. The space of all d-dimensional $H^{1/2}$-shapes is given by

$$\mathcal{B}^{1/2}(\Gamma_0, \mathbb{R}^d) := \mathcal{H}^{1/2}(\Gamma_0, \mathbb{R}^d)/\sim,$$

where

$$\mathcal{H}^{1/2}(\Gamma_0, \mathbb{R}^d)$$
$$:= \{w \colon w \in H^{1/2}(\Gamma_0, \mathbb{R}^d) \text{ injective, continuous; } w(\Gamma_0) \text{ Lipschitz shape}\}$$

and the equivalence relation $\sim$ is given by

$$w_1 \sim w_2 \Leftrightarrow w_1(\Gamma_0) = w_2(\Gamma_0), \text{ where } w_1, w_2 \in \mathcal{H}^{1/2}(\Gamma_0, \mathbb{R}^d).$$

A theorem in [42] provides the space $\mathcal{B}^{1/2}$ with a diffeological structure. A diffeological space is a generalization of a manifold. The difference between manifolds and diffeological spaces is clarified in [42]. Roughly speaking, a manifold of dimension n is getting by glueing together open subsets of $\mathbb{R}^n$ via diffeomorphisms. In contrast, a diffeological space is formed by glueing together open subsets of $\mathbb{R}^n$ with the difference that the glueing maps are not necessarily diffeomorphisms and that n can vary. However, note that manifolds deal with charts and diffeological spaces deal with plots. For an introduction into diffeological spaces, we refer to [18]. Note that, so far, there is no theory for shape optimization on diffeological spaces. However, from a theoretical point of view, a diffeological space is very attractive in shape optimization. It can be supposed that a diffeological structure suffices for many differential-geometric tools used in shape optimization techniques. The appearance of a diffeological space in the context of shape optimization can be seen as a first step or motivation towards the formulation of optimization techniques on diffeological spaces. It can be observed that Algorithm 3 works in $\mathcal{B}^{1/2}$ (cf. [35, 37]). For more details about this approach and in particular the implementation details we refer to these publications.

Gradient flows and consequently geodesics and retractions depend on the metric which is chosen on a shape space. In the following, we have a look on geodesics in $\mathcal{B}^{1/2}$. We will see that the geodesics in $\mathcal{B}^{1/2}$ with respect to g^S are the geodesics discussed in [14]. The elastic deformation energy given in [14] and defined in the sequel is induced by the Steklov-Poincaré metric g^S due to the following reasoning: We get a representation of the shape gradient in terms of g^S by solving (4.6). In particular, we obtain a tangent vector arising from the solution of (4.6), i.e., a deformation inducing shape morphings. The scalar product $g^S(\cdot,\cdot)$ on the surface can be reformulated equivalently as a symmetric and coercive bilinear form $a(\cdot,\cdot)$ for volume representations (cf. (4.8)). Thus, it is sufficient to consider such an inner product to measure distances—in other words, to get geodesics—between shapes in $\mathcal{B}^{1/2}$ with respect to g^S. If we choose the inner product as the weak form of the linear elasticity equation, i.e.,

$$a(U, V) = \int_X \sigma(U) : \epsilon(V)\, dx \qquad (U, V \in H_0^1(X, \mathbb{R}^d)), \tag{4.9}$$

we get an elastic deformation energy and a corresponding distance measure as in [14]. Here σ is the stress tensor and ϵ is the strain tensor in linear elasticity.

Now, we consider the shape space $\mathcal{B}^{1/2}(\Gamma_0, \mathbb{R}^d)$, more precisely, elements of it. We recall once again that—in this setting—shapes are given by the images of injective and continuous $H^{1/2}$-deformations of Lipschitz shapes Γ_0. These $H^{1/2}$-deformations come from H^1-deformations of open and bounded domains $X \subset \mathbb{R}^d$ containing Γ_0. Roughly speaking, the elements of $\mathcal{B}^{1/2}(\Gamma_0, \mathbb{R}^d)$ arise from H^1-deformations acting on Lipschitz sets $\mathcal{X}_0 \subset X$ with boundaries Γ_0. Thus, a variation Γ of Γ_0 is associated with a mapping

$$U \colon [0, 1] \to H_0^1(X, \mathbb{R}^d),\ t \mapsto U(t), \tag{4.10}$$

where the variable $t \in [0, 1]$ represents geometrically the coordinate along a path of transport fields $U(t) \in H_0^1(X, \mathbb{R}^d)$. In this setting, $\mathcal{X}(t) \subset X(t)$ describes the deformed object and $\Gamma(t)$ characterizes the deformed shape at time $t \in [0, 1]$. Therefore, a smooth path

$$\gamma \colon [0, 1] \to \mathcal{B}^{1/2}(\Gamma_0, \mathbb{R}^d),\ t \mapsto u(t) \tag{4.11}$$

in this shape space is associated with a family $(U(t))_{t\in[0,1]}$ of deformations, where $u(t) = U(t)|_{\Gamma_0}$.

Remark 4.4 Note that a path (4.11) is given by a curve of injective and continuous deformation fields, i.e.,

$$\gamma \colon [0, 1] \to \big(\Gamma_0 \to \mathbb{R}^d\big),\ t \mapsto \big(\theta \mapsto u(t, \theta)\big). \tag{4.12}$$

More precisely, a shape variation Γ $(= \Gamma(t))$ of a prior shape Γ_0 is given by the image of an injective and continuous deformation u of Γ_0 at time $t \in [0, 1]$, i.e., $\Gamma = u(t, \Gamma_0)$. Since Γ_0 is fixed, we write $\Gamma = u(t)$ $(= \gamma(t))$ to be in line with (4.11).

In our setting, such deformations U arise from (4.6) as already stated above. Thus, we can choose the inner product as the weak form of the linear elasticity equation (4.9).

In a first step to a definition of a suitable shape distance measure and in particular geodesics, we give the definition of a *volume elastic deformation energy* (cf. [14]).

Definition 4.5 Let $\lambda \geq 0$ and $\mu > 0$ denote the Lamé parameters involved in the stress tensor σ and strain tensor ϵ in linear elasticity. Moreover, let $U \in H_0^1(X, \mathbb{R}^d)$ denote a vector field defined on X. The elastic energy of the deformation U on X is given by

$$E(U) = a(U, U) = \int_X \sigma(U) : \epsilon(U)\, dx. \tag{4.13}$$

In the following, we consider deformation energies on the shape itself. The energies should deform a shape $\Gamma \in \mathcal{B}^{1/2}(\Gamma_0, \mathbb{R}^d)$ in normal direction. This means that we have to define an elastic energy of a surface deformation $\phi \in H^{1/2}(\Gamma)$. For this purpose, we define a special kind of trace operator. Let Γ be smooth enough to admit a normal vector field n. Moreover, let $\mathrm{tr}\colon H_0^1(X,\mathbb{R}^d)\to H^{1/2}(\Gamma,\mathbb{R}^d)$ be the trace operator on the Sobolev spaces for vector valued functions restricted to Γ. The trace operator

$$\mathrm{tr}_n\colon H_0^1(X,\mathbb{R}^d)\to H^{1/2}(\Gamma),\ U\mapsto \langle \mathrm{tr}(U),n\rangle \tag{4.14}$$

defined in [14] is continuous and surjective. This means that for every $\phi\in H^{1/2}(\Gamma)$ there exists an $U\in H_0^1(X,\mathbb{R}^d)$ such that $\mathrm{tr}_n(U)=\phi$, as shown in [14].

Now, we are able to define the *surface elastic deformation energy* as the infimum of all domain elastic energies (2.3), which deform a shape Γ in normal direction. Note that X depends on Γ in our setting. To clarify this, we write X_Γ in the following.

Definition 4.6 Let $\Gamma\in\mathcal{B}^{1/2}(\Gamma_0,\mathbb{R}^d)$. The elastic energy of a surface deformation $\phi\in H^{1/2}(\Gamma)$ is defined as

$$E_\Gamma(\phi):=\inf_{\substack{U\in H_0^1(X_\Gamma,\mathbb{R}^d)\\ \mathrm{tr}_n U=\phi}} E(U). \tag{4.15}$$

With this surface elastic deformation energy, which is firstly defined in [14], we obtain an elastic shape distance on $\mathcal{B}^{1/2}(\Gamma_0,\mathbb{R}^d)$ with respect to the Steklov-Poincaré metric g^S. For a shape variation along a path $\gamma\colon[0,1]\to\mathcal{B}^{1/2}(\Gamma_0,\mathbb{R}^d)$, which is induced by the deformation family $(U(t))_{t\in[0,1]}$, the path length L is given by

$$L(\gamma):=\int_0^1\sqrt{E_{\gamma(t)}(\dot\gamma(t))}\,dt \tag{4.16}$$

and the energy E is given by

$$E(\gamma):=\int_0^1 E_{\gamma(t)}(\dot\gamma(t))\,dt, \tag{4.17}$$

where $\dot\gamma=\frac{\partial\gamma}{\partial t}$ denotes the temporal variation of γ at time $t\in[0,1]$. The temporal variation $\dot\gamma(t)$ is the velocity of $\gamma(t)$ normal to $\gamma(t)$. This is a tangent vector at $\gamma(t)\in\mathcal{B}^{1/2}(\Gamma_0,\mathbb{R}^d)$ and, thus, a continuous surface deformation $\phi(t)\in H^{1/2}(\Gamma(t))$.

With all this, we can define a geodesic path.

Definition 4.7 A geodesic path between two shapes $\Upsilon,\tilde\Upsilon\in\mathcal{B}^{1/2}(\Gamma_0,\mathbb{R}^d)$ is a curve $\gamma\colon[0,1]\to\mathcal{B}^{1/2}(\Gamma_0,\mathbb{R}^d)$ with $\gamma(0)=\Upsilon$ and $\gamma(1)=\tilde\Upsilon$ which locally minimizes the length L or, equivalently, the energy E defined in (4.16)–(4.17).

Definition 4.7 establishes the relation between the geodesics in $\mathcal{B}^{1/2}$ with respect to g^S and the geodesics discussed in [14] and, thus, we obtain a distance measure between shapes. The authors propose an energy of infinitesimal deformation (cf. surface elastic deformation energy) of shapes in $B_e(S^d, \mathbb{R}^{d+1})$. This energy is based on the elastic deformation energy (cf. volume elastic deformation energy), i.e., on linear elasticity. A shape metric is derived from it. Since geodesics are locally minimizers of the length or, equivalently, of the energy, they also obtain a definition of geodesics in defining the energy of infinitesimal deformation of shapes. However, they do not formulate such a definition, but it should be mentioned that it would be almost the same as Definition 4.7. The difference is that we work with the shape space $\mathcal{B}^{1/2}(\Gamma_0, \mathbb{R}^d)$ and we consider the Steklov-Poincaré metric g^S. This leads us to the above definition. In contrast, in [14], the shape space $B_e(S^d, \mathbb{R}^{d+1})$ is considered and such a definition would be obtained in defining these energies without considering a special metric. However, they give an alternative interpretation of the elastic deformation energy based on differential geometry. Moreover, the authors show that the perturbation of the metric on a shape, which is deformed by an infinitesimal deformation, is a special case of the elastic deformation energy. In our case, the above energies are justified by (4.8), which allows to reformulate the scalar product $g^S(\cdot, \cdot)$ on the surface equivalently as a symmetric and coercive bilinear form $a(\cdot, \cdot)$ for volume representations. If we choose the inner product as the weak form of the linear elasticity equation, we obtain the elastic deformation energies and a corresponding distance measure as in [14]. Note that we also get a connection to [43–45, 47] by establishing the correlation to [14].

Finally, it should be mentioned that there are several open questions from a theoretical point of view. As already stated in [14], the existence of minimizers of the length (or the energy), i.e., geodesics, is an open problem. In a finite dimensional setting, the theorem of Hopf and Rinow ensures the existence of geodesics on manifolds which are complete as metric spaces. However, in our infinite dimensional setting, this theorem cannot be applied.

Acknowledgements This work has been partly supported by the German Research Foundation within the priority program SPP 1962 "Non-smooth and Complementarity-based Distributed Parameter Systems: Simulation and Hierarchical Optimization" under contract number Schu804/15-1 and the research training group 2126 "Algorithmic Optimization".

References

1. P. Absil, R. Mahony, and R. Sepulchre. *Optimization Algorithms on Matrix Manifolds*. Princeton University Press, 2008.
2. L. Ambrosio, N. Gigli, and G. Savaré. Gradient flows with metric and differentiable structures, and applications to the Wasserstein space. *Rendiconti Lincei – Matematica e Applicazioni*, 15(3–4):327–343, 2004.

3. M. Bauer, P. Harms, and P. Michor. Sobolev metrics on shape space of surfaces. *Journal of Geometric Mechanics*, 3(4):389–438, 2011.
4. M. Bauer, P. Harms, and P. Michor. Sobolev metrics on shape space II: Weighted Sobolev metrics and almost local metrics. *Journal of Geometric Mechanics*, 4(4):365–383, 2012.
5. M. Beg, M. Miller, A. Trouvé, and L. Younes. Computing large deformation diffeomorphic metric mappings via geodesic flows of diffeomorphisms. *International Journal of Computer Vision*, 61(2):139–157, 2005.
6. J.-D. Benamou and Y. Brenier. A computational fluid mechanics solution to the Monge-Kantorovich mass transfer problem. *Numerical Mathematics*, 84(3):375–393, 2000.
7. J.-D. Benamou, Y. Brenier, and K. Guittet. The Monge-Kantorovitch mass transfer and its computational fluid mechanics formulation. *International Journal for Numerical Methods in Fluids*, 40(1–2):21–30, 2002.
8. F. Bookstein. *Morphometric Tools for Landmark Data: Geometry and Biology*. Cambridge University Press, 1997.
9. T. Cootes, C. Taylor, D. Cooper, and J. Graham. Active shape models – their training and application. *Computer Vision and Image Understanding*, 61(1):38–59, 1995.
10. M. Delfour and J.-P. Zolésio. *Shapes and Geometries: Metrics, Analysis, Differential Calculus, and Optimization*, volume 22 of *Advances in Design and Control*. SIAM, 2nd edition, 2001.
11. M. Droske and M. Rumpf. Multi scale joint segmentation and registration of image morphology. *IEEE Transactions on Pattern Analysis and Machine Intelligence*, 29(12):2181–2194, 2007.
12. S. Durrleman, X. Pennec, A. Trouvé, and N. Ayache. Statistical models of sets of curves and surfaces based on currents. *Medical Image Analysis*, 13(5):793–808, 2009.
13. S. Durrleman, X. Pennec, A. Trouvé, P. Thompson, and N. Ayache. Inferring brain variability from diffeomorphic deformations of currents: An integrative approach. *Medical Image Analysis*, 12(5):626–637, 2008.
14. M. Fuchs, B. Jüttler, O. Scherzer, and H. Yang. Shape metrics based on elastic deformations. *Journal of Mathematical Imaging and Vision*, 35(1):86–102, 2009.
15. J. Glaunès, A. Qui, M. Miller, and L. Younes. Large deformation diffeomorphic metric curve mapping. *International Journal of Computer Vision*, 80(3):317–336, 2008.
16. B. Hafner, S. Zachariah, and J. Sanders. Characterization of three-dimensional anatomic shapes using principal components: Application to the proximal tibia. *Medical and Biological Engeneering and Computing*, 38(1):9–16, 2000.
17. D. Holm, A. Trouvé, and L. Younes. The Euler-Poincaré theory of metamorphosis. *Quarterly of Applied Mathematics*, 67(4):661–685, 2009.
18. P. Iglesias-Zemmour. *Diffeology*, volume 185 of *Mathematical Surveys and Monographs*. American Mathematical Society, 2013.
19. D. Kendall. Shape manifolds, procrustean metrics, and complex projective spaces. *Bulletin of the London Mathematical Society*, 16(2):81–121, 1984.
20. M. Kilian, N. Mitra, and H. Pottmann. Geometric modelling in shape space. *ACM Transactions on Graphics*, 26(64):1–8, 2007.
21. A. Kriegl and P. Michor. *The Convient Setting of Global Analysis*, volume 53 of *Mathematical Surveys and Monographs*. American Mathematical Society, 1997.
22. S. Kurtek, E. Klassen, Z. Ding, and A. Srivastava. A novel Riemannian framework for shape analysis of 3D objects. In *Proceedings of the IEEE Conference on Computer Vision and Pattern Recognition*, pages 1625–1632, 2010.
23. S. Kushnarev. Teichons: Solitonlike geodesics on universal Teichmüller space. *Experimental Mathematics*, 18(3):325–336, 2009.
24. H. Ling and D. Jacobs. Shape classification using the inner-distance. *IEEE Transactions on Pattern Analysis and Machine Intelligence*, 29(2):286–299, 2007.
25. P. Michor and D. Mumford. Vanishing geodesic distance on spaces of submanifolds and diffeomorphisms. *Documenta Mathematica*, 10:217–245, 2005.
26. P. Michor and D. Mumford. Riemannian geometries on spaces of plane curves. *Journal of the European Mathematical Society*, 8(1):1–48, 2006.

27. P. Michor and D. Mumford. An overview of the Riemannian metrics on spaces of curves using the Hamiltonian approach. *Applied and Computational Harmonic Analysis*, 23(1):74–113, 2007.
28. P. Michor, D. Mumford, J. Shah, and L. Younes. A metric on shape space with explicit geodesics. *Rendiconti Lincei – Matematica e Applicazioni*, 9:25–57, 2007.
29. W. Mio, A. Srivastava, and S. Joshi. On shape of plane elastic curves. *International Journal of Computer Vision*, 73(3):307–324, 2007.
30. J. Nocedal and S. Wright. *Numerical Optimization*. Springer, 2nd edition, 2006.
31. D. Perperidis, R. Mohiaddin, and D. Rueckert. Construction of a 4D statistical atlas of the cardiac anatomy and its use in classification. In J. Duncan and G. Gerig, editors, *Medical Image Computing and Computer Assisted Intervention*, volume 3750 of *Lecture Notes in Computer Science*, pages 402–410. Springer, 2015.
32. W. Ring and B. Wirth. Optimization methods on Riemannian manifolds and their application to shape space. *SIAM Journal on Control and Optimization*, 22(2):596–627, 2012.
33. V. Schulz and M. Siebenborn. Computational comparison of surface metrics for PDE constrained shape optimization. *Computational Methods in Applied Mathematics*, 16(3):485–496, 2016.
34. V. Schulz, M. Siebenborn, and K. Welker. Structured inverse modeling in parabolic diffusion problems. *SIAM Journal on Control and Optimization*, 53(6):3319–3338, 2015.
35. V. Schulz, M. Siebenborn, and K. Welker. Efficient PDE constrained shape optimization based on Steklov-Poincaré type metrics. *SIAM Journal on Optimization*, 26(4):2800–2819, 2016.
36. E. Sharon and D. Mumford. 2D-shape analysis using conformal mapping. *International Journal of Computer Vision*, 70(1):55–75, 2006.
37. M. Siebenborn and K. Welker. Computational aspects of multigrid methods for optimization in shape spaces. *SIAM Journal of Scientific Computing*, 39(6):B1156–B1177, 2017.
38. M. Söhn, M. Birkner, D. Yan, and M. Alber. Modelling individual geometric variation based on dominant eigenmodes of organ deformation: Implementation and evaluation. *Physics in Medicine and Biology*, 50(24):5893–5908, 2005.
39. J. Sokolowski and J. Zolésio. *Introduction to Shape Optimization*, volume 16 of *Computational Mathematics*. Springer, 1992.
40. A. Trouvé and L. Younes. Metamorphoses through Lie group action. *Foundations of Computational Mathematics*, 5(2):173–198, 2005.
41. M. Vaillant and J. Glaunès. Surface matching via currents. In G. Christensen and M. Sonka, editors, *Information Processing in Medical Imaging*, volume 3565 of *Lecture Notes in Computer Science*, pages 381–392. Springer, 2005.
42. K. Welker. Suitable spaces for shape optimization. *arXiv:1702.07579*, 2017.
43. B. Wirth. *Variational Methods in Shape Spaces*. PhD thesis, Friedrich-Wilhelms-Universität Bonn, 2009.
44. B. Wirth, L. Bar, M. Rumpf, and G. Sapiro. Geodesics in shape space via variational time discretization. In M. Figueiredo, J. Zerubia, and A. Jain, editors, *Energy Minimization Methods in Computer Vision and Pattern Recognition*, volume 5681 of *Lecture Notes in Computer Science*, pages 288–302. Springer, 2009.
45. B. Wirth, L. Bar, M. Rumpf, and G. Sapiro. A continuum mechanical approach to geodesics in shape space. *International Journal of Computer Vision*, 93(3):293–318, 2011.
46. B. Wirth and M. Rumpf. A nonlinear elastic shape averaging approach. *SIAM Journal on Imaging Sciences*, 2(3):800–833, 2009.
47. B. Wirth and M. Rumpf. Variational methods in shape analysis. In O. Scherzer, editor, *Handbook of Mathematical Methods in Imaging*, pages 1363–1401. Springer, 2011.
48. J.-P. Zolésio. Control of moving domains, shape stabilization and variational tube formulations. *International Series of Numerical Mathematics*, 155:329–382, 2007.

Printed in the United States
By Bookmasters